AF901566

eBook Bonus.

With the purchase of this book, you also receive access to its electronic version. Use the following personal code to download the electronic pages.

80181-4uugl-
26r11-k00da

Take advantage of the eBook Bonus on your computer, smartphone, or tablet. Go to **www.hanserpublications.com/ebookbonus** and register with your personal access code.

Download of this book as an eBook is subject to legal regulations, which can be found at www.hanserpublications.com/ebookbonus. Requirements: Internet connection, Adobe® Reader®, and account registration.

Kohlgrüber (Ed.)

Co-Rotating Twin-Screw Extruders: Fundamentals

Klemens Kohlgrüber (Ed.)

Co-Rotating Twin-Screw Extruders: Fundamentals

HANSER

Hanser Publishers, Munich Hanser Publications, Cincinnati

The Editor:

Dr.-Ing. Klemens Kohlgrüber, Kürten, Germany

MIX
Papier aus verantwortungsvollen Quellen
FSC® C016439
FSC www.fsc.org

Distributed in the Americas by:
Hanser Publications
414 Walnut Street, Cincinnati, OH 45202 USA
Phone: (800) 950-8977
www.hanserpublications.com

Distributed in all other countries by:
Carl Hanser Verlag
Postfach 86 04 20, 81631 Munich, Germany
Fax: +49 (89) 98 48 09
www.hanser-fachbuch.de

Library of Congress Control Number: 2019949172

Editor: Dr. Mark Smith
Production Management: Jörg Strohbach
Coverconcept: Marc Müller-Bremer, www.rebranding.de, Munich
Coverdesign: Stephan Rönigk
Typesetting: Kösel Media GmbH, Krugzell
Printed and bound by Druckerei Hubert & Co GmbH und Co KG BuchPartner, Göttingen
Printed in Germany

ISBN: 978-1-56990-747-4
E-Book ISBN: 978-1-56990-748-1

Preface

The twin-screw extruder is of great importance in various industrial sectors, such as in the plastics, food, and pharmaceuticals industries. The editor published a book on this subject in late 2007 as both English- and German-language editions, the former of which was called simply "Co-Rotating Twin-Screw Extruders". In the meantime a considerably extended and updated 2nd German edition of the book (*Der gleichläufige Doppelschneckenextruder*) was published in 2016. The preface of this German edition translated into English is appended below. This current English edition comprises about half of that greatly expanded German edition, with a focus on the basics of co-rotating twin-screw extruders. In particular, the following main points are described:

- Historical development.
- Process comprehension, especially compounding.
- Geometry of twin-screw screws and new patents for them.
- Material properties of polymers.
- Transport, pressure, and torque (power) behavior.

The editor would like to thank all the section authors, especially for their English translations. My thanks also go to Mr. Thomas König, who has clarified technical terms and also carried out an overall review. In particular, I would like to thank Dr. Smith from Carl Hanser Verlag, who managed this English edition and supported the publisher extraordinarily well!

Klemens Kohlgrüber, August 2019

Preface to the Second German Edition

The 50th anniversary of the "twin-screw compounder (ZSK)" was the occasion for the first edition of this book. Therefore, only authors of the companies Bayer (licensor, Chapter 1) and Werner & Pfleiderer (today Coperion, licensee) were involved. The elaboration of the first edition took place under considerable time pressure because, after the first idea for this book, it should appear on the occasion of the Plastics and Rubber Fair "K 2007".

For the present edition it was my intention as editor to incorporate especially the following improvements and extensions:

- The participation of different companies and universities.
- A greater involvement of technical topics.
- Naturally the consideration of the further developments that have been made in the meantime (concerning screw geometries, calculation approaches, applications, ...).
- The basics of the extruder technique and the process descriptions by means of models should be described in more detail.
- Especially application-oriented practical examples should be incorporated to a larger extent.
- The contributions should be better coordinated.

This has succeeded now in many points of the present second edition. The reader may decide himself on the qualitative improvements. The extent has grown because of the number of contributions and by the more detailed depiction of the basics. The book should now be readable for apprentices in technical professions and simultaneously represent a benefit for experts due to the described applications. Some chapters are partly overlapping; this has been done intentionally. Due to different authors with different explanations regarding the same facts, some topics will become clearer. When coordinating the contributions I have tried to ensure that largely the same denominations and formula symbols have been used. The description of a topic and the interpretation of findings have been the focus of the respective author. In particular cases, a fact can be seen differently by different authors, for example the evaluation regarding usefulness of models (for more details please see Section 1.4). For this reason I refrained from the original intention to write a summary for each contribution. This could lead to an assessment being "counterproductive" in the sense of cooperation.

I would like to take this opportunity to offer heartfelt thanks to all authors for their contributions! I thank Mr. Lechner for the coordination of the contributions of Coperion.

My thanks go to all those who contributed with their comments on improvements and detailed definitions. Furthermore I would like to thank my daughter Kristina for the review of my contributions.

Here my special thanks are due to Ms. Wittmann of the publisher Hanser! She always accompanied the "book project" from the preparation phase until the end and gave valuable contributions for designing the book.

Klemens Kohlgrüber, May 2016

The Authors

The Editor

Dr. Klemens Kohlgrüber completed a metalworking apprenticeship, after which he obtained two years of professional experience. He then undertook further education in Cologne to become a mechanical engineering technician, and then studied in Wuppertal to become a mechanical engineer, followed by a licentiate degree and doctorate from the RWTH Aachen University (each in Germany). From 1986 to 2015 he was employed at Bayer AG, in roles including leading the group on high-viscosity, mixing, and reactor technology. In parallel and over many years he has lectured on compounding/preparation of polymers to master's students in chemistry at the University of Dortmund, Germany. Also for many years, he has led the working group on high-viscosity technology at the *Forschungsgesellschaft Verfahrenstechnik* (German Research Association for Process Engineering) and was a member of the Association of German Engineers (VDI) advisory board on plastics preparation/compounding technology. He leads annual VDI seminars on the topic of extruders.

The Coauthors

Section	Author	
1.2	Martin Ullrich †	Formerly of Bayer Technology Services GmbH
1.3	Dr. Reiner Rudolf	Covestro Deutschland AG
1.5, 2.1	Dr. Thomas König	Covestro Deutschland AG
2.2	Dr. Ralf Kühn	Coperion GmbH
2.3, 4.7	Dr. Michael Bierdel	Covestro Deutschland AG
3.1, 3.4	Dr. Jens Hepperle	Bayer Crop Science AG
3.2	Dr. Jürgen Flecke	Covestro Deutschland AG
3.3	Dr. Heino Thiele	Formerly of BASF
3.5	Dr.-Ing. habil. Kalman Geiger	Formerly of the University of Stuttgart
3.5	Dr.-Ing. Gerhard Martin	Kunststoff Prozess Technik GmbH
4.5	Dr. Ulrich Liesenfelder	Covestro Deutschland AG
4.6	Dr. Carsten Conzen	Covestro Deutschland AG
4.6	Prof. Dr. Olaf Wünsch	University of Kassel

Order according to chapter structure.

Contents

1 Introduction

1.1 Technical and Economic Importance of Extruders

Klemens Kohlgrüber

1.1.1 Extruder Types and Terms

Screw machines are used for many process technology tasks. Normally the application takes place in continuous processes in which a screw machine can execute several process tasks simultaneously. It is a "multifunctional" machine. Although screw machines are able to do far more than extrude, mostly the term extruder is used. In the older German use of language also the terms "press" and "kneader" have been used. Corresponding to the old rubber screw presses, screw machines/extruders for plastics have initially been named plastic screw presses. This has been expressed for example by the title "Screw Presses for Plastics" of the first edition of the book of Gerhard Schenkel in 1959. The second edition of 1963 was renamed to "Plastic Extrusion Technology" [1]. Consistent with the current book title "Co-Rotating Twin-Screw Extruders" both terms, screws and extruders, have been "incorporated" into the book at hand.

Werner & Pfleiderer acquired licenses from Bayer for twin-shaft, exactly self-wiping, closely intermeshing co-rotating screw machines (see Section 1.2). They were named "ZSK", and this term was for a long time a synonym for this screw type. The term "ZSK" of Werner & Pfleiderer (today Coperion) is according to the former staff member and author Heinz Herrmann an abbreviation for *Zweiwellige Knetscheiben-schneckenpresse* ("twin-shaft screw compounder" in German; [2], p 179). Today the term is mostly shorted to "twin-(shaft)-screw kneader".

For this machine type many synonyms are in use, for example:

- Co-rotating twin screws (tightly intermeshing or non-intermeshing)
- Co-rotating extruder

- Co-rotating, closely intermeshing twin-shaft screw
- Co-rotating twin-screw extruder
- Co-rotating double-screw extruder
- Co-rotating twin-shaft extruder

The closely intermeshing twin-shaft screw with co-rotating shafts occupies a dominant position among the "extruders" and is applied in a variety of processes. An important application is found in the production, compounding, and processing of plastics. The co-rotating screws are also used in other industry sectors, e.g. the rubber and food industry.

1.1.2 Screw Machines and Plastics

The history of the plastics is very short, compared with the history of other materials (e.g., wood, metal, ceramic). The tremendous growth is very clearly illustrated in Figure 1.1.

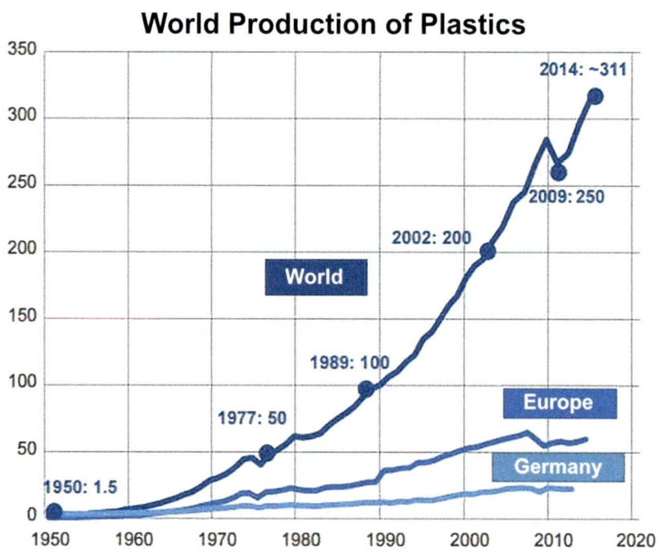

Figure 1.1 Diagram relating to the development of plastics worldwide during the last decades (ordinate: million tons) [Plastics Europe Deutschland e. V.]

What is the connection between the extruder and plastics production?

The success of the plastics industry is closely connected to the success of the extruders. Initially plastics were exclusively compounded discontinuously. This causes, however, economic limits at increasing production quantities. Furthermore, larger quality variations of the material were caused by discontinuous compounding (batch processing). Therefore, in the 1960s the continuous compounding process by means of extruders was developed. Meanwhile, nearly every plastic "passes through" an extruder during compounding and/or processing.

Figure 1.2 shows a typical plastic production comprising the reaction (or synthesis), compounding at the material producer and manufacturer of compounds, and processing to the semi-finished product or finished product.

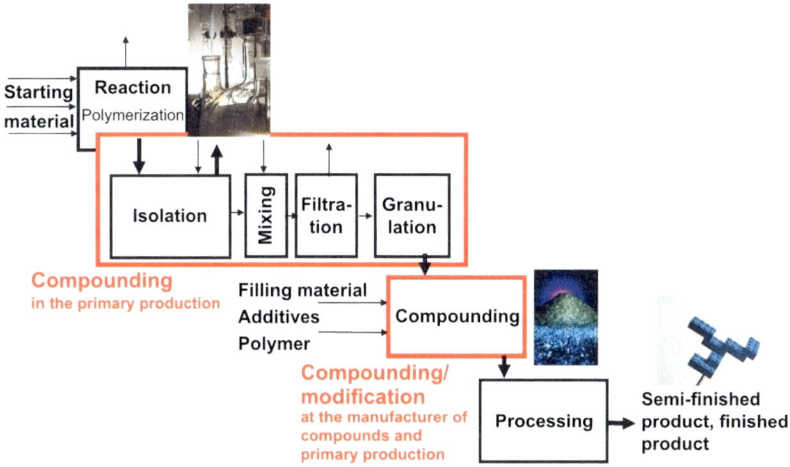

Figure 1.2 Plastic production with the compounding steps at the producer of raw materials and manufacturer of compounds

During compounding and processing of plastics the extruder plays a decisive role for melting, mixing, and dispersing. Therefore, two process steps shall be mentioned in the following section.

1.1.3 Economic Core Function of an Extruder in the Plastics Industry

1. **Melting of granulates or powders**
2. **Incorporation of additives or other polymers**

Why is the extruder a very important machine regarding the first point?

Plastics are usually treated in granular form (sometimes also as powder) between the polymer producer and the compounder, and between the compounder and processor. Only extruders are economically dominant for the continuous melting of granulate.

Why is the extruder a very important machine regarding the second point?

No polymer works without additives after the polymerization reaction. Normally, additives must be incorporated already at the raw material producer so that the polymer has a sufficient stability. In this context it is necessary to refer to the possible damage mechanisms (see Section 3.4). Furthermore, the polymers will be modified by additives. Only by this will they become ready for use and better able to be processed. The compounder also generates many new products with "tailor-made properties" by the addition of additives and production of blends. History has shown that new plastics mainly originate in this way, not by the development of new basis polymers. Conclusion: extruders are used very successfully for the compounding of existing products and development of new products. The following Figure 1.3 of Coperion GmbH gives an impression of the sold quantities of compounding machines.

Polyolefins	814
Technical plastics	4,350
Masterbatch	1,160
Long-fiber reinforced plastics	36
Temperature and shear sensitive plastics (PVC, cables, TPE)	1,659
Powder coatings, toner	602
Direct extrusion/calender	160
Chemistry and reaction technology	543
Food extrusion	270
Others	766
In total	**10,360**

More than 25% of all installed compounder systems worldwide come from Coperion.

Figure 1.3 Importance of compounding extruders [Coperion GmbH]

KraussMaffei Berstorff GmbH has also built "several thousand twin-screw extruders" as per information on the occasion of the Chinaplast trade show in April 2016.

The core function, the incorporation of additives, requires the extruder to have good mixing and dispersing abilities. This is one of the strengths of the co-rotating twin-screw extruder. In the range of the product discharge, a pressure build-up is required and normally the product must be cooled via the housing walls. The extruder type can perform these functions only with low effectivity. This is due to the poor pump effectiveness (Chapter 4) and the restricted cooling ability of heavy machines.

Compounding is described in detail in Section 1.3. Besides the description of the functional zones and process variables, practical information regarding the layout of the compounding machines are given.

1.1.4 Extruder Types and Advantages of Closely Intermeshing Co-Rotating Screws

There are several types of screw machines or extruders, which can be classified according to their number of shafts (Figure 1.4).

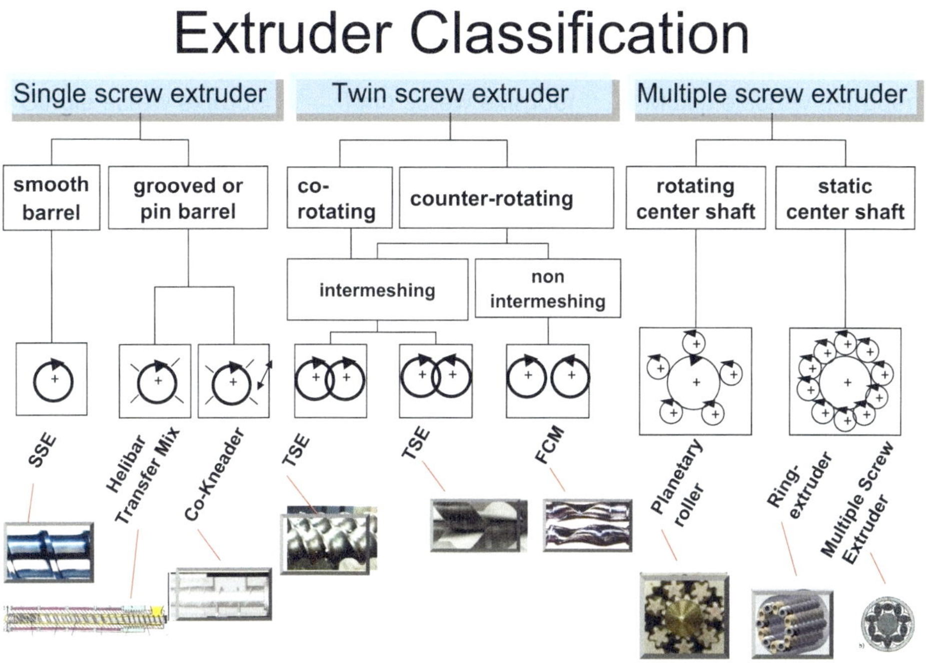

Figure 1.4 Classification of screw machines/extruders by the number of shafts

Single-shaft (single-screw) machines with a smooth-bore housing (barrel) as well as those with grooves and/or pins in the housing are utilized in plastic processing primarily for melting and pressure build-up. As the mixing ability of single-screw extruders is limited, co-rotating twin-screws (with two shafts) are often used for compounding tasks. This advantage is mentioned above as a core function.

For special applications a conical co-rotating execution has been developed. Furthermore, there are multi-shaft extruders, which partly have the geometry of two-shaft extruders, for example the ring extruder of the company Extricom. Multi-shaft extruders are presented and compared in detail elsewhere [3].

Co-rotating twin-screws are built using a modular design and can thus be adapted easily to handle a variety of processing requirements and product characteristics. A further essential advantage of the closely intermeshing co-rotating screw is – in contrast to the single-shaft machines – that the flights mesh tightly except for the necessary clearance. The screws, and thus the machine, are designated as kinematically "self-cleaning". Compared to a normal single-screw machine, where the flights scrape the inside of the housing (while maintaining a certain clearance between the screw and housing), the flights in a closely intermeshing twin-screw arrangement also clean each other. Conceptually, the twin-screw arrangement can thus be understood as a primary screw and a "cleaning screw".

On the market, there are very small laboratory extruders for product developments with small throughputs of $x \cdot 10$ g/h up to production extruders with $x \cdot 10$ t/h (factor $x < 10$). Because of the small product requirement, the smaller machines are especially suitable for development tasks with very expensive materials.

1.1.5 First Closely Intermeshing Co-Rotating Screws

The first closely intermeshing co-rotating screws were built by Bayer using their own design and featured a vertical arrangement of screws. Figure 1.5 and Figure 1.6 show such a screw arrangement built by Bayer for chemical reactions.

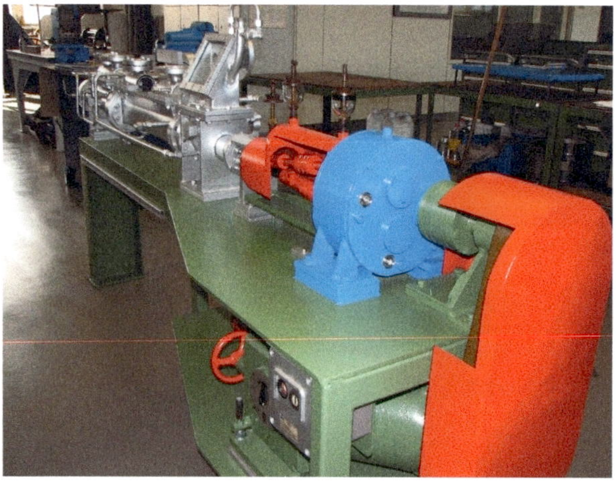

Figure 1.5
Historical chemical laboratory screw from Bayer with vertical arrangement of screws

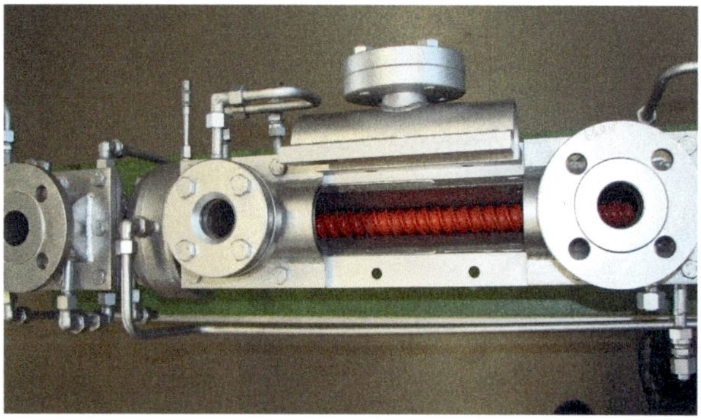

Figure 1.6
Screw shown in Figure 1.5, view from above

The screw shown in Figure 1.5 and Figure 1.6 has been restored by Coperion and can be visited there.

As a laboratory and trial machine, the machine shown in Figure 1.5 and Figure 1.6 has no casing to provide easy access and be readily convertible. In contrast, the first machines of the company formerly called Werner & Pfleiderer (today Coperion) were nearly fully encased in "the fashion of the day"; see Figure 1.7, which shows two ZSK machines from the 1950s. The first co-rotating screw type ZSK went into operation in 1957. Also, Werner & Pfleiderer initially built machines with vertical screw arrangement, as the sectional drawing in Figure 1.8 shows. The figure also shows that the actual processing section is very small in relation to the drive unit. At that time, the machine brochure promoted the "oversized drive" (with reliable operation as the benefit).

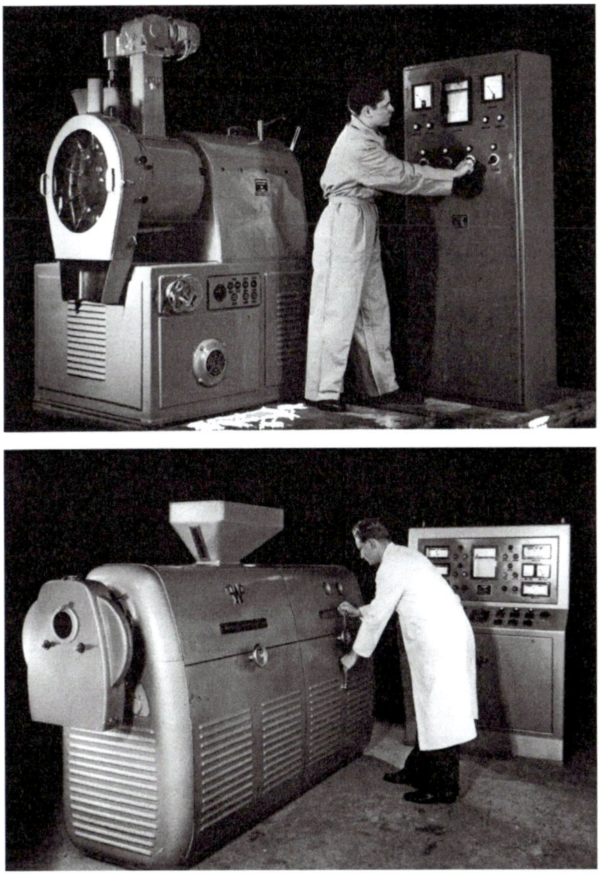

Figure 1.7 Werner & Pfleiderer's ZSK machine from the 1950s. Original figure caption: "In the 1950s WP, together with major chemical companies, made a significant contribution to the burgeoning age of plastics. The top figure shows a plasticizing unit for producing flexible PVC, the bottom figure the first twin-screw extruder for compounding plastic pallets"

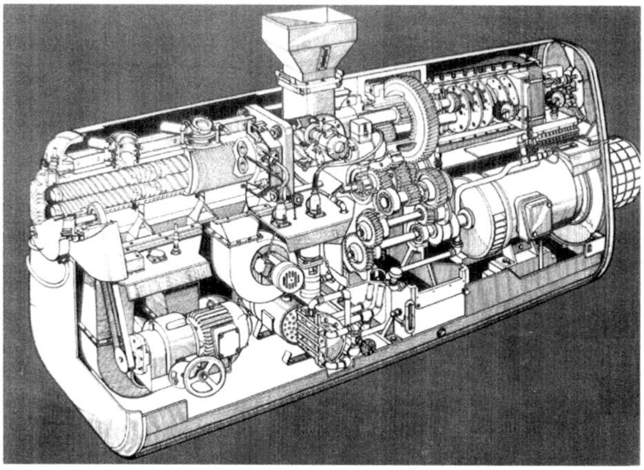

Figure 1.8 Section through a historic ZSK (Coperion GmbH)

In fact, the available torques were high for machines built then, but over the course of time it was possible to increase them even further. As the following sections will show, besides the torque also the speeds could be increased considerably.

1.1.6 Details of Twin-Screws

Section 1.2 presents a detailed summary of additional historical developments of co-rotating twin-screw extruders. Here, the modular approach to screw configuration will be specifically detailed. The shaft with screw flights is no longer manufactured “in one piece”, but rather consists of a core shaft with slipped-on screw elements and kneading elements. Section 2.2 provides an overview of the many screw elements used and their principle of operation. An extensive survey of patents relating to screw elements and screw geometries can be found in Section 2.3.

Section 2.1 presents a very detailed description of how to create the basic geometries for conveying and kneading elements. For closely intermeshing profiles, the geometric cross-section of these screw elements depends on only three characteristics: the number of flights, the diameter, and the distance between shaft centers (Section 2.1.2). Conveying occurs at the screw profile with a pitch. Together with the necessary clearances, six specifications are sufficient for the determination of the basic geometry: number of flights, diameter of housing, distance between the shaft centers, the flight pitch, the clearance between the screws themselves, and the clearance between the screw and housing. For the clearance between the screws themselves a “clearance strategy” must still be fixed (Section 1.2.2.4.2). The so-called longitudinal cut equidistance represents a good compromise.

1.1.7 Objective of the Book

For the optimal design of a co-rotating screw for adaption to the task and product, a comprehensive knowledge of the machine (what can the machine perform?) and process (how does the product behave in the machine?) is required. Besides the specific knowledge in the areas product, process technology, and mechanical engineering, especially the knowledge of the interplay of the areas is necessary (Figure 1.9). Just this necessary basic knowledge is what this book conveys. It can be used for screw design and scale-up/-down.

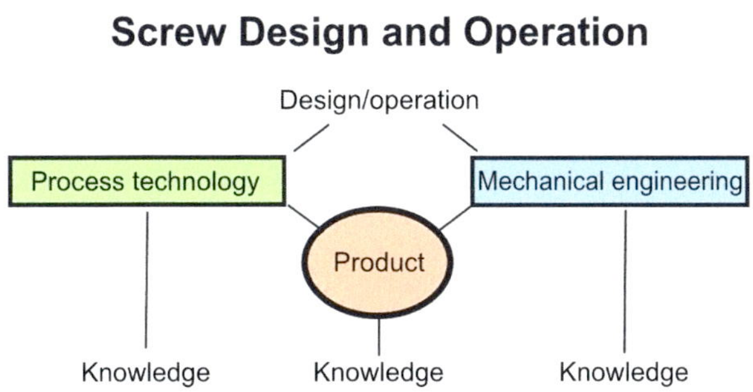

Figure 1.9 Necessary basic knowledge of the individual areas and their interaction

In Figure 1.9 the term product has been chosen for the material to be processed in the extruder. This term needs further clarification, because for machine builders "their product" is of course the machine. The term in Figure 1.9 stands as a synonym for material, pellets, granulate, polymer, food, extrudate, blend, etc.

In accordance with the three states of matter, the "material" in the extruder can be solid, liquid, and gaseous. For instance, in the melting area of a compounding extruder there is solid granulate, melted granulate as highly viscous liquid, and maybe drawn-in air or auxiliary gases from the feed section. Gaseous materials occur also when removing volatile components, e. g. solvents and monomers. Solid materials (granulate, powder, etc.) are dosed and can occur also as extrudate, e. g. at crystallization in special crystallization screws. Very highly viscous materials with simultaneously low-viscous water are utilized in so-called (single-shaft) **dewatering extruders** in the rubber industry. For most applications of the screw machines, however, the process zones, also named functional areas, are supplied with highly viscous liquids and blends. These applications are the main focus of this book.

1.1.8 Summary

The co-rotating twin-screw extruder is, due to its mode of operation and modular design, a multifunctional machine. Its success in terms of the number of units was and is closely connected with the economic success of plastics. The economic core function of extruders in the plastics industry is:

- Melting of granulates or powders

and additionally of co-rotating twin-screw extruders:

- Incorporation of additives or other polymers.

The disadvantages of a poor pump efficiency to the pressure build-up and the restricted product cooling effectivity in large extruders, both relevant especially in the range of the product outlet at the end of the extruder, are accepted.

1.1.9 Prospects

A forecast of the author regarding further developments and applications:

- Further development of new self-cleaning screw geometries differing from the basic "Erdmenger" profile
- Incremental increases of the "usual further developments": maximal length of screw L/D, product volume (expressed by D_a/D_i), and torque M_d/A^3 (A = center distance)
- Increased applications in the food and pharmaceutical industries
- Sporadic additional applications in the primary plastic production for reactions in mass without solvents (at short reaction times)

References for Section 1.1

[1] Schenkel, G., *Kunststoff-Extrusionstechnik,* Hanser (1963)

[2] Herrmann, H., *Schneckenmaschinen in der Verfahrenstechnik,* Springer, Berlin, Heidelberg, New York (1972)

[3] Kohlgrüber, K., Zwei- und Mehrwellenextruder, annual VDI-Forum with speakers from machine manufacturers and the chemical industry *(https://www.vdi-wissensforum.de/weiterbildung-kunststoff/zwei-und-mehrwellenextruder/)*

■ 1.2 Historical Development of Co-Rotating Twin-Screw Extruders

Martin Ullrich

1.2.1 Preface and Recognition of Bayer Scientists

Klemens Kohlgrüber

Since the publication of the first edition the author of the section "Historical Development of Co-Rotating Twin-Screw Extruders", Mr. Martin Ullrich, sadly passed away. Also Mr. Juri Pawlowski, who provided fundamental identification numbers and interrelations through his works on similitude theory of extruders, died at the end of 2013. In memory and as recognition of the work of these scientists I therefore would like to add some remarks.

In Section 1.2.2.1, Ullrich briefly goes into the matter concerning the Bayer developments for the "co-rotating screw system". Professor J. L. White, who died in 2009, describes in his book *Twin Screw Extrusion* quite extensively the research from Bayer (formerly IG Farben) in this area. This is expressed by some sub-chapter titles from [1] (selection):

- The Meskat-Erdmenger-Geberg, L. G. Farbenindustrie Wolfen Works Twin Screw Development Program
- Meskat Bayer (Dormagen) Program
- Erdmenger Bayer (Leverkusen) Program
- The Werner-Pfleiderer ZSK System Erdmenger
- Later Bayer Developments

As a complement, reference is made to the "timetable of the history of development of screw machines" ([2], pp. 38–47).

The basic patent of the threaded screw comes from Meskat and Erdmenger and dates back to 1944 (Section 1.2.2.1.2). Meskat later became mayor of Dormagen, and Erdmenger managed the "Screw Group" at Bayer in Leverkusen until 1976. Then Ullrich took over this group, which he later renamed "High-Viscosity Technology". After Ullrich's retirement in 2000, I became head of the group and was able to expand the group, in both staff and thematic, and renamed it to "High-Viscous, Mixing and Reactor Technology".

The tightly meshing self-cleaning geometries for co-rotating screws as per the former patents are often called "Erdmenger Geometries". This reflects also the above sub-chapter title of [1] ZSK System Erdmenger. A recognition for Erdmenger is given in Section 1.2.2.2.4. Here, I would like to refer briefly to the works of Pawlowski and Ullrich.

Juri Pawlowski made many valuable contributions, especially in his theoretical works at Bayer concerning the similitude theory in physical-technical research, on which he wrote a book [3]. A valuable development was the derivation of dimensionless process factors and dimensionless conveying parameters (profile parameter $A_1, \ldots$) for single-shaft screws [3]. The conveying parameters of the single-shaft screw were measured then. Later, we calculated these very exactly and were surprised at the extremely good measured values [4]. On another topic, the representative shear rate, Mr. Pawlowski sent me an e-mail on January 12, 2013 (when he was 93 years old):

> *"The application of the representative shear rate results i. a. in an unfounded and an inadmissible similarity theory of reduction of the pi-area which is required by a chemical engineering process. But there are some more simple problems of the creeping flow at which the 3rd invariant of the deformation speed tensor div(v) is equal to zero (e. g., parallel plate-, tube-, and Couette flow) and the representative values (viscosity, shear rate, shear stress) act as mathematically exact solutions.*
>
> *And still another incidental remark: Several authors are seeking to approximate the flow curves of non-Newtonian materials mathematically. But this is of little use with regard to the similarity theory: 1. There are legions of rheological material equations of different mathematical structure and 2. The methods based on the similarity theory apply likewise to all flow curves, regardless of their possible approximation."*

In his book, Pawlowski shows that in non-Newtonian materials "without memory" only two dimensionless process factors of the extruder arise and the other material parameters arise isolated dimensionless, for example at creeping flow (Equation 6.3.2 in [3]).

$$\Delta p^{*} = f_{\Delta p}\left(\dot{V}^{*}, n \cdot \theta, \frac{L}{D}, \pi_{rheological,i}\right) \tag{1.1}$$

Here, the dimensionless process factors comprising two dimensioned properties are pressure and extruder speed. θ is a time constant in the rheological description and $\pi_{\text{rheological,i}}$ are dimensionless material constants which do not depend on process parameters. With the Carreau approach for viscosity (Section 3.1), often used in this book, the dimensionless pressure can be formed with the zero viscosity. The product of speed and rheological time constants is also named Deborah number. $\pi_{\text{rheological}}$ is then the flow exponent. Detailed information regarding Equation (1.1) will follow in Section 1.4.

Pawlowski worked mostly on a theoretical basis and also published his results, whereas Ullrich worked very closely with polymer production companies and documented his developments in the field of the co-rotating screw system mainly in internal Bayer reports. The following Figure 1.10 shows a summary of a report of 1973. Hence follows:

- The "dimensionless description according to the theory of similarity" of the conveying, pressure, and capacity characteristic for single-shaft extruders introduced by Pawlowski has been applied by Ullrich to twin-screw extruders.
- The linear dependence of dimensionless process key figures for single-shaft screws theoretically proven by Pawlowski has been shown by Ullrich experimentally for twin-screw extruders.
- The conveying parameters (= profile parameters) A_1, A_2, B_1, and B_2 have been determined using a newly developed measurement equipment via the axis intercept equation.
- In this report several screw pitches have been analyzed and further examinations with varied screw clearance, kneading discs, and partly filled screws have been announced.

Das Förderverhalten gleichsinnig rotierender, kämmender Doppelschnecken bei Newtonschen Flüssigkeiten

Zusammenfassung

Das Förderverhalten von Schneckenmaschinen bei Newtonschen Flüssigkeiten hat einen Druckaufbau- und einen Leistungsaspekt und kann nach Dr.Pawlowski mit 3 dimensionslosen Kennzahlen in 2 linearen Zusammenhängen vollständig dargestellt werden. Ein beliebiges Schneckenprofil wird in dieser Darstellung durch 4 experimentell zu ermittelnde Parameter gekennzeichnet.

In Versuchen mit einer zweiwelligen, kämmenden Meßschnecke (Schneckentyp der ZSK-Serie) wurde die Eignung dieser Betrachtungsweise auch für mehrwellige Schnecken nachgewiesen (s. Abb. 2-9). Weiterhin wurden die wichtigen Profilparameter für verschiedene Schneckensteigungen bestimmt, so daß die fördertechnischen Aspekte von ZSK-Schnecken jetzt auch ohne Experimente berechnet werden können (s. Beispiel Pkt. 7).

Weitere Untersuchungen über das Förderverhalten von teilgefüllten Schnecken, Knetscheiben und Schnecken mit variiertem Spiel sollen sich anschließen.

Figure 1.10 Extract from a report of Martin Ullrich dated September 20, 1973 [5]. English translation below

The Conveying Characteristic of Co-Rotating, Intermeshing Twin-Screws at Newtonian Liquids

Summary

The conveying characteristic of screw machines at Newtonian liquids has a pressure build-up and a capacity aspect and can as per Dr. Pawlowski depicted totally with 3 dimensionless key figures in 2 linear interrelationships. An arbitrary screw profile will be marked in this representation by 4 parameters to be identified experimentally.

During tests with a two-shaft, intermeshing measuring extruders (screw type of the ZSK series), the suitability of this approach has been proven also for multi-shaft screws (see Fig. 2-9). Furthermore, the important profile parameters for different screw pitches have been determined, so that the conveying aspects of ZSK screws can be calculated now without experiments (see example point 7).

Further tests relating to the conveying characteristic of partly filled screws, kneading discs and screws with varied clearance shall follow.

The tests announced in the last point were also executed with the newly developed experimental apparatus (Figure 1.11). Using a two-sided bearing of the screw shafts and a floating bearing of the drive, the torques can be measured very exactly by means of a spring balance. The housings are partly made of Plexiglas and as model liquid mostly high-viscous transparent silicone oil was used. This allows experiments at room temperature. By means of temperature measurements a product heating by liquid friction (dissipation) was determined and taken into account in the evaluation (viscosity reduction). All sensors were "mechanical" and did not have electronic components, which were not available at that time. Today the equipment is still operational and is used occasionally.

Figure 1.11
Experiment apparatus for determination of the conveying characteristic of twin-screws

For decades a staff member of Mr. Ullrich executed measurements with this measuring screw. The linear relations between dimensionless pressure and dimensionless capacity in dependence of the dimensionless throughput pointed out theoretically by Pawlowski have been confirmed also for twin-screw extruders by Ullrich based on numerous measurements. The linear dependence for Newtonian materials has been very generally proven by Böhme for creeping flows [6].

From the numerous results [7], the reciprocal values of the maximal achievable pump efficiency for some extruder geometries are depicted as an example in Figure 1.40. The related terms can be found in Figure 1.42.

In the area of polycarbonate primary production Ullrich introduced a "key technology" with the co-rotating screws. There existed a division of work with Dieter Löhr. Ullrich developed the production screws for the polycarbonate primary production (also polycarbonate manufacturing) and Löhr was, as head of the compounding technical center, responsible for the compounding. A description of the contemporary Bayer compounded products and basics thereof was provided by Löhr in the *Kunststoffhandbuch* (Handbook of Plastics) [8].

Due to the collaboration of Ullrich with the polymer production, people from production did not want to publish further developments in the field of co-rotating screws. The collaboration with the extruder manufacturer Werner & Pfleiderer (Bayer licensee) was effected in great secrecy. Internally, the works of Ullrich were very much appreciated and thus, a company-wide Bayer award, the Otto-Bayer-Medal, was presented to Ullrich in 1988, as for the scientist Pawlowski three years earlier. In my later farewell speech on the retirement of Mr. Ullrich, I honored symbolically the "screw developers" Erdmenger, Pawlowski, and Ullrich (Figure 1.12) and named a relation for co-rotating screws, which is not presented in this book, after Ullrich [9].

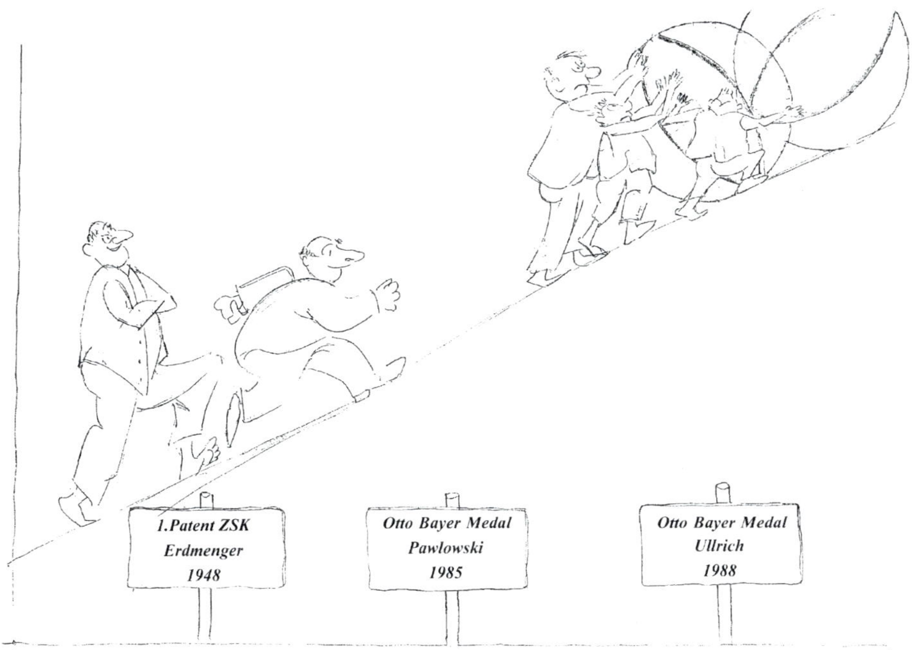

Figure 1.12 "Advancement" of the Bayer screw developers [8]

Ullrich has recorded his innovations, besides in the internal Bayer reports, in 47 patents. Thereby, the accomplishments of Mr. Ullrich became known beyond Bayer, and Werner & Pfleiderer proposed and arranged for the honoring of Mr. Ullrich with the "Heinz Herrmann Award" (Figure 1.13). This award should have been presented him on the occasion of the ANTEC conference in the USA in 2010. However, because of his ill health, Mr. Ullrich was no longer able to travel, although he had prepared an acceptance speech. From this:

> *"At the end of my speech I would like to thank the SOCIETY OF PLASTIC ENGINEERS for the outstanding honor you gave me with the bestowal of the HEINZ HERRMANN TWIN-SCREW EXTRUSION AWARD. This special recognition is one of the greatest delights of my professional life."*

Society of Plastics Engineers

ANTEC 2010

In acceptance of the

Heinz Hermann Twin-Screw Extrusion Award

to the late Mr. Martin Ullrich

Figure 1.13 Award for Martin Ulrich

Regrettably, he died before the date of the award ceremony. Together with an American co-worker, Mr. Vitkovski, I transmitted as a "thank you" a video message that was presented at the ANTEC.

At the retirement celebration for Ullrich from Bayer in 2000 I honored the three Bayer screw scientists Erdmenger, Pawlowski, and Ullrich by the following Figure 1.14. Mr. Erdmenger was already dead then. Mr. Pawlowski was very pleased about this drawing on January 3, 2013.

Figure 1.14 "Screw developers": Erdmenger, Pawlowski, and Ullrich [8]

At this point I would like to thank Mr. and Mrs. Löhr for creating Figure 1.12 and Figure 1.14 for that occasion [9]!

References for Section 1.2.1

[1] White, L.J., Kim, E.K., *Twin Screw Extrusion,* 2nd ed., Hanser (2010)

[2] Herrmann, H., *Schneckenmaschinen in der Verfahrenstechnik,* Springer (1972)

[3] Pawlowski, J., *Die Ähnlichkeitstheorie in der physikalisch-technischen Forschung,* Springer, (1971)

[4] Beck, D., Kohlgrüber, K., Berechnung der Wärmeübertragung in Schneckenmaschinen, CIT 66 (1994), No. 8, pp. 1073–1076

[5] Ullrich, M., Das Förderverhalten gleichsinnig rotierender, kämmender Doppelschnecken bei newtonschen Flüssigkeiten, internal report, Bayer, September 20, 1973

[6] Böhme, G., Theoretische Betrachtungen über schleichende Strömungen. In Festschrift, Universität-GH Essen (1995), pp. 27–40

[7] Ullrich, M., Das Förderverhalten von Doppelschnecken, internal report, Bayer, February 9, 1989

[8] Löhr, D., Herstellen von Compounds, in: *Kunststoffhandbuch,* Becker, G.W., Braun, D., Eds, Vol. 3/4, pp. 299–314

[9] Kohlgrüber, K., booklet on the speech at the retirement from Bayer of Martin Ullrich, October 26, 2000

1.2.2 Historical Development of Co-Rotating Twin-Screw Extruders

Martin Ullrich

This section covers the engineering-related history of the twin-screw, or more precisely the co-rotating twin-screw extruder, where both screw shafts rotate in the same direction. In the following, we will simply refer to them as co-rotating extruders (instead of co-rotating twin-screw extruders).

The best place to start with is a brief account of the origin of all multiple screws, namely the single-screw extruder. Its inventor, Archimedes [1] (approximately 2250 years ago), used it to transport water overcoming differences in elevation. The same principle is still used today in Egypt, Holland, and in many water purification plants.

1.2.2.1 Early Developments

The single screw as an “extrusion apparatus” was developed in the second half of the 19th century and was used intensively in industrial and heavy engineering applications. It was used in three major industries:

- Pottery industry: ceramic compounds
  - extrusion, shaping
- Rubber industry: natural rubber, gum
  - plastification, extrusion,
  - profile production

- Food industry: oily fruits, oil seeds
  - extracting biological oils,
  - separation of material using "strainer screws",
  - meat processing by meat grinder

Product feed in a single screw initially appears somewhat strange. While every metal molecule remains in the same cross-sectional plane, the product is nevertheless conveyed axially. Following is an attempt at an explanation: when considering the screw and the product, the screw rotates without changing position, although the product does not rotate but slides axially, in other words, it is axially conveyed. This so-called "theoretical" conveying does not exist in practice, however, because the product is not a solid body but a highly viscous fluid with a rheological character.

The adhesion and friction characteristics of the plastic material determine the intensity of the flow. In the case of Newtonian fluids this is half of the theoretical conveying (at constant pressure) and even less with counterpressure (extrusion), even down to zero. In the latter case, the product rotates with the shaft and throughput ceases.

This weakness of the single screw, particularly the fact that there is no cleaning of the shaft and the strong dependence of conveying on rheological properties, motivated inventors to seek solutions to these problems. The co-rotating extruder was therefore initially proposed as a self-cleaning mechanism. Six patent citations [2–7] over a 70-year period (1869–1939) show that the co-rotating extruder remained very much at the forefront of engineers' minds. J. L. White [44] provides an extensive, thorough description of the developments and patent situation in this field.

In the early 1940s, a systematic investigation into the co-rotating extruder system began at the IG plant in Wolfen, Saxony-Anhalt. It involved the combined physical, mathematical, engineering, and mechanical expertise of a team composed of W. Meskat, A. Geberg, R. Erdmenger, and their staff. The team was commissioned to develop a reliable "mechanical apparatus" for chemical processes with highly viscous products.

The work was continued at Bayer AG in Leverkusen with strongly process-orientated groups [8] under the new "applied physics" (AP) organizational structure introduced by K. Riess and implemented by K. Sigwart in 1948. R. Erdmenger founded and led one of these engineering groups, which was composed of 10 to 15 specialists and employees, until 1976. The group was later given the name "High Viscosity Technology".

This team, as part of the chemical industry, was of course primarily involved with solving problems in the area of high-viscosity engineering, particularly in developing chemical processes for the Bayer AG. The mechanical aspect was developed as required and to varying degrees of intensity.

With respect to screw geometry, we turn back the clock to Wolfen, where the team was searching for the perfect mechanical apparatus for high-viscosity technology. It had to function despite the adhesive, frictional, and antifrictional properties of the product, cope with various material consistencies, and overcome rheological changes caused, for instance, by reactions in the machine.

The desired self-cleaning function led to the development of the intermeshing twin screw. The counter-rotating screw was discarded, because it tended to get blocked by solids and was a poor mixer; attention focused instead on the intermeshing, co-rotating extruder.

1.2.2.1.1 Basic Geometry

A. Geberg addressed the geometric kinematic problem with a mathematical equivalent view. He discovered the fact that the co-rotation of two shafts around their fixed axes is the kinematic equivalent of the "movement without rotation" of one shaft around another fixed shaft (Figure 1.15). In the case of this so-called "movement without rotation", which happens when the profiles are touching, all mass points of the moved screw move in circles with radii equivalent to the centerline distance (Figure 1.15).

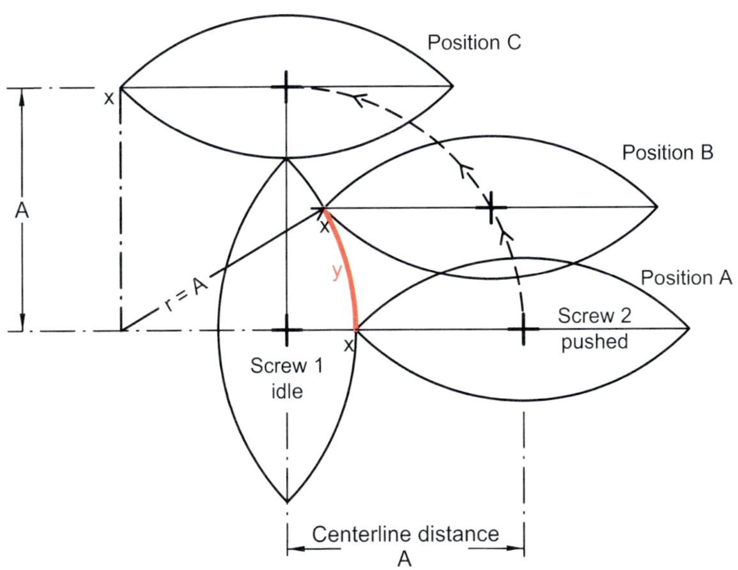

Figure 1.15 Kinematic equivalent view "movement without rotation" (screw with tip width 0)

Because the – mathematically precise – system is intended to be fully wiping, the central shaft can be a wax blank that is shaped to its corresponding contour by the metal moved screw. The moved screw (Figure 1.15) with its metal tip *x* then forms the flank arc *y* (bold) in the fixed wax shaft. As all mass points of the moved screw describe circles with a radius equal to the centerline distance, including the tip *x*, the flank arc *y* of the wax screw must also be an arc with a radius equal to the centerline distance of the two screw shafts: an astonishingly simple solution.

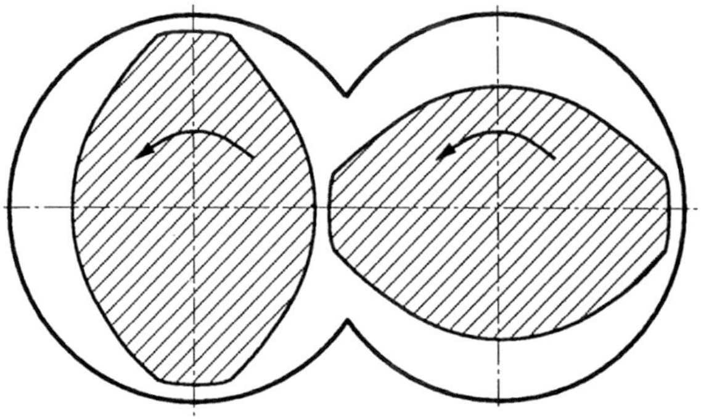

Figure 1.16 Screw with tip width, cross-section

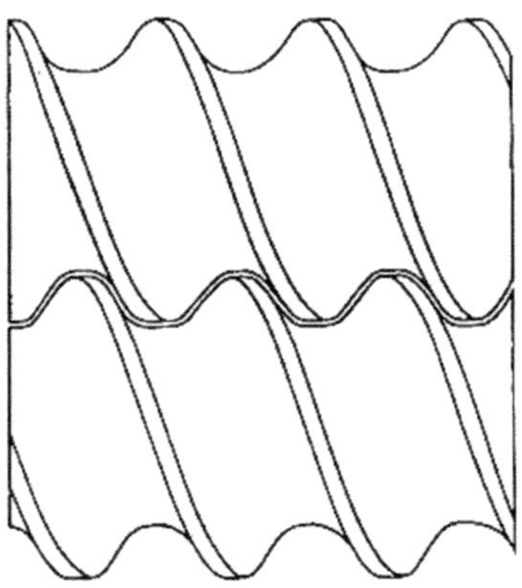

Figure 1.17
Axial view of screw

Real screws do not have points in position *x*. They have specific tip widths (Figure 1.16), which have previously been omitted to clarify the kinematics (Figure 1.15). It helps here to determine the kinematics in cross-section, then to advance the resulting cross-section profiles axially, and finally to apply a twist to obtain the longitudinal section contour and the desired three-dimensional screw (Figure 1.17).

A. Geberg supplemented his investigations by determining the basic geometries of screws in practical applications with varied parameters: number of threads and channel depth and their dependent variables, tip angle (Figure 1.18), and free cross-sectional area that can be filled with product (Figure 1.19).

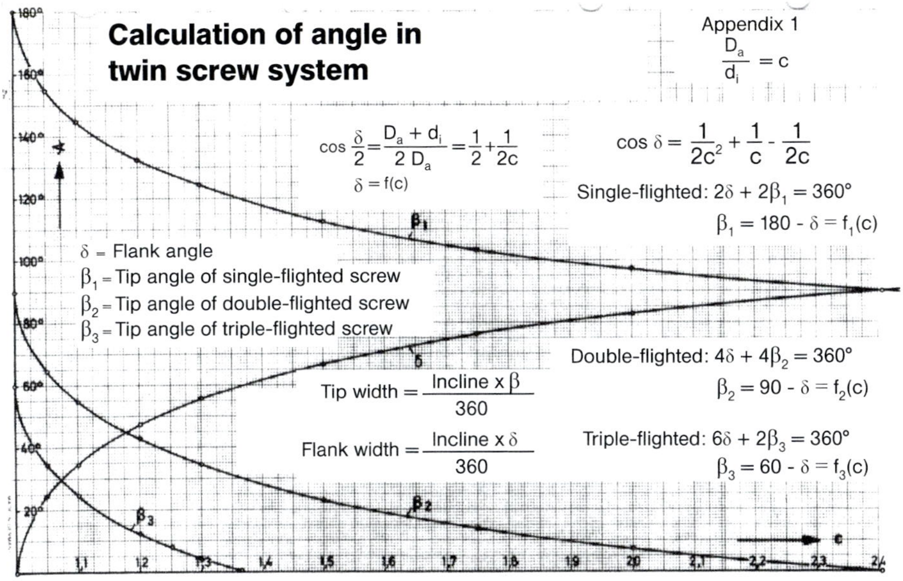

Figure 1.18 Characteristic angle diagram [11]

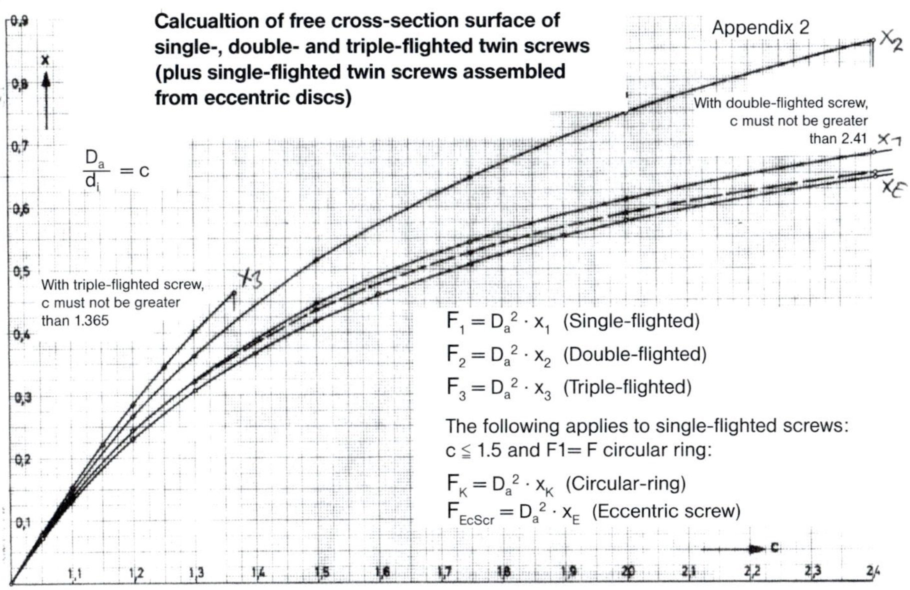

Figure 1.19 Characteristic cross-section area diagram [11]

In 1978, during his retirement, R. Erdmenger published a paper [9] looking back over his career and the significant role he played in the development of the co-rotating extruder. In this paper he cites the work of W. Meskat, A. Geberg, and R. Erdmenger (1944) on the basic geometry in Wolfen [10], extracts from R. Erdmenger and S. Neidhardt (1948) in Leverkusen [11], and his own tests on laboratory screws in Leverkusen in 1948/51 [12].

1.2.2.1.2 Basic Patents

The bibliography includes 11 patents [13–23] with inventor, year of application, patent number, and one of the characteristic abstracts describing the fundamental subject of the patent.

They concern co-rotating extruders with a self-cleaning profile, all relevant types of (single-, double-, and triple-flighted) mixing and kneading discs, modular designs, mixed-technology, backwards-pumping screw principle, and special designs for dewatering and evaporating.

It is interesting to note that

- the 20 years in which these 11 patents were issued is far shorter than the preliminary period mentioned above, and
- the material of the co-rotating extruder system is now precisely mathematically founded, is used in process engineering, and is being advanced by engineers.

1.2.2.1.2.1 Basic Patent of Threaded Screws

In view of its significance, an extract from the basic patent for the co-rotating threaded screw [13] is reproduced here (Figure 1.20 and Figure 1.21):

Erteilt auf Grund des Ersten Überleitungsgesetzes vom 8. Juli 1949
(WiGBl. S. 175)

BUNDESREPUBLIK DEUTSCHLAND

AUSGEGEBEN AM 12. JANUAR 1953

DEUTSCHES PATENTAMT

PATENTSCHRIFT

Nr. 862 668

KLASSE **39a** GRUPPE 10 15

F 3192 XII / 39 a

Dr. Walter Meskat, Dormagen und
Dipl.-Ing. Rudolf Erdmenger, Bergisch Gladbach
sind als Erfinder genannt worden

Farbenfabriken Bayer, Leverkusen-Bayerwerk

Vorrichtung zum Verkneten, Gelatinieren und Verpressen von plastischen Massen

Patentiert im Gebiet der Bundesrepublik Deutschland vom 7. Juli 1944 an
Der Zeitraum vom 8. Mai 1945 bis einschließlich 7. Mai 1950 wird auf die Patentdauer nicht angerechnet
(Ges. v. 15. 7. 51)
Patentanmeldung bekanntgemacht am 27. September 1951
Patenterteilung bekanntgemacht am 20. November 1952

PATENTANSPRÜCHE:

1. Vorrichtung zum Kneten, Gelatinieren und Verpressen von plastischen Massen, bei der zwei oder mehr oder ein- oder mehrgängige Schneckenwellen, die sich in einem dicht an den äußeren Umfang der Kämme anschließenden Gehäuse in gleichgerichtetem Sinne drehen, verwendet werden, dadurch gekennzeichnet, daß die Gewindenut einer Schnecke durch den eingreifenden Kamm der Gegenschnecke oder Gegenschnecken an den Flankenberührungsstellen vollständig oder nahezu vollständig längs einer Raumkurve abgedichtet ist, wobei das Gewindeflankenprofil so ausgebildet ist, daß die Flankenfläche in ihrer Verschneidung mit beliebigen ebenen Flächen senkrecht zur Schneckenachse Kreisbogenstücke ergibt, die den Radius r = Achsabstand der beiden Schneckenwelen besitzen und den Kernkreis der Schneckenwelle tangieren.

2. Vorrichtung nach Anspruch 1, dadurch gekennzeichnet, daß die Gewindekammbreite einer Schneckenwelle stetig zunimmt, wobei sich die zugehörige Gewindenutbreite in der anderen Welle entsprechend vergrößert.

Figure 1.20 Basic patent for threaded screws – title page and patent claims

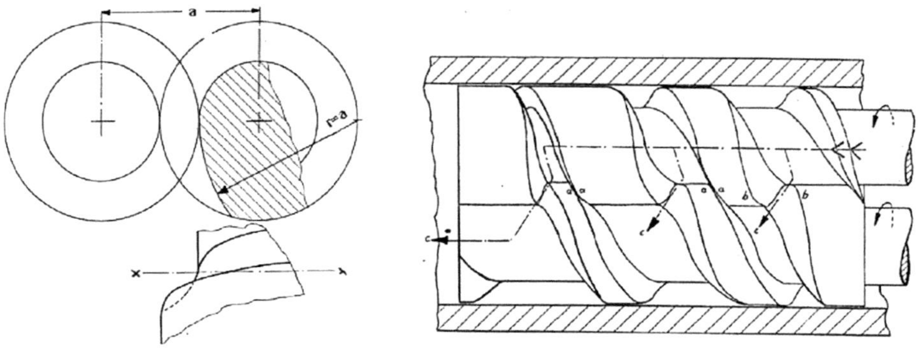

Figure 1.21 Basic geometry (illustration 3 according to patent)

The following items are of interest and worthy of mention:

- After the application in 1944, this patent was issued in West Germany in 1953 via a *"bridging act"* and was given a five-year post-war extension of validity (23 instead of 18 years, i. e., valid until 1967).
- The rapidity with which the language of technology changes is evident from the title given in 1944: "Vorrichtung zum Verkneten, Gelatinieren und Verpressen von plastischen Massen" (Apparatus for kneading, gelatinizing, and compressing of plastic masses).
- Just 20 years later, in the 1960s, the wording was as follows: "Vorrichtung zum Kneten, Plastifizieren und Extrudieren von Polymeren" (Apparatus for kneading, plastification, and extrusion of polymers).
- The first and the primary claims are extremely telling: Double-or multiple-shaft co-rotating extruders in single or multiple-flighted versions with the sealing profile, i. e., defined with the flank profiles in cross-section by circles with radius equal to the centerline distance (see Section 1.2.2.1.1).
- The second and final claim with the varied tip width along the axis with correspondingly varied groove widths of the neighboring shaft (for the purpose of increasing the mixing effect) has lost its significance with the subsequent invention of simpler kneading discs (see below).

In his paper [9], R. Erdmenger added material movements with explanations to Fig. 1 of the patent [13] (Figure 1.22): a is the purely axial movement (= theoretical conveying, see Section 1.2.2.1 above), b is the drag flow of a Newtonian fluid ($b < a$), and c is the minimal flow of a material adhering to the shafts ($c < b$).

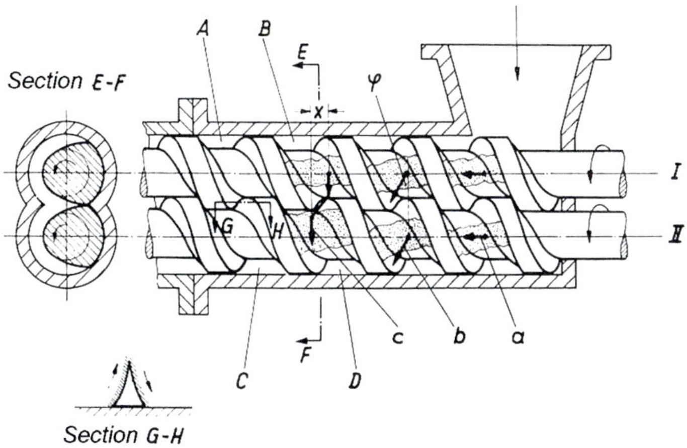

Figure 1.22 Multiple shaft co-rotating extruder system with full intermeshing and mutual cleaning of the shafts. There is a continuous connection between *A, B, C,* and *D,* etc.
The possible material movements in principle are as follows:
a) Purely axial shift of a rigid sliding body (e.g., wax)
b) Dragging of deforming substances (e.g., stringy liquids) in the direction of the arrow (angle φ), shearing in the gap, section G ÷ H.
c) Purely tangential conveying of materials adhered to the screws with smaller axial components (x) at transition from shaft I to shaft II.

1.2.2.1.2.2 Basic Patents for Kneading Discs, DBP [16], USP [17], DBP [20]

Kneading discs are prismatic bodies formed by axial displacement of the cross-section contour discussed above but without rotation common with screw threads (see Section 1.2.2.1.1).

Several kneading discs arranged on a core shaft create kneading elements (or kneading blocks) with special effects. The geometric variables of these kneading elements are the thickness of the individual discs and their offset angles with respect to one another, viewed in cross-section, and the resulting rotation of the individual kneading discs with the effect of a spiral staircase. This results in kneading, crushing, and shearing effects combined with active conveying (right-hand rotation), neutral, or reversing effects (left-hand rotation).

The same principles apply for the cross-sections of kneading discs as for the cross sections of their corresponding screw threads (see Section 1.2.2.1.1). The geometrical form is determined by number and depth of flights.

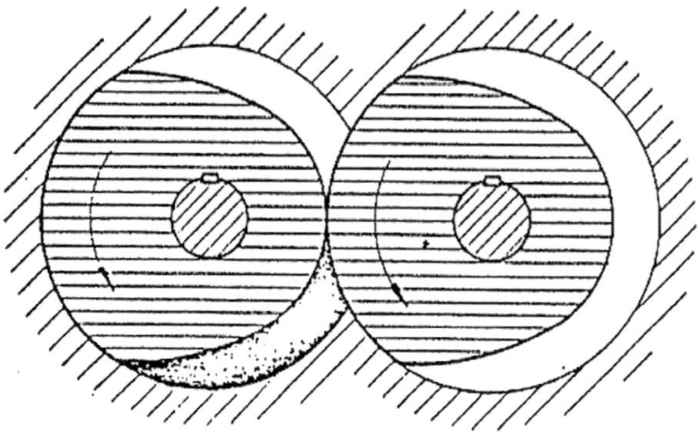

Figure 1.23 Single-flighted kneading discs

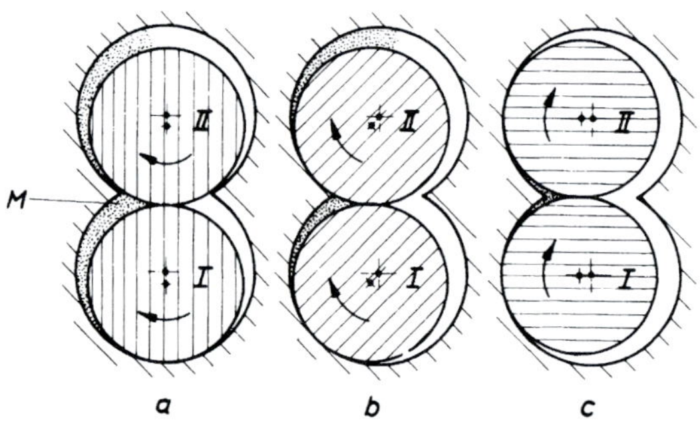

Figure 1.24 Operating method of eccentric, synchronous, co-rotating, circular discs: *a, b, c* three operating positions: mass *M* is crushed and sheared in the key during the quarter turn around their rotation axis described by each disc (*I* and *II*).

Figure 1.23 represents the normal case of a single-flighted kneading disc. Figure 1.24 shows the interesting special case of the single-flight, self-cleaning circular discs.

Figure 1.25 also shows the normal double-flighted kneading disc, also with Erdmenger's commentary. The practical limit for double-flighted elements is shown in Figure 1.26 with a small tip angle and large product volume (see Section 1.2.2.1.1).

Figure 1.27 explains the triple-flighted kneading disc and its effect according to R. Erdmenger [9].

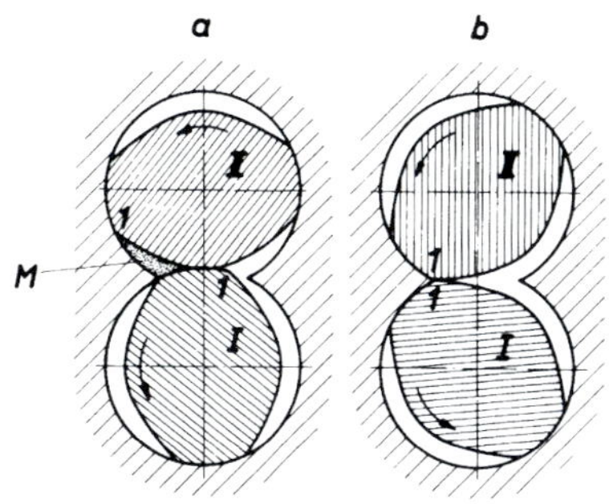

Figure 1.25 Double-flighted kneading discs operation of blunted lens-shaped discs
a) Commencement of kneading of deformable material (*M*)
b) End of the kneading phase
The mass *M* is crushed twice per revolution due to the co-rotation of discs *I* and *II* (approach of points 1–1) and escapes vertically to the indicated plane

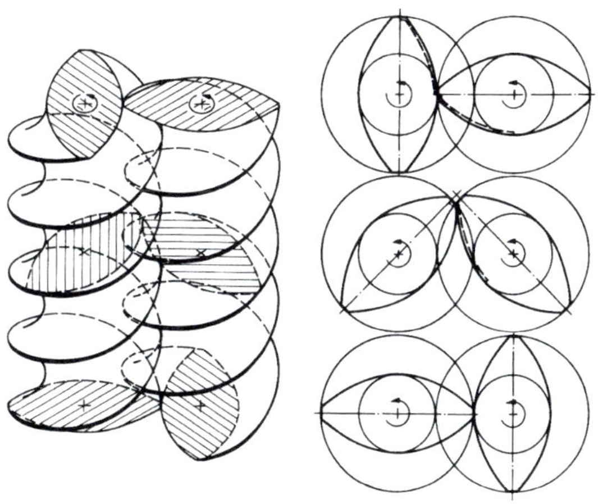

Figure 1.26 Practical limit for double-flighted elements

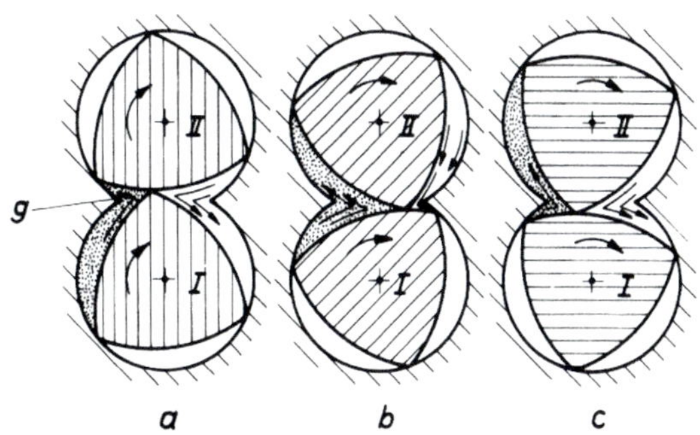

Figure 1.27 Operating method of 3-point kneading discs: *a, b, c* three operating positions. The mass *M* is forced to flow around the sharp barrel ridge *G* three times per revolution by the synchronous rotation of the co-rotating discs *I* and *II* wherein it is sheared to an extreme degree.

1.2.2.1.2.3 Basic Patents for Modular Design

Further patents describe the modular principle for different screw and kneading elements threaded and clamped to a core shaft. The benefits are striking:

- Formation of different functional zones along the machine with optimized technical processes.
- Variation of the screw geometry for tests and at start up.
- Adjustment to subsequent process changes.
- Typifying and standardization of screw and kneading element production.
- Simplification and cost savings in spare parts acquisition.

These benefits resulted in the USP [21] with priority 1959 for R. Erdmenger. The modular design is illustrated in Figure 1.28 below with a combination of different consecutive screw and kneading zones and in Figure 1.29 as a modular assembly diagram.

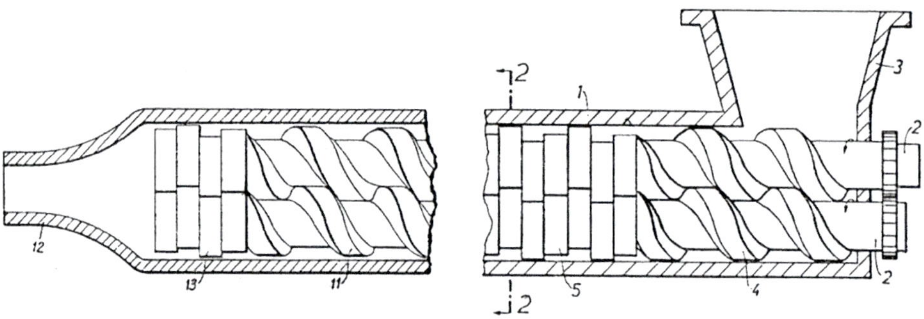

Figure 1.28 Modular principle

Figure 1.29 Modular assembly diagram

W. Meskat und J. Pawlowski obtained a German patent (DBP) [18] by 1950 for a co-rotating extruder with sections featuring localized backward-feed with respect to the primary feed direction. This was achieved by backward pumping screws, i.e., by right-to-left hand rotation combinations in the screw geometry whereby specific mixing effects are achieved. Figure 1.30 shows this apparatus in detail.

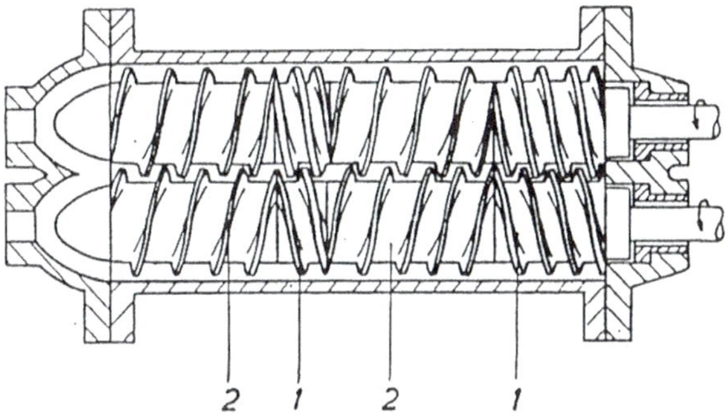

Figure 1.30 "Backward pumping screw" with left/right hand thread

These three groups of patents, namely, the widely used *threaded screws,* the *kneading elements,* and the *modular design* principle paved the way for the development of a highly effective new machine type for process engineering and plastics technology.

At this point it is also worth mentioning the (modified) comment made by H. Herrmann [36]:

"W. Meskat, A. Geberg and R. Erdmenger's idea of a co-rotating and closely intermeshing twin screw of 1944 was readdressed and turned into a complete and well-founded geometrical solution. Since this time, it has been possible to talk of *close meshing* and *self-cleaning* co-rotating extruder machines."

1.2.2.2 Pioneering Period

R. Erdmenger had worked together with A. Geberg since 1943 in W. Meskat's twin-screw core team in Wolfen, and he pursued this work intensively after 1945 for Bayer AG in Leverkusen, achieving success and acclaim with his team. His work comprised three fundamental components:

1. The necessary in-house *machine development* for co-rotating extruders.
2. The use of these machines in *chemical processes* and/or the realization of innovative chemical processes using co-rotating extruders.
3. The *issue of the license* for this machinery system in the face of increasing demand.

1.2.2.2.1 Machine Development

With respect to the first point it is obvious that laboratory screws, developmental experiments, and prototype product machines were developed with the workshop technology available in-house. Here in particular, Erdmenger's university training in mechanical engineering and his leadership were an essential driving force on the road to success. The team produced the $DA/DK = 32/24$ mm laboratory screw as a Bayer house model around 1948.

Figure 1.31 illustrates another example of a highly specialized machine from the pioneering period. Produced around 1955, it has a degressive pitch in the transport direction, which, surprisingly, was cut from the solid feedstock.

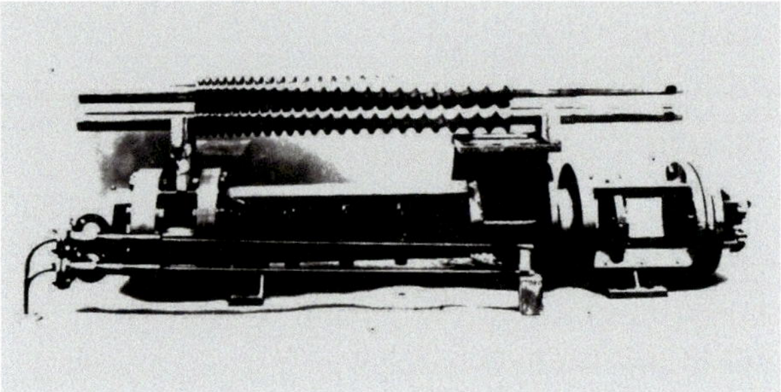

Figure 1.31 Pioneering screw, circa 1955, with degressive pitch

1.2.2.2.2 Use in Chemical Processes

The application of co-rotating extruders in chemical processes only stood a chance, if the new components of the co-rotating extruder made the product

- *more economically viable,* or
- to a *higher quality,* or
- *more environmentally friendly.*

This is taken into account in an engineering sketch drawn by R. Erdmenger himself in 1953 of a chemical screw process (Figure 1.32). From today's point of view, the last two illustrations display the historical patina and archaic simplicity of engineering times long past. The same also applies for Figure 1.33 which shows a somewhat rustic experimental center from 1957.

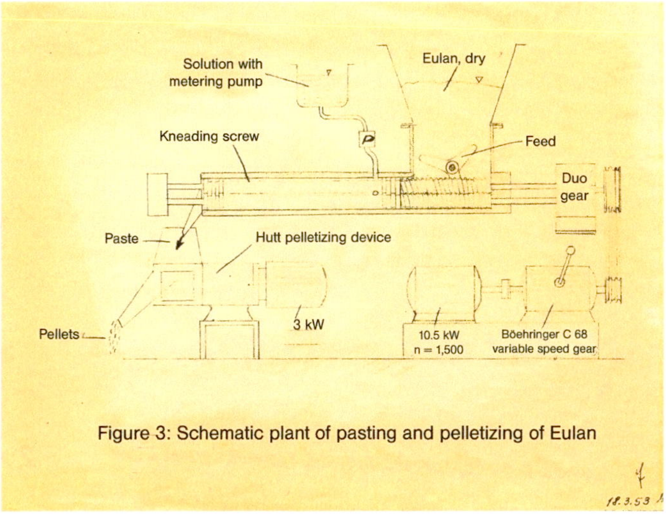

Figure 1.32 Pioneering process according to R. Erdmenger 1953

Figure 1.33 Experimental center, circa 1957

1.2.2.2.3 Licensing

Licensing of the new technology came to fruition in 1953. After intensive preliminary work and negotiations with several mechanical engineering firms, the noted company Werner und Pfleiderer in Stuttgart obtained the exclusive worldwide license (as is so often annoying for the inventor, nine years had passed since the initial invention in 1944). Although the licensee today operates under the name "Coperion Werner & Pfleiderer GmbH & Co. KG, Stuttgart",[1] we will refer to them as "Werner & Pfleiderer" in the following.

[1] The company name has changed to Coperion since this chapter was written.

A key reason for the issue of the license was that in-house production in chemical workshops was no longer possible for screw diameters beyond 120 mm. At the same time, there was a demand for higher quality machinery.

1.2.2.2.4 Recognition for R. Erdmenger

The development period of the co-rotating extruders, which was shaped by R. Erdmenger (he is portrayed in Figure 1.35), is considered the pioneering period for this technology today.

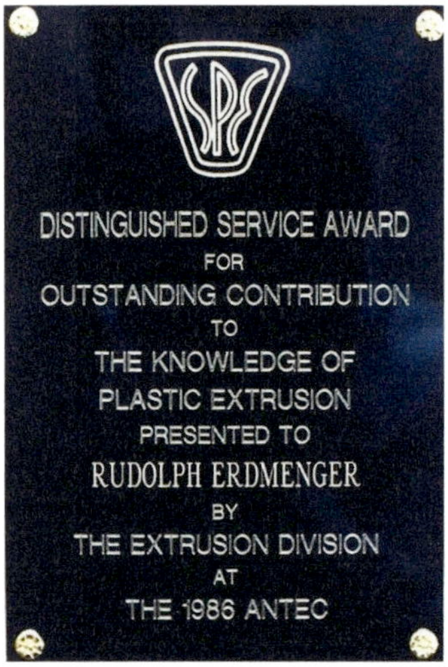

Figure 1.34 SPE's "Distinguished Service Award" for R. Erdmenger in 1986

Figure 1.35 Portrait of Rudolf Erdmenger

After his retirement, Erdmenger received two awards in the US:

- the "Distinguished Service Award" of the Society of Plastics Engineers "SPE" (Figure 1.34), and
- introduction into the University of Akron/Ohio's "Hall of Fame".

1.2.2.3 New High-Viscosity Technology with Co-Rotating Extruders

Soon after the patents cited above were issued, publications about the new types of co-rotating extruders and their applications began to appear in increasing numbers. The secondary literature also cites further publications [24–35] from the period between 1951 and 1974, many of which were authored by the licensors: Bayer, i. e., predominantly R. Erdmenger and the licensee: Werner und Pfleiderer, i. e., H. Herrmann. These publications increasingly pushed process engineering and the various uses of the co-rotating extruder.

1.2.2.3.1 Screw Machines in Process Engineering

It is to the great credit of H. Herrmann that, after the licensing process and alongside his primary developmental role at Werner & Pfleiderer, he edited the standard work on this topic, published in 1972 [36]. In this work, two further important points are worth emphasizing: on the one hand, the mechanical engineering description and historical classification of all known screw types and on the other hand, the appreciation of the co-rotating extruder along with their process engineering application areas and operating methods.

1.2.2.3.2 Similarity Theory for Screw Machinery

J. Pawlowski extended the knowledge base and deepened understanding of the physical functions and operating methods of screw machinery considerably through the development of the *similarity theory* and its application to these machines [37–39]. His work's significant consequences:

- Seven independent, dimension-specific values for filled screws were reduced to just three dimensionless parameters for throughput, pressure, and power (Figure 1.36).
- The dependencies of these dimensionless parameters for Newtonian fluids can be simplified to two linear equations (Figure 1.37) for the target values (measured on a co-rotating extruder) of pressure (Figure 1.38) and power (Figure 1.39). These straight lines are fixed by their axial sections A_1, A_2 for pressure and B_1, B_2 for power. The value of these "profile parameters" depends on the "internal geometry" of the screw, i.e., on the shape of the profile (see 1.2.2.4.1) and on the screw type (single shaft, multiple shaft, co-rotating, counter-rotating). The "external geometry" (d, L) is already incorporated into the three dimensionless parameters (Figure 1.36).

$\dot{V}$	*Volumetric flow*
n	*Screw speed*
Δp	*Pressure differences*
η	*Viscosity*
P	*Drive power*
d	*Screw barrel internal diameter*
L	*Screw length*

$$\frac{\dot{V}}{n d^3} \qquad \frac{\Delta p d}{\eta n L} \qquad \frac{P}{\eta n^2 L d^2}$$

Figure 1.36 Reduction of dimension specific factors (7 → 3) according to J. Pawlowski [37], (for Newtonian media)

$$\frac{1}{A_1} \cdot \frac{\dot{V}}{nd^3} + \frac{1}{A_2} \cdot \frac{\Delta p d}{\eta n L} = 1$$

$$\frac{1}{B_1} \cdot \frac{\dot{V}}{nd^3} + \frac{1}{B_2} \cdot \frac{P}{\eta n^2 L d^2} = 1$$

Profile parameter: A_1, A_2, B_1, B_2:
Axial sections of straight-line equations

Figure 1.37 Linear pressure and power characteristic of filled screws

The four profile parameters are determined experimentally in a far less time-consuming process. The measurements for *co-rotating extruders* from [40] in Figure 1.38 and Figure 1.39 show that Pawlowski's similarity method is valid in terms of flow engineering [37] even for this complex screw type.

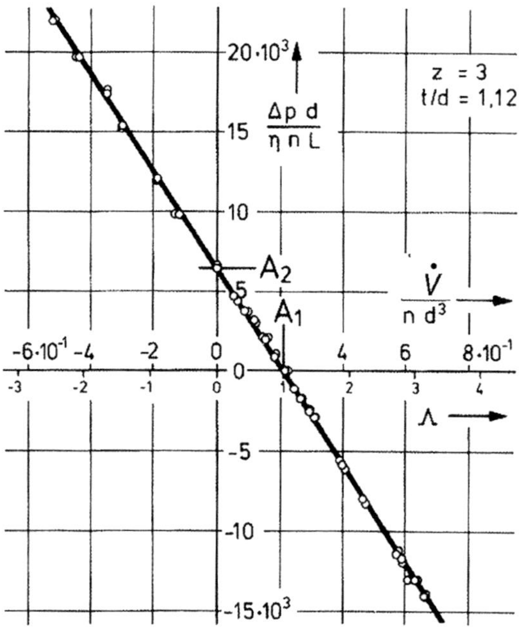

Figure 1.38 Pressure measurements on a co-rotating extruder

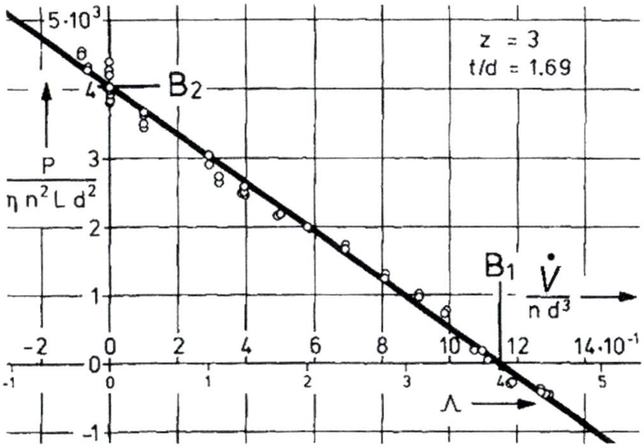

Figure 1.39 Power measurements on a co-rotating extruder

The same data, determined by not too complex means, can also be used to answer further questions, e. g. relating to the efficiency of the pressure build-up. Figure 1.40 illustrates the "product stress" (the reciprocal value of the pump efficiency) comparing counter-rotating and co-rotating extruders [40], i. e., two substantially different machines.

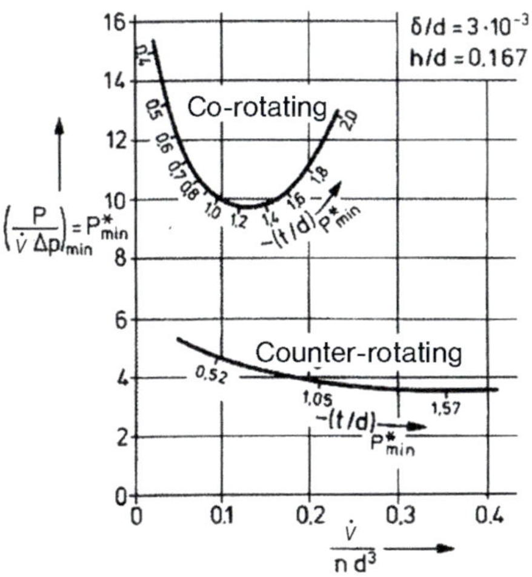

Figure 1.40
Energetic optimization, i. e., "minimal stress" for co-rotating and counter-rotating screws

1.2.2.3.3 Versatile High-Viscosity Processes

The high-viscosity processes developed during and after the pioneering period and the applications discovered for high-viscosity machines [8] are as striking as they are varied.

Table 1.1 Applications of High-Viscosity Machines

Process-orientated tasks	Product groups
Mixing, compounding, kneading, pasting, dissolving	Plastics, dyes, fibers, silicon polymer
Expelling dewatering	Rubbers
Evaporating, releasing, residual degassing	Plastics, rubbers, adhesives
Polymerization, polycondensation, polyaddition, and other reactions	Thermoplastic polyurethanes, silicone polymers
Crystallization, melting	Pesticides, organic matter
Extruding, forming, pelletizing	Plastics, rubbers

The process engineering tasks are classified according to the product groups for which they are used. There are four main areas: material mixing, material separation, reaction, and phase transformation. The common factor of all the processes in Table 1.1 is of course the high-viscosity phase and consequently the specific "difficulties" of the machinery (technical design, dimensioning with respect to forces, torques, drive power, pressures).

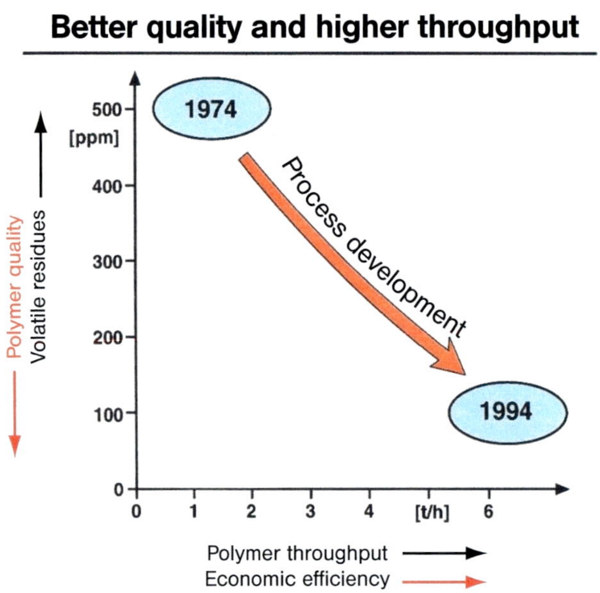

Figure 1.41
Goals of process optimization

Along with the work within the parameters of the basic operations, the process engineer has to investigate important processes in-depth to develop and optimize those processes (Figure 1.41) [8]. This involves meeting the constantly increasing market demand for higher quality while maintaining economic viability, in other words increasing throughput. From a physical point of view, however, the residual content of volatile materials increases dramatically as the throughput increases. Therefore, specific process engineering measures are required to reach the goal of increasing throughput by a factor of six while reducing the residual content by a factor of five (Figure 1.41).

1.2.2.4 Special Developments from Bayer-Hochviskostechnik (High Viscosity Technology Group)

1.2.2.4.1 Extended Kinematics, Profile Geometries

During the pioneering time of co-rotating extruders, predominantly graphical methods were applied for the basic geometry illustrated in Section 1.2.2.1.1 and the technical design of new machines, but the advances in automated computing technology soon proved to be ideal for the design of co-rotating extruders. Here, the internal geometry of the screw (Figure 1.42) is determined by six independent geometrical factors:

1. Internal barrel diameter d
2. Centerline distance a
3. Tip-wall clearance δ
4. Screw-screw clearance S
5. Screw pitch t
6. Number of flights Z

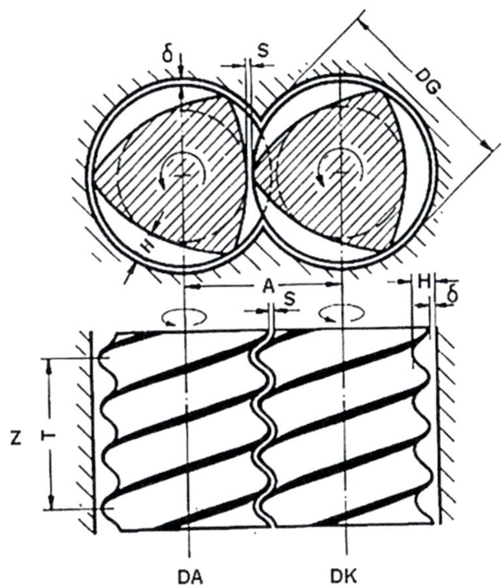

Figure 1.42 Internal geometry of the co-rotating extruder with dimensioning

The clearances δ and S are not required for the fully-wiping screw (see Section 1.2.2.4.2). This leaves four independent factors. Other six-fold combinations of independent geometrical values can also be used.

The application of the basic geometry by computer for a screw described in this way (see Section 1.2.2.1.1) results in the calculation of the profile curves in cross-section and in longitudinal section and, for example output for manufacture. Process engineering target values, such as the usable product volume and all surfaces, are also determined.

1.2.2.4.2 Clearance Strategies

"No screw without clearances": there are five reasons for this motto:

- The natural surface roughness of the screw and barrel material with clearances $S = 0$ and $\delta = 0$ leads to "cold welding" and therefore to erosion and blockages. This is unacceptable and clearances are therefore required.
- All production methods involve inherent production tolerances that must be ensured by the application of respectively larger clearances.
- The quality of the axial bearings of the two shafts in the duo-gear as well as that of the rotation angle allocation leads to tolerances and clearances in the screw profile.
- Occasional irregularities in the heating process or torsional stress on the two shafts can lead to mutual shifting and thus require clearances.
- Process engineering aspects can also require particular clearances if, for example, intensive shearing in the tip zones affects product quality.

These different reasons for the need for clearances led to different strategies in their implementation:

The clearances δ (tip–wall) and S (tip–base of neighboring screw) should be set as trivial because they are radially effective.

- In the pioneering period of the co-rotating extruder, technical designs corresponded exactly with the basic geometry, with the result that the two shafts tended to move apart somewhat in the actual machines *("centerline distance enlargement")*. Ideally S was obtained at the base of the screw while at the outside of the flank, particularly with small pitches, varying flank clearance occurred, sometimes much smaller than the required S (with a consequential danger of collision).
- A considerably better clearance strategy was realized graphically or, as described in Section 1.2.2.4.1, the *"planar offset"*. In the longitudinal section profile curve of the fully wiping screw ($S = 0$) calculations are carried out based on a $S/2$ orthogonally-shifted longitudinal section profile curve (Figure 1.43) and are retroceded to the cross-section profile. The disadvantages of method 1: "centerline distance enlargement" are largely reduced by the clearance strategy 2: "planar offset".

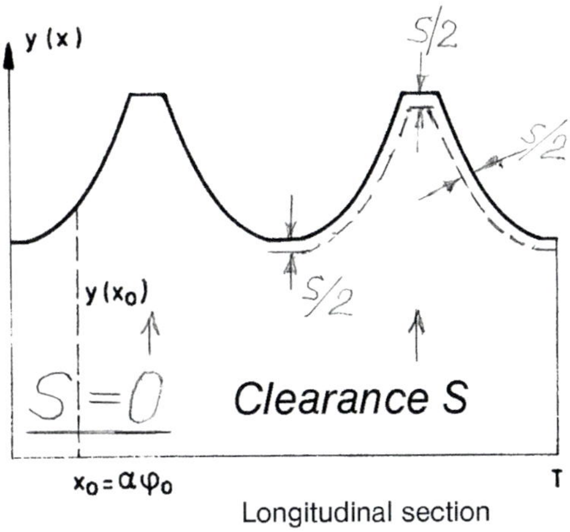

Figure 1.43 Planar offset

- The corresponding screws cut against one another through a spatial curve. Therefore, the effective clearance orientated to the spatial curve is relevant as it is in view of the self-wiping effect and the product stress. The result is the "*spatial offset*" with the aforementioned advantages.

The clearance strategies are compared in Table 1.2.

Table 1.2 Evaluation of the clearance strategies

Clearance Strategy	Grade	Flank effect
Fully wiping screw $\delta = 0$, $S = 0$	1	"Eroding"
"Centerline distance enlargement" $S \neq$ constant	2	Danger of collision
"Planar offset" $S \approx$ constant	3	Good compromise
"Spatial offset" $S \equiv$ constant	4	Perfect solution

The clearance considerations above apply for constant screw geometry along the axis and particularly for a constant pitch. Modular screws with elements with different pitches create the following dilemma:

With constant flank clearances along the entire machine, the contact points of different element pitches exhibit slight differences in their cross-section contours. On the other hand, applying congruent cross-section contours and therefore identical element contact points to avoid this results in varying flank clearances.

See Chapter 4 and Section 1.4 for an overview of computational programs for process screws. See [41] for details about flow simulation using "CFD".

1.2.2.5 Developments after Licensing

After the license was awarded to Werner & Pfleiderer (Section 1.2.2.2.3) and after an initial warm-up phase, machinery and process engineering underwent dramatic development. This took place during the years of industrial development and the simultaneous advances in the areas of polymer chemistry and plastics engineering that were taking place at the same time (e. g., compounding technology). Five major developments characterize this boom:

1. The increase in production quantity of co-rotating extruders; by the beginning of the 1960s, Werner und Pfleiderer had delivered the 100th ZSK 83 model of this machine.
2. The modular technology was developed both for the internal screw geometry (Figure 1.44) and for whole machines (Figure 1.45).
3. The trend towards large machines (Figure 1.46) began soon after, leading to today's ZSK 380 "mega-compounder" with a throughput of up to 50 t/h.
4. Modified "light duty" screws were designed for special chemical applications (for example see Figure 1.47).
5. Approximately 5000 co-rotating extruders of all types and sizes have been produced so far by Werner & Pfleiderer.

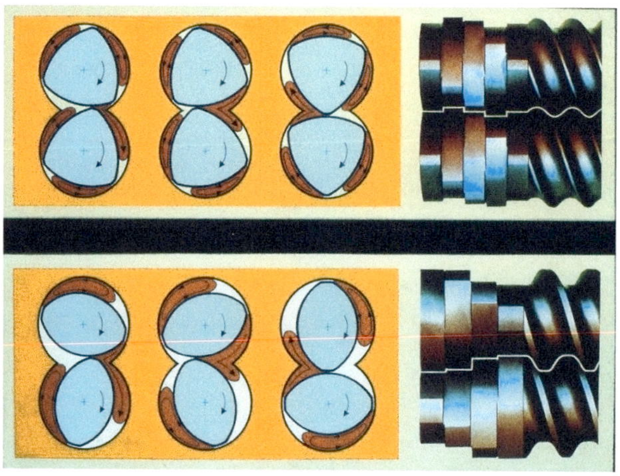

Figure 1.44 Modular technology in screws (Courtesy: Werner & Pfleiderer)

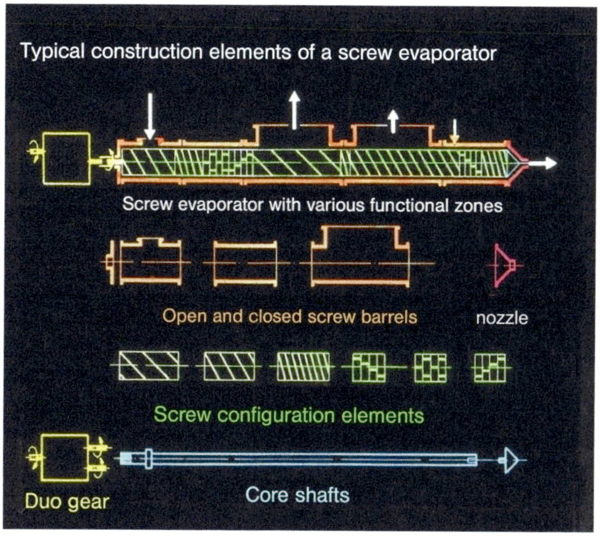

Figure 1.45 Overall machine comprising typical modular components

Figure 1.46 Large co-rotating extruder: Megacompounder ZSK 340 (Courtesy: Werner & Pfleiderer)

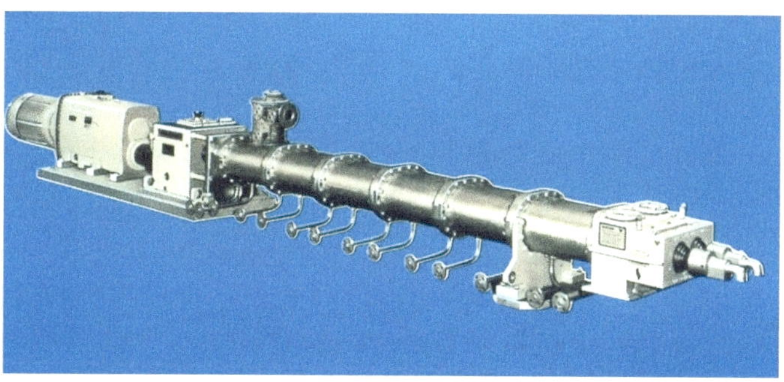

Figure 1.47 Light duty co-rotating extruder: ZDSW 160 (Courtesy: Werner & Pfleiderer)

It took a well-established mechanical engineering company to turn this development effort into a successful engineering and commercial enterprise. Even the individual areas of research and development, testing, project planning, technical design, production, assembly, commissioning, sales, and customer service now required teams with management and technical expertise, while during the pioneering period a small group developed in this field (Section 1.2.2.2).

Another feature of the post-licensing period is the increase in published university research and publications from mechanical and chemical engineering companies. This includes H. Herrmann's standard work of 1972 [36], referred to in Section 1.2.2.3.1, along with other selected examples [42, 43].

1.2.2.6 Developments after Expiration of the Primary Patents

Once the co-rotating extruder had become the standard element for plastics conditioning and processing, compounding in other words, and the patents referred to in Section 1.2.2.1.2 had expired (approximately 1970), the co-rotating principle was adopted by numerous mechanical engineering companies. Today, approx. 50 companies around the world can supply small- or medium-sized (approx. < 120 mm) screws. However, just five are able to supply large screws of 120 mm to 380 mm (estimated values).

The "patent vacuum" has also prompted new applications from several areas. Apparatus patents are among the foremost of these, i. e., numerous variants of innovative screw, mixing, and kneading elements from mechanical engineering companies, while process patents employing co-rotating extruders are proposed by the chemical industry.

We are also witnessing a considerable increase in new specialized literature. The most important of these is the standard work "Twin Screw Extrusion" by J. L. White [44]. It is a rich source of information, both about the historical development and the state-of-the-art technology, as well as about patent literature and specialized publications. A few examples [45, 46] from the wide-ranging specialized literature on the subject of co-rotating extruders should be mentioned here.

Plastics technology and co-rotating extruders have undergone dramatic developments which, to a certain extent, have been synchronous and mutually supportive. Plastics technology has profited from the increasing potential of the co-rotating extruder, while polymer technology has itself been a hothouse for advances in co-rotating extruders. We can therefore conclude: There is hardly a kilogram of plastic today that has not encountered a co-rotating extruder somewhere on its journey between polymerization and processing.

Figure 1.48 shows the dynamic development of the co-rotating extruders with the increase in volumetric throughput over time. It also includes the milestones in the development and process engineering operations and their respective starting points.

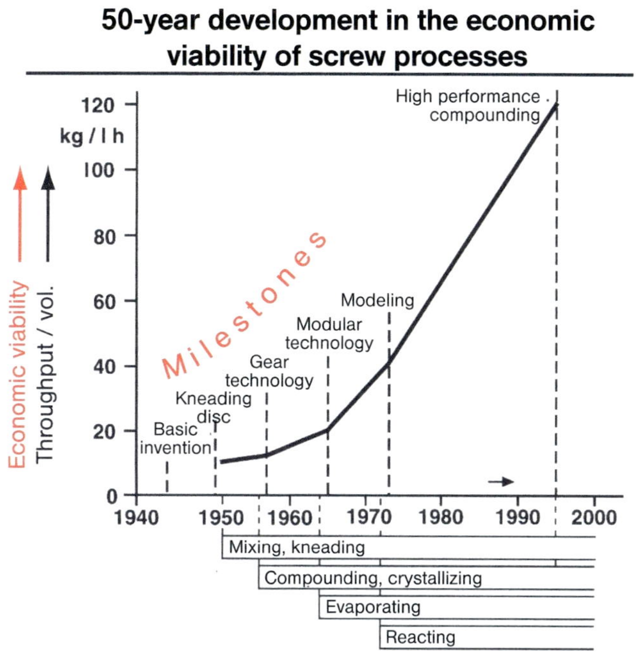

Figure 1.48 Dynamic development of the co-rotating extruders

References for Section 1.2.2

[1] Personen Lexikon 1 (A–L), Chronik-Verlag, p. 76

[2] Coignet, F., US Patent 93035 (1869)

[3] Wünsche, A., German Patent 131392 (1901)

[4] Easton, R. W., GB Patent 109663 (1916)

[5] Nelson, W. K., US Patent 1868671 (1931)

[6] Pease, F. F., US Patent 2048286 (1933)

[7] Colombo, R., Italian Patent 370578 (1939)

[8] Wagner, W., 50 Jahre Verfahrenstechnik im Bayer-Konzern 1948–1998 – Rückschau und Ausblick. Bayer-Jubiläumsschrift der ZT-Technische Entwicklung

[9] Erdmenger, R., Schneckenmaschinen für die Hochviskostechnik, Bayer AG (1978)

[10] Meskat, W., Geberg, A., Erdmenger, R., IG internal reports (1944)

[11] Erdmenger, R., Neidhardt, S., Bayer internal reports (1948)

[12] Erdmenger, R., Bayer internal reports (1948/51)

[13] Meskat, W., Erdmenger, R., DBP 862668 Gleichdrall-Doppelschnecke mit Dichtprofil. Basic patent for the threaded screw. Priority: 1944 (transferred to West Germany: 1953). Extracts from this key patent are included in the text and its importance emphasized

[14] Meskat, W., Erdmenger, R., DBP 872732, Gleichdrall-Dreifachschnecke mit Dichtprofil zum Abpressen. Priority: 1944 (transferred to West Germany: 1953)

[15] Erdmenger, R., DBP 815641, Gleichdrall-Doppelschnecke mit Sperrbarrieren zum Mischen (1949)

[16] Erdmenger, R., DBP 813154, Gleichdrall-Knetscheiben mit Dichtprofil, eingängig und zweigängig sowie Sonderfall Kreis-Exzenter. Basic patent for the kneading element (1949)

[17] Erdmenger, R., USP 2670188 as above: DBP 813154, however, without double-flighted kneading element (1950)

[18] Meskat, W., Pawlowski, J., DBP 949162 Gleichdrall-Doppelschnecke mit Dichtprofil und "Bremsgewinde" (rechts-links-Kombination) zum Mischen (1950)

[19] Erdmenger, R., Neidhardt, S., DBP 915689 Gleichdrall-Doppelschnecke mit Dichtprofil und "Rückwärtsentgasung" über "Seitenschnecke" (1951)

[20] Erdmenger, R., DBP 940109, (1953) = USP 2814472, (1954), Gleichdrall-Knetscheiben mit Dichtprofil, dreigängig, Knetelemente, Erweiterung

[21] Erdmenger, R., USP 3122356, Gleichdrall-Doppelschnecke mit Dichtprofil und modularer "Baukasten-Bauweise" (1959)

[22] Erdmenger, R., Oetke, W., DBP 1111154, Vierwellenschnecke mit Dichtprofil für Verdampfungsprozesse (1960)

[23] Erdmenger, R., USP 3254367, Gleichdrall-Doppelschnecke mit Dichtprofil und "Schulter" im Schneckenkanal (1963)

[24] Riess, K., Meskat, W., Schneckenmaschinen für die Verfahrenstechnik zähflüssiger und plastischer Massen, *Chem.-Ing.-Tech.* (1951), 23 (9/10), pp. 205–212

[25] Riess, K., Erdmenger, R., Der heutige Stand der Technik kontinuierlicher Press- und Knetverfahren in der chemischen und verwandten Industrie, *VDI-Z.* (1951), 95 (19/20), pp. 633–638

[26] Erdmenger, R., Stoffzerkleinernde Mischprozesse, *Chem.Ind.* (1953), pp. 513–518

[27] Riess, K., Probleme kontinuierlicher Verfahren in der Verbrauchsgüterindustrie, *Chem.-Ing.-Tech.* (1955), 27, pp. 457–464

[28] Erdmenger, R., Zur Entwicklung von Schneckenverdampfern, *Chem.-Ing.-Tech.* (1962), 34, pp. 751–754

[29] Herrmann, H., Entwicklungstendenzen bei Aufbereitungsmaschinen für die Kunststoff-industrie, *Chem. Ind.* (1962), pp. 603–606

[30] Erdmenger, R., Mehrwellenschnecken in der Verfahrenstechnik, *Chem.-Ing.-Tech.* (1964) 36 (3), pp. 175–185

[31] Herrmann, H., Die zweiwellige Knetscheiben-Schneckenpresse ZSK 160, *Kunststoff und Gummi* (1964), pp. 217–223

[32] Herrmann, H., Das Aufbereiten von Kunststoffen in Schneckenmaschinen, *Kunststoffe* (1965), pp. 845–849

[33] Herrmann, H., Mischungsverhältnisse und Selbstreinigung in einwelligen und zweiwelligen Schneckenmaschinen, *Chem.-Ing.-Tech.* (1966), pp. 25–29

[34] Erdmenger, R., Ullrich, M., Untersuchungen an einem Schneckenverdampfer, *Chem.-Ing.-Tech.* (1970), 42(1), pp. 1–5

[35] Erdmenger, R., Entgasen hochviskoser Flüssigkeiten, *Masch.markt* (1974), 80 (1, 2–6, and 10), pp. 148–152

[36] Herrmann, H., *Schneckenmaschinen in der Verfahrenstechnik*, Springer-Verlag, Berlin, Heidelberg, New York (1972)

[37] Pawlowski, J., *Die Ähnlichkeitstheorie in der physikalisch-technischen Forschung* Springer-Verlag, Berlin, Heidelberg, New York, (1971)

[38] Pawlowski, J., *Transportvorgänge in Einwellen-Schnecken* Verlag Salle u. Sauerländer, Frankfurt/Main (1990)

[39] Pawlowski, J., *Veränderliche Stoffgrößen in der Ähnlichkeitstheorie*, Verlag Salle u. Sauerländer, Frankfurt/Main (1991)

[40] Ullrich, M., two papers in VDI-K-proceedings: Optimierung des Compoundierprozesses durch Rezeptur- und Verfahrensverständnis, VDI-Verlag, Düsseldorf (1997)

[41] Kohlgrüber, K., Beck, D., two papers in VDI-K-proceedings: Optimierung des Compoundier-prozesses durch Rezeptur- und Verfahrensverständnis, VDI-Verlag, Düsseldorf (1997)

[42] Werner, H., Das Betriebsverhalten der zweiwelligen Knetscheibenpresse vom Typ ZSK bei der Verarbeitung von hochviskosen Flüssigkeiten, Technical University Munich, Dissertation (1976)

[43] Herrmann, H., Burkhardt, U., Vergleichende Funktionsanalyse dicht kämmender Gleichdrall- und Gegendrall-Doppelschnecken, *Kunststoffe* (1978), 68 (1), pp. 19–26

[44] White, J. L., *Twin Screw Extrusion*, Hanser, Munich (1990)

[45] VDI-K conference proceedings: Optimierung des Compoundierprozesses durch Rezeptur- und Verfahrensverständnis, VDI-Verlag, Düsseldorf (1997)

[46] VDI-K conference proceedings: Der Doppelwellenextruder, Grundlagen und Anwendungen, VDI-Verlag, Düsseldorf (1998)

1.3 General Overview of the Compounding Process: Tasks, Selected Applications, and Process Zones

Reiner Rudolf

1.3.1 Compounding Tasks and Requirements

The properties of polymers need constant refinement to meet the constantly increasing demands on plastic parts. Because the development of new polymers is steadily decreasing [1], deliberate modification of the properties of a base polymer by means of additives and/or blending with other polymers is becoming increasingly important. This is the main task of compounding.

In Europe, approximately 7 million tons of compounds and 1 million tons of masterbatches were consumed in 2010. The European compounding and masterbatch market is shown in Figure 1.49.

The requirements of the compounding process are characterized by

- Increasing throughputs
- Decreasing batch sizes
- Increasing product variety
- Increasing quality requirements,
- Increasing filler contents
- New products (e. g., nano-filler materials, modified and blended polymers)

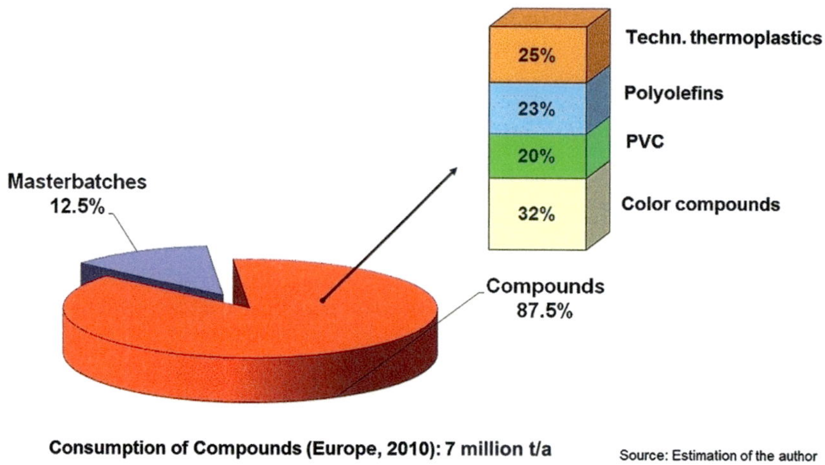

Figure 1.49 Compound and masterbatch market in Europe 2010

Today, compounding is predominantly carried out with co-rotating twin-screw extruders with constantly increasing available drive powers, torques, and screw speeds.

In particular, increasing throughput means increasing the screw speed, which in turn leads to

- Higher thermal and mechanical polymer stress ($q = \eta \cdot \dot{\gamma}^2$; $\tau = \eta \cdot \dot{\gamma}$)
- Shorter available processing times $\left(t_V = f\left(n, \dot{M}\right)\right)$

This increases the risk of loss of product quality due to a decrease in the molar mass, thermal inhomogeneity, and unmelted recipe components. Hence, the extruder and screw concepts need to be constantly optimized and adapted to avoid quality loss.

Typical compounding tasks include

- Reinforcing polymers
  - Incorporation of fibers (glass, carbon, or natural fibers)
- Modification of thermoplastics to achieve high-impact resistance
  - Blending with rubber elastic components
- Improving the dimensional stability and compression strength of polymers
  - Incorporation of inorganic fillers, glass beads
- Improvement of flow, mold release, and flame-retardant behavior of polymers
  - Incorporation of low-viscosity substances
- Production of polymer blends
  - Mixing compatible or incompatible polymer melts
- Coloring polymers
  - Incorporation of pigments, masterbatches, or liquid dyes

- Improvement of chemical/physical stability of polymers
  - Incorporation of low-viscosity stabilizers, antistatic agents.

The consumption of mineral fillers in Western Europe for thermoplastics processing alone was 1.3 million tons in 2015 [2]; a breakdown is shown in Figure 1.50.

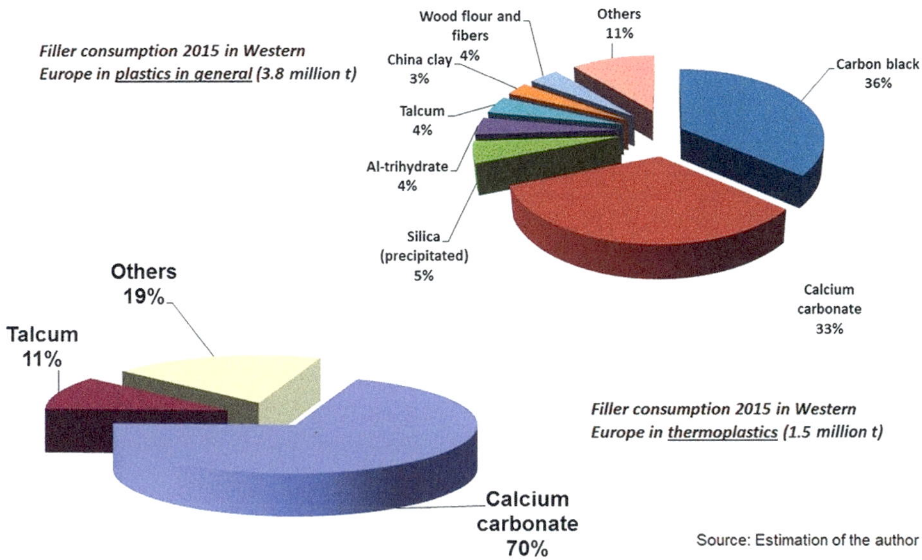

Figure 1.50 Consumption of mineral fillers for thermoplastic processing in Western Europe 2015

1.3.2 Tasks and Design of the Processing Zones of a Compounding Extruder

The various processing zones of a twin-screw extruder are arranged in series, as illustrated in Figure 1.51. Each processing step is linked to the next. Therefore, the different steps cannot be considered independently from each other. If, for example, the speed or throughput is changed, this inevitably affects all process zones. As will be explained below, this can be advantageous for one process zone, but it can have negative effects for another.

Furthermore, the process zones cannot be strictly separated. Dispersion processes, for example, already take place in the plasticizing zone; the incorporation of fibers dosed into the melt takes place not only in the designated dispersing zone but also in the discharge zone and in partially-filled screw channels.

Analyzing the unit operations in the extruder to a satisfactory degree is only possible when sufficient information about the rheological and thermodynamic properties of the polymers and polymer compounds is available. However, providing these data is very difficult, particularly in zones with considerable changes in state. Therefore, it is frequently necessary to refer back to model investigations and trials [3].

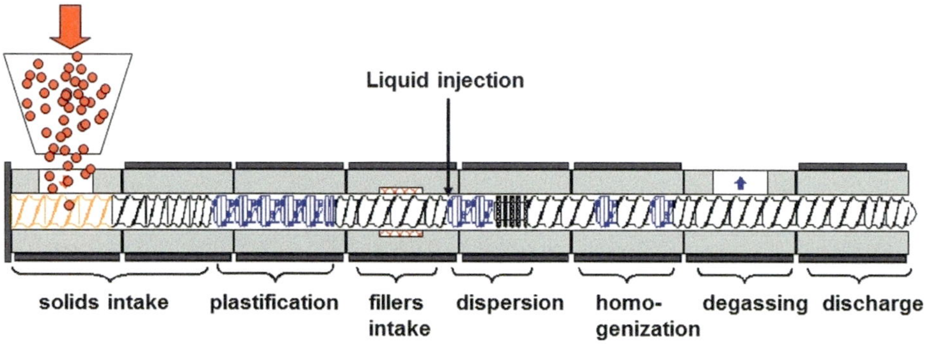

Figure 1.51 Processing zones of a twin-screw extruder

Figure 1.52 compares the independent setting parameters of the compounding process with the process values. Other than with a single-screw extruder, the throughput rate can be chosen independently of screw speed.

The process parameters in the extruder are variable in axial as well as radial direction. As a rule, however, only integral values along the entire extruder can be measured, e.g. the melt temperature in the emerging melt strand.

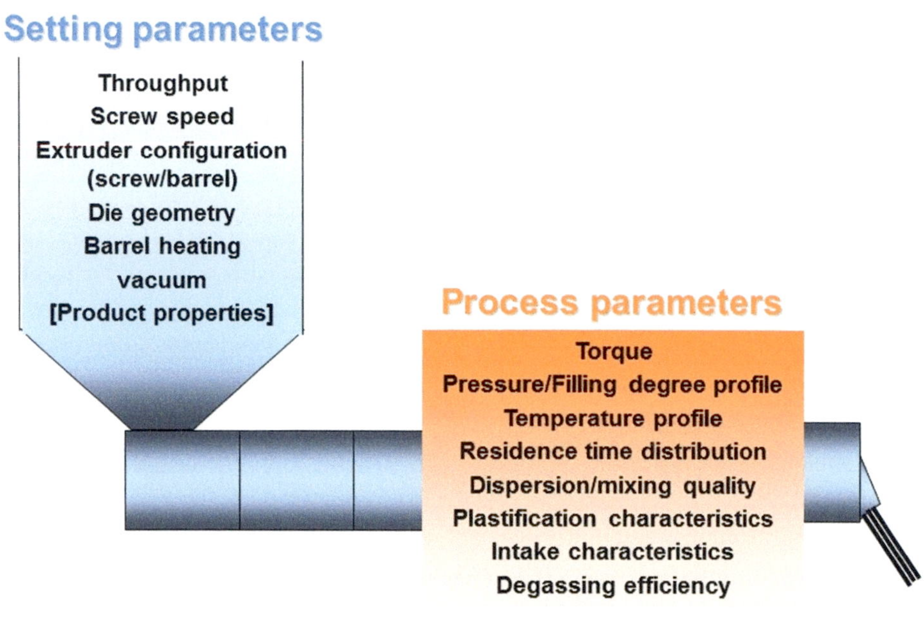

Figure 1.52 Parameters of the compounding process

Depending on the product to be processed and its requirements, the throughput of a compounding process can be limited by the available volume, torque, and/or the maximum tolerable product temperature.

In the following, we will describe each processing zone in a co-rotating twin-screw extruder.

1.3.2.1 Intake Zone

Tasks of the intake zone are:

- Convey and compress solids
- Remove any air that is drawn in

Key setting parameters are:

- Screw speed
- Throughput
- Free screw volume
- Pitch
- Bulk density of the product
- Friction between product and barrel wall

All intake processes are limited by volume [4].

The feed capacity depends on the free screw volume (channel depth, number of threads), the pitch, the screw speed, and the bulk density of the product. The product can only be conveyed if there is friction between the product and the cylinder wall.

Self-wiping screw elements with the maximum possible number of threads give the largest free cross-section (see Figure 1.19 in Section 1.2.2.1.1). Thus, triple-flighted elements should be used in the intake zone of twin-screw extruders with $D_o/D_i < 1.3$, and double-flighted elements in the case of $D_o/D_i > 1.3$. An intake zone length of 4 to 6 D is generally adequate for effective product feed.

The pitch should be 1.5 D directly below the feed hopper. A reduction of the pitch in the feed zone should be avoided to avoid accumulation of the product, as this can lead to a reduction in throughput. At the end of the intake zone, the compression of the solid can cause local pressures of up to 200 bar which, depending on the polymer, may lead to a greater or lesser degree of wear. It is therefore recommended that the elements at the end of the intake zone and at the beginning of the plastification zone are wear protected or have a reduced diameter.

An increase in the free cross-section and hence a 10 to 15% increase in the resulting maximum feed capacity is possible by using (non-self-wiping) pushing flight elements.

When conveying solids with a low bulk density, high fill rates in the intake zone can cause fluidization, if the air that is drawn in cannot escape from the extruder. This can be avoided by a partially filled downstream plastification zone (e.g., with only conveying kneading blocks and no re-conveying elements) and a vent opening installed downstream of the plastification zone.

The product is heated via the barrel wall and heating depends on the available heat exchanger surface and the residence time in the intake zone. Because the surface/volume ratio decreases as the screw diameter increases, product preheating in the intake zone is not relevant in large machines. Effective preheating prior to melting is therefore only possible in large machines before the product is fed into the extruder [5].

The intake barrel is cooled to prevent low melting point components, e.g. waxes, from adhering to the barrel wall or to the feed hopper, because this would severely affect the intake characteristics.

An exact prediction of the conveying behavior or the conveying capacity in the intake zone is currently only possible by means of trials. The knowledge of material characteristics such as bulk density, flowability, and particle size distribution also do not allow a clear conclusion on the feeding behavior. However, conveying of simple, granular materials can already be simulated very well today. It is expected that conveying behavior of mixtures of granules and powders, as they often occur in compounding, can also be sufficiently simulated in the future.

1.3.2.2 Plastification Zone

Tasks of the intake zone are:

- Melting of polymers
- Pre-dispersion of fillers

Key setting parameters are:

- Screw speed
- Throughput
- Screw geometry
- Heat flux
- Specific heat capacity
- Heat conduction
- Melting enthalpy
- Particle size

Comprehensive investigations of the polymer melting process in co-rotating twin-screw extruders by Bastian/Gabor [6] yielded the following results:

- The melting process is influenced to a high degree by screw speed and throughput; in other words, the degree of filling and the residence time

- The required melting length depends substantially on the melting enthalpy of the polymer, the viscosity of the melt, and the screw geometry
- Significant melting starts at the point, where the screw is initially fully filled
- Varying the barrel temperature in the plastification zone makes hardly any difference to the melting process.

Melting can be accelerated by [6]:

- A high degree of filling
- Low melting temperatures
- Small pellet sizes
- A low degree of crystallinity
- Low melting enthalpy
- High melting viscosities
- Low flow activation energies (= temperature influence on viscosity variation)
- Narrow kneading discs

In addition, for polymer blends, the ratio of melt viscosities of the blend partners $r = \eta_{dispers} / \eta_{contin}$ has a major influence on the melting process. In this case, $\eta_{dispers}$ is the viscosity of the blend partner present in the dispersed phase and η_{contin} is the viscosity of the blend partner present in the continuous phase. The smaller the value of *r*, the sooner the melting process begins, but also the longer it takes [6]. In unfavorable process conditions, as yet unplasticized particles may be encountered in the compounded pellets (see Figure 1.53).

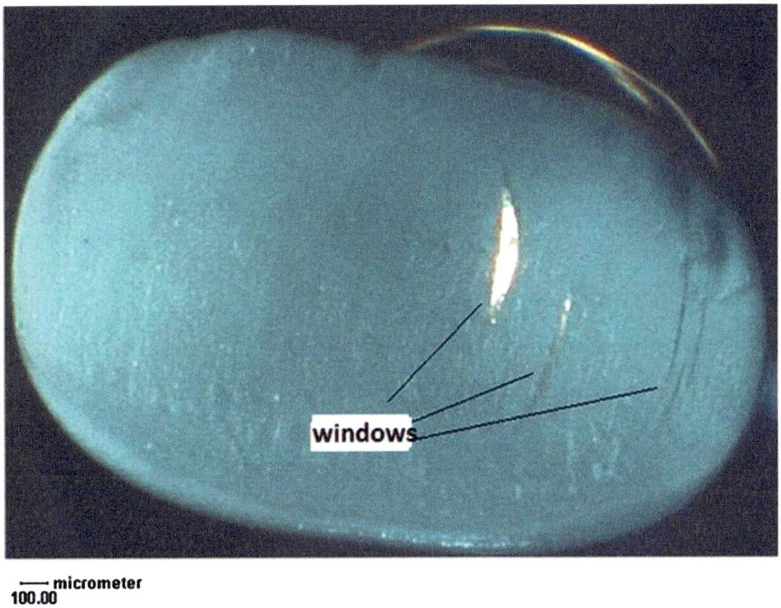

Figure 1.53 Unplasticized particles (so-called windows) within a pellet

The addition of low-viscosity additives can impede the formation of the melt film that is essential for melting. On the other hand, high-viscosity or solid additives can cause increased energy dissipation due to internal and external friction, which accelerates the formation of the melt film but can also lead to overheating of the viscous phase.

Along with the actual melting, part of the key tasks of dispersion and homogenization also takes place in the plastification zone.

Depending on the subsequent process steps, achieving 100% melting at the end of the plastification zone may not be mandatory. Solid particles still present after the plastification zone can be melted in subsequent zones, e.g. in the dispersion or discharge zones. Figure 1.54 illustrates a typical melting profile along a plastification zone.

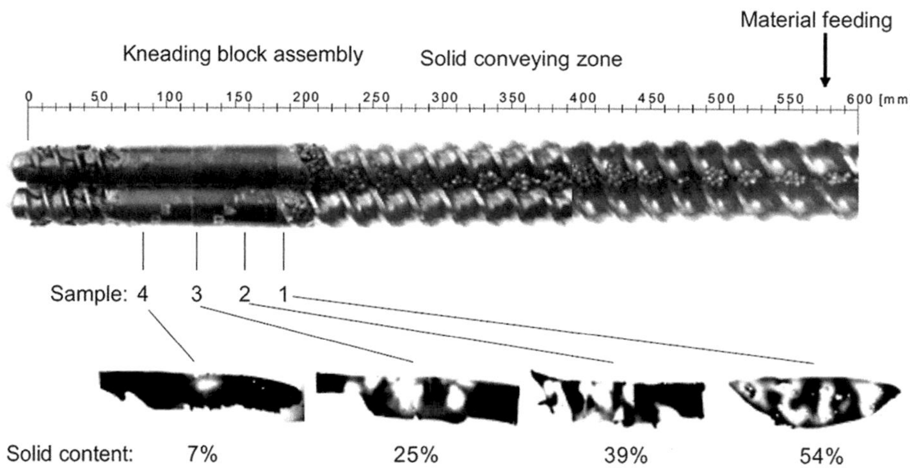

Figure 1.54 Melting profile along a plastification zone [7]

The energy required to melt the polymer components is mainly transmitted via the screw shafts; the heat flow through the barrel wall serves only to form a melt film on the wall [3]. This melt film is important to create an adhesion of the polymer on the wall, thus generating a shear gradient [8]. To achieve this, the temperature of the barrel wall must be higher than the softening point of the polymer.

Up to 80% of the overall mechanical energy input in the twin-screw extruder takes place in the plastification zone. Unfortunately, the calculation of the melting process is currently only rudimentary.

Typically, double- or triple-flighted kneading blocks are used to facilitate melting. The influence of the kneading block geometry on the process parameters is shown using a double-flighted kneading block (see Figure 1.55).

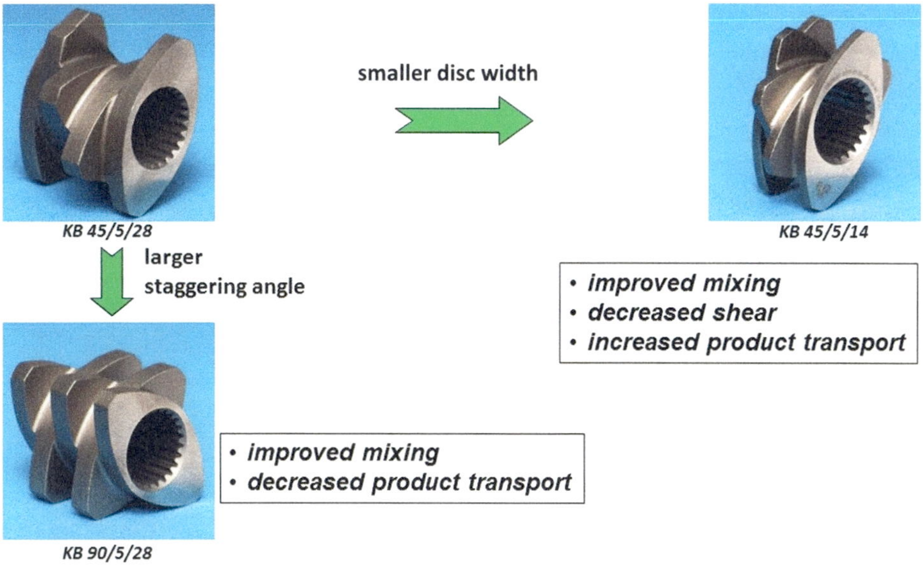

Figure 1.55 Geometric values and process variables of a double-flighted kneading block

A larger staggering angle leads to:

- A reduction of the downstream conveying action of the melt and thus to
- An improved mixing performance because more leakage streams are generated and the residence time increases.

A narrower kneading disc results in:

- Reduced shear in the kneading gap,
- Increased downstream conveying action of the melt, and
- More efficient mixing due to the generation of more leakage streams (see Figure 1.56).

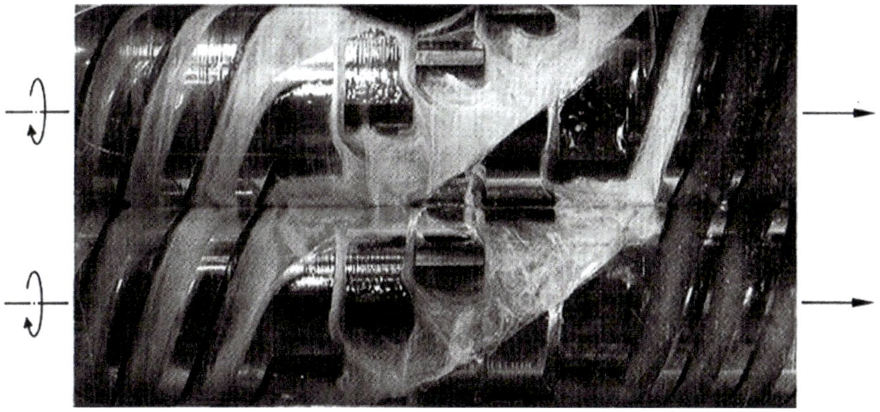

Figure 1.56 Melt streams in a kneading block zone: broad main stream in the feed direction and narrow leakage flow [9]

Melting with triple-flighted kneading blocks results in less energy dissipation compared to double-flighted kneading blocks. However, the melting length required is longer [6].

Melting is accelerated in completely filled zones. It is therefore beneficial to install a re-conveying element at the end of the plastification zone to cause back-pressure in the melt. Kneading blocks (re-conveying or neutral), re-conveying screw elements, or slightly re-conveying mixing elements are generally used for this purpose. However, re-conveying screw elements can cause high pressure and temperature peaks and should be avoided as far as possible (see Chapter 4). For complex melting operations, e.g. with high/low-viscosity multicomponent mixtures, sieve discs can also be used to create back-pressure, provided that the higher energy input generated by these discs is tolerable.

If the feedstock polymer is a fine powder, or if it is necessary to incorporate low bulk-density fillers in large quantities, the plastification zone should be operated partially filled to prevent fluidization in the intake zone. Partial filling can be achieved when the plastification zone only consists of forward pumping kneading blocks and there are no re-conveying elements at the end of the plastification zone. In this case, however, the melting length must be considerably longer.

Melting in a fast-running, high-performance extruder can result in temperature peaks if conventional kneading blocks are used. These peaks can damage the polymer. To prevent these temperature peaks, less closely intermeshing screw elements have been developed. They reduce the shear peaks to provide a gentler compounding action. Examples include Multi-Process Elements (MPE) (Berstorff), triple-flighted eccentric kneading blocks with reduced external diameters, and shoulder-type kneading blocks (e.g., Extricom, Coperion). Scheel has shown, for example, that it is possible to achieve melting of polyamides with an MPE plastification zone with 10% lower energy input than with a conventional kneading block plastification zone (see Figure 1.57) [10]. This means a reduced melt temperature with the same throughput and/or a higher throughput at the same melt temperature.

The energy input via the screw elements and kneading discs results in splaying forces, particularly in the plastification zone, which can cause wear to the cylinders (see Figure 1.58). This can be reduced by an appropriate design of the plastification zone and by heating the cylinder wall, which results in the formation of a melt film [5].

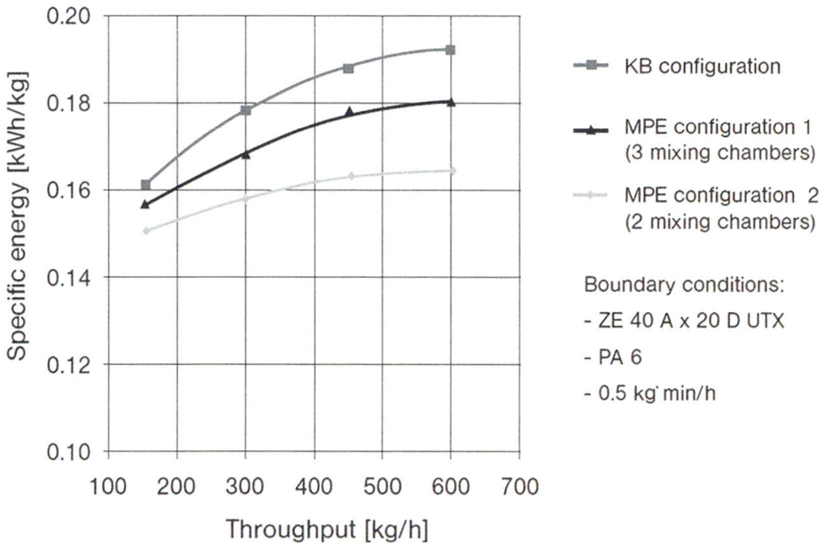

Figure 1.57 Energy input as a function of throughput for polyamide melting with kneading block and MPE plastification zones [10]

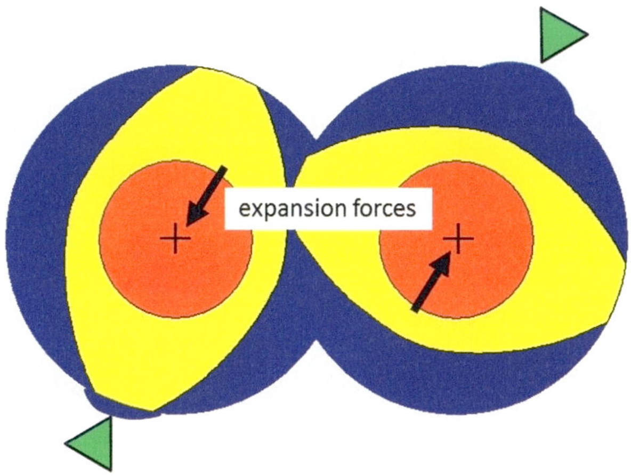

Figure 1.58 Cylinder wear caused by expansion forces [5]

1.3.2.3 Melt Conveying Zone

Tasks of the melt conveying zone are:

- Convey melt

Key setting parameters are:

- Screw speed
- Throughput

- Pitch
- Material properties

Melt conveying zones are generally partially filled, except for the back-pressure zones upstream of pressure consuming elements.

Figure 1.59 shows an example of the filling degree as a function of the pitch at two different screw speeds (400 rpm and 600 rpm) and a constant throughput of 2500 kg/h for a twin-screw extruder ZSK 92 Megacompounder with 92 mm outer screw diameter.

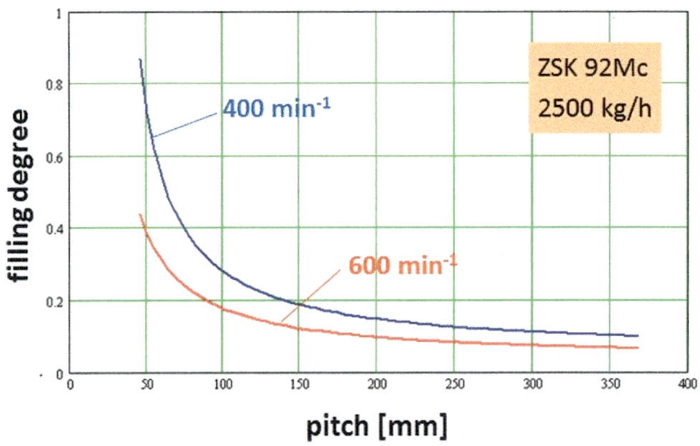

Figure 1.59 Filling degree as function of pitch and screw speed

The melt conveying zones serve to transport the product from one process zone to the next and to feed fillers. Leakage flows over the screw tip and in the intermeshing zone also cause some slight back-mixing. As little energy as possible should be dissipated in the melt conveying zones.

For simple melt conveying, screw elements with a pitch of 1 D are commonly used. If other components are to be added (e.g., via a side feeder), the pitch should be 1.5 D to increase the feed capacity.

As will be shown in subsequent chapters, melt conveying can be calculated very well provided that sufficient material data is available.

1.3.2.4 Distributive Mixing Zone

Tasks of the distributive mixing zone are:

- Distribution of solids or fluids in the melt
- Homogenization of the melt temperature

Key setting parameters are:

- Screw speed
- Throughput
- Screw geometry

The aim of the mixing process is to achieve homogeneous distribution of solids or liquids within the melt with the shortest mixing lengths and minimum energy input.

Due to the laminar flow of polymer melts in an extruder, mixing is entirely the result of kinematic distributing processes related to the extruder operating characteristics. This can be described in terms of the ratio of drag and pressure flows. The actual homogenization process (distribution process) depends entirely on the intensity and direction of these flow portions and not on the viscosity of the melt.

The mixing performance depends significantly on the dimensionless kinematic parameter of flow $\Lambda = \frac{1}{A_1} \cdot \frac{\dot{V}}{n \cdot D^3}$ (Figure 1.60). A_1 indicates the maximum throughput of the screw element per revolution that the machine can achieve when it is completely filled and operated without back-pressure (known as inherent throughput, see Section 4.1 for a more detailed description of the dimensionless representation). $\dot{V}$ represents the volumetric flow rate, n is the screw speed, and D the barrel diameter. Figure 1.60 shows the mixing quality as a function of Λ for silicone oil with Newtonian viscosity behavior into which iron powder has been mixed [11]. In principle, this behavior also applies to shear thinning materials.

The higher the screw speed and/or the lower the throughput, the better is the mixing performance. The figure also shows that re-conveying elements ($\Lambda < 0$) provide significantly better mixing performance than conveying elements ($\Lambda > 0$).

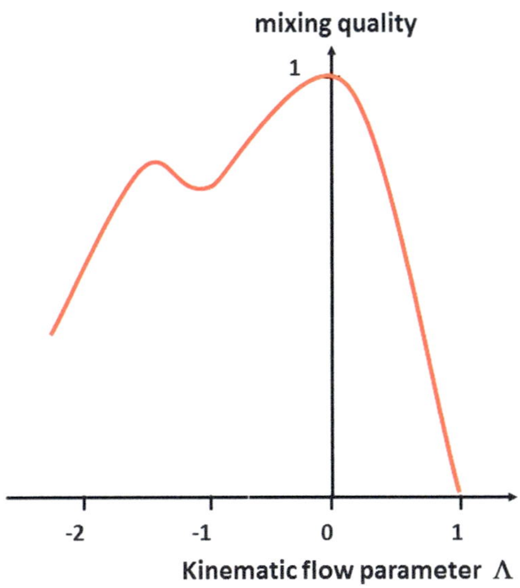

Figure 1.60 Influence of the kinematic parameter of flow Λ on the mixing quality [11]

Mixing is improved when the melt streams are subjected to displacement in addition to shear deformation. Special mixing elements have been designed for this purpose, e.g. toothed blocks or toothed mixing elements. Kneading blocks with narrow discs also provide a better mixing performance than screw elements (advantages compared to broad discs: better mixing performance because the greater number of discs per length results in more melt stream splitting; less shear due to less wedge flow between screw flank and barrel see also Figure 1.56). Mixing/homogenization ideally takes place in melt-filled zones.

If a large quantity of a low-density powdered component is to be incorporated, the air that is drawn in with the powder must be removed to prevent fluidization. In principle, there are three methods to achieve this:

1. Partially filled mixing zone with conveying kneading blocks so that the air can escape in the direction of flow through a downstream vent opening.
2. Feeding of the powder at two or more positions in the extruder (e.g., one at the first intake and the rest via a side feeder).
3. Use of pre-compacted powder to reduce the amount of air drawn in.

1.3.2.5 Dispersive Mixing Zone

Tasks of the dispersive mixing zone are:

- Breaking down solids, polymer particles, and fluid droplets

Key setting parameters are:

- Screw speed
- Throughput
- Screw geometry
- Viscosity

Like the degree of distribution, the degree of dispersion is dependent on the kinematic parameter of flow Λ. However, along with the axial and radial velocities in the screw channel, the dispersion process is also dependent on viscosity and is therefore affected by the shear stresses in the screw channel (Figure 1.61). The figure shows the dispersion degree of TiO_2 agglomerates as a function of Λ for silicone oils of differing viscosities for an isothermally-operated single-screw extruder [12]. In principle, the same applies to twin-screw extruders.

The dispersion effect is dependent on the shear stress as well as on the duration of the load. Which of the two parameters has the biggest influence depends on the material to be dispersed [13]. On the basis of tests on a single-screw extruder, Martin has demonstrated that the dispersion of carbon black agglomerates is only possible by exceeding a minimum shear stress and a minimum load time [14] (Figure 1.62). According to Potente, there is also a maximum shear stress above which an agglomerate spontaneously breaks down [18].

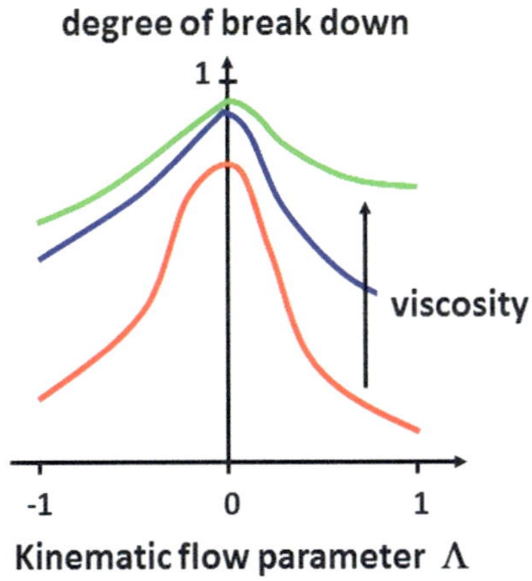

Figure 1.61 Influence of kinematic parameter of flow Λ and viscosity on the dispersion degree of TiO_2 agglomerates in silicone oil (trials with a single-screw extruder) [12]

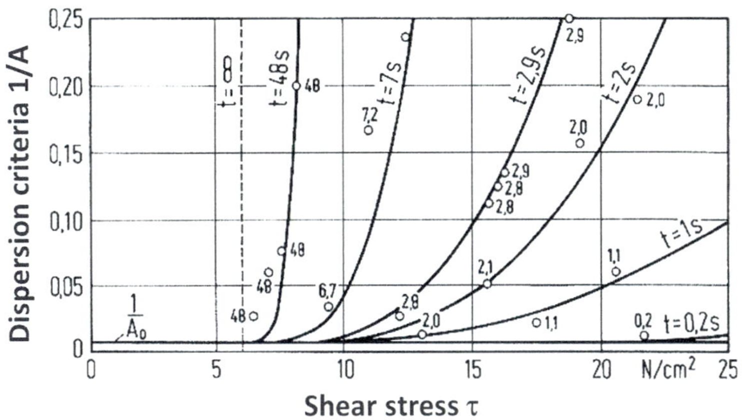

Figure 1.62 Dispersion rate of carbon black agglomerates in a PE melt depending on the shear stress and stress duration (trials with a single-screw extruder) [14]

Together with a sufficient residence time in the shear zone, effective dispersion is also encouraged by

- High screw speed
- Low throughput
- Narrow shear gaps
- High melt viscosity

Wide-disc kneading blocks achieve the most effective dispersion because they generate the highest shear stresses. This also leads to high energy input, however, which increases the melt temperature and reduces the viscosity of the predominantly shear-thinning polymer melts, which in turn leads to a reduction of shear stress. Therefore, dispersive mixing is an optimization problem.

The purely mechanical dispersion of the agglomerates can be assisted by the addition of dispersion aids (e.g., amide, polypropylene, or polyethylene waxes; metal carboxylates; organic fatty acids and their esters), which wet the particle surfaces, thus reducing the interaction between them (Figure 1.63) [16]. This significantly lowers the shear stresses required to disperse the agglomerates.

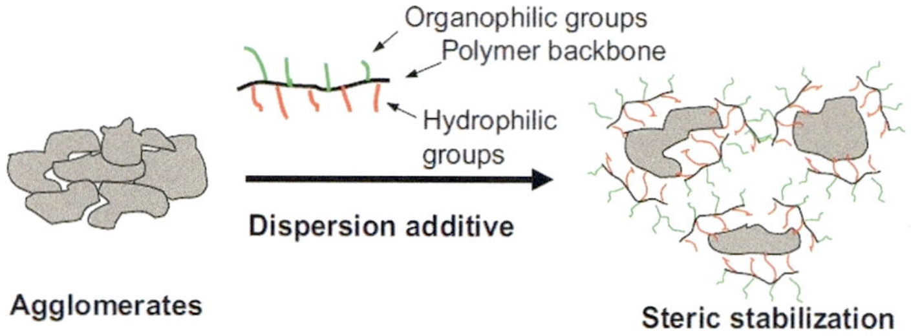

Figure 1.63 Effect of dispersion additives

1.3.2.6 Devolatilization Zone

Tasks of the devolatilization zone are:

- Removal of water, residual monomers, and solvents

Key setting parameters are:

- Screw speed
- Throughput
- Available screw volume
- Vacuum

Due to the axially open screw channels in the co-rotating twin-screw extruder, the devolatilization zone must be sealed by completely filled zones on either side of the devolatilization opening to prevent, for example, extraction of components that have not yet been incorporated from upstream or downstream zones. Re-conveying or neutral kneading blocks neutral re-conveying or mixing elements are used to achieve the melt-filled zones. Generally, screw elements with 1.5 *D* pitches are used in the devolatilization zone itself. These are operated partially filled to provide as large a product surface as possible for devolatilization and also to prevent product from discharging into the vent.

The best devolatilization efficiency can be achieved with

- High melt temperature
  - Low equilibrium concentration, high diffusion constant
- Low degree of filling caused by a large available screw volume (large channel

depth, large number of threads, long zone) and a low throughput/screw speed ratio
  - Large devolatilization surface
- High screw speed
  - Frequent surface renewal
- Addition of entraining agent
  - Low partial pressure
- High vacuum
  - Low equilibrium concentration.

The vent opening should be as large as possible to avoid high gas speeds, which can lead to product being drawn into the vent.

In practice, three vent inserts are used to ensure effective devolatilization on the one hand and to prevent product from discharging into the vent on the other hand (Figure 1.64):

- Form A: for polymer melts that do not adhere to the screw (e.g., PVC, polyolefins)
- Form B: for polymer melts that adhere to the screw (e.g., PA, PET, PC)
- Form C: for large volumes of gas in polymer melts that adhere to the screw (e.g., PA, PET, PC)

Alternatively, vents with vent stuffers can be used to prevent product discharging. However, this will reduce the devolatilization efficiency due to the reduction of the available devolatilization cross-section.

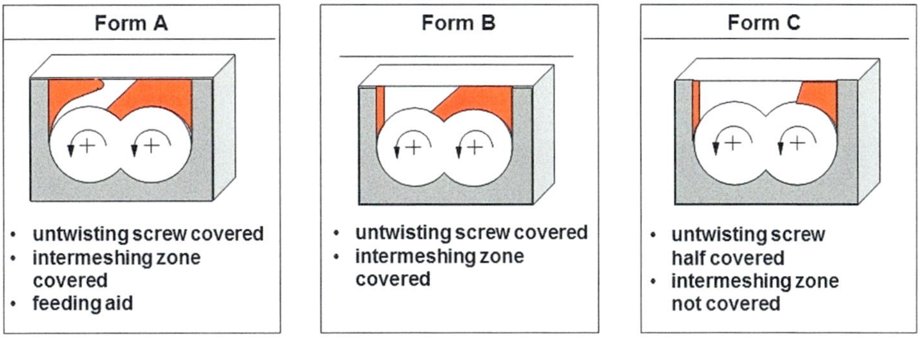

Figure 1.64 Vent inserts (according to Coperion [13])

1.3.2.7 Pressure Build-Up Zone

Tasks of the pressure build-up zone are:

- Build up pressure

Key setting parameters are:

- Screw speed
- Throughput
- Pitch
- Gap between screw and barrel

Pressure build-up zones are located in the extruder in the discharge zone and upstream of each backward conveying screw element, e.g. mixing elements, re-conveying, or neutral kneading blocks.

The pressure required to overcome the backward conveying elements has to be generated in the pressure build-up zones. The pressure consumed by the tool and any subsequent units (e.g., melt filters) has to be generated in the discharge zone. There is a greater reflux of melt over the screw tips in the pressure build-up zone resulting in more intense melt mixing.

Today, the pressure build-up zone can be calculated with sufficient accuracy provided that adequate material data are available. The aim of the design is generally to generate the required pressure consuming as little energy as possible. Figure 1.65 shows the energy input in the melt-filled pressure build-up zone on the basis of preset values for pressure build-up, throughput, and screw speed. The energy inputs are compared for single-flighted and double-flighted ($Z = 1$ or 2) conveying elements with different pitches T (as multiple of barrel diameter D) and screw/barrel clearances (δ). The end of the pressure build-up length is characterized by the end of the curve. It is apparent that with single-flighted conveying elements the required pressure can be generated with considerably less energy input and a shorter pressure build-up length. However, it should be noted that single-flighted conveying elements have a tendency to pulsate, because there is only one product flow in comparison to three product flows in the case of double-flighted elements. Metering fluctuations can therefore more rapidly lead to throughput fluctuations than in the case of double- or triple-flighted elements.

For the same example, Figure 1.66 shows that pressure build-up ability and energy input depend not only on screw geometry but also on the process parameters (in this case: screw speed), i.e., in the end on the kinematic parameter of flow Λ. For a pitch of $T = 0.3\ D$, the energy input decreases and the pressure build-up length increases as the screw speed decreases. For a pitch of $T = 0.2\ D$, however, the energy input increases as the screw speed decreases. Pressure build-up is no longer possible above $n = 145$ rpm at a throughput of 100 kg/h, because Λ is greater than 1 (see also Section 4.2).

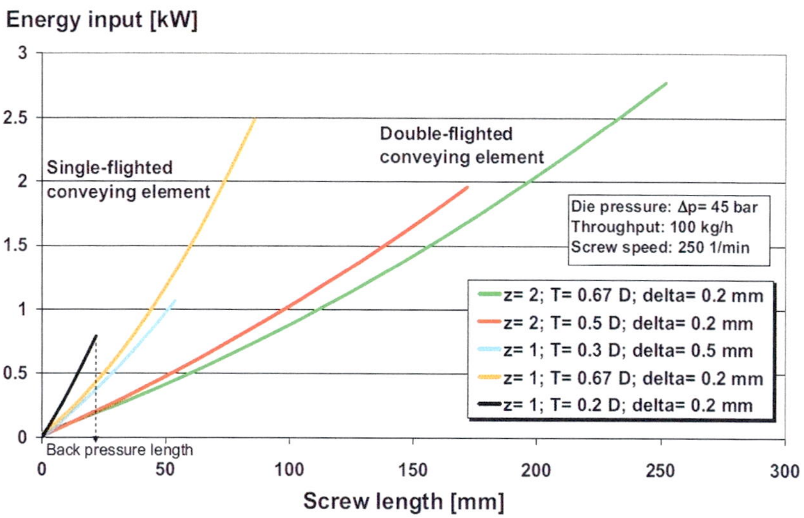

Figure 1.65 Energy input in the pressure build-up zone for different conveying elements

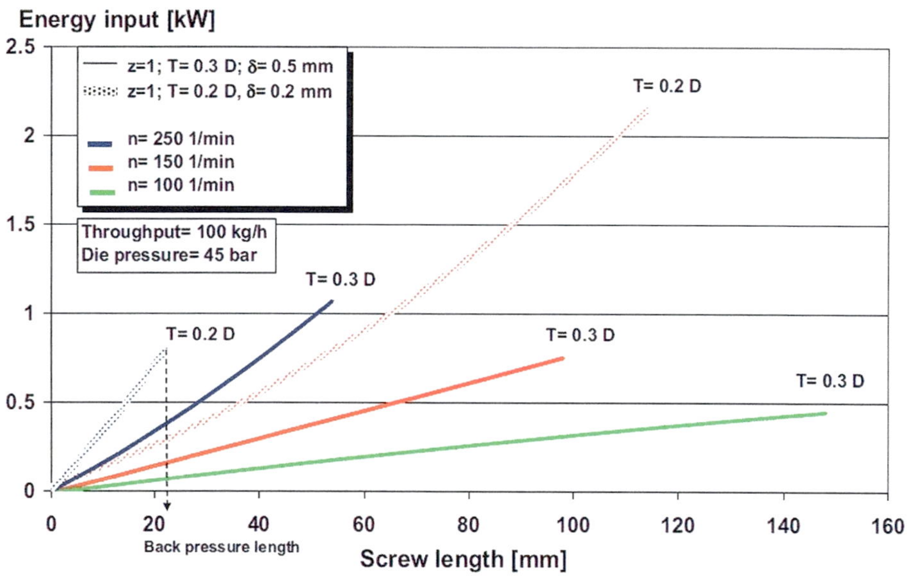

Figure 1.66 Influence of screw speed and pitch on the energy input in the pressure build-up zone

To prevent excessive energy input in the discharge zone, pressure build-up for very high pressures and/or temperature-sensitive polymers can also be generated by a downstream gear pump that has a considerably greater pumping efficiency than the co-rotating twin-screw extruder (see Figure 1.67 and Section 4.3.2).

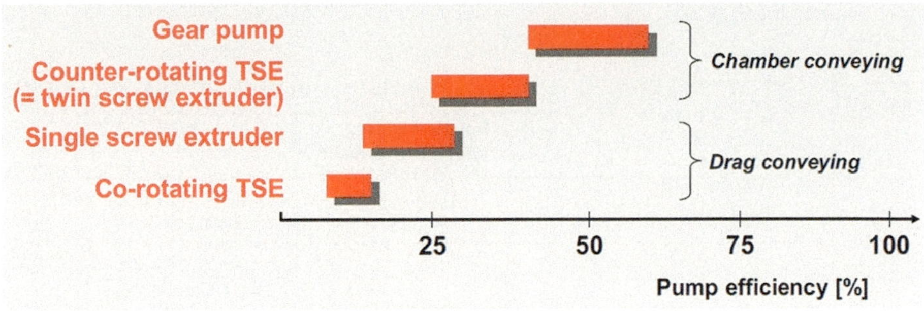

Figure 1.67 Pumping efficiency of different conveying units

1.3.3 Characteristic Process Parameters

1.3.3.1 Specific Energy Input

The specific energy input is the most important characteristic value. It is characteristic to each product and applies irrespectively of the size of the extruder. Generally, only measurable value is the specific mechanical energy input e via the screw shafts. This is calculated using the following Equation (1.2):

$$e = \frac{2 \cdot \pi \cdot n \cdot M_D}{\dot{M}} \tag{1.2}$$

where n is the screw speed, M_D is the torque of the screw shafts, and $\dot{M}$ is the throughput.

The heat flow across the barrel wall is generally only measurable on lab extruders. Figure 1.68 shows the specific mechanical energy input for different volume and/or torque-limited processes.

Figure 1.69 to Figure 1.71 show the relationship between screw speed, throughput, specific mechanical energy input, and melt temperature. It is apparent that melt temperature and specific mechanical energy input increase as screw speed increases and decrease as the throughput increases (Figure 1.69 and Figure 1.70), while the temperature increase in relation to the screw speed is greater than the temperature decrease in relation to the throughput. Therefore, according to Equation (1.2), the melt temperature ultimately rises as the specific energy input rises (Figure 1.71).

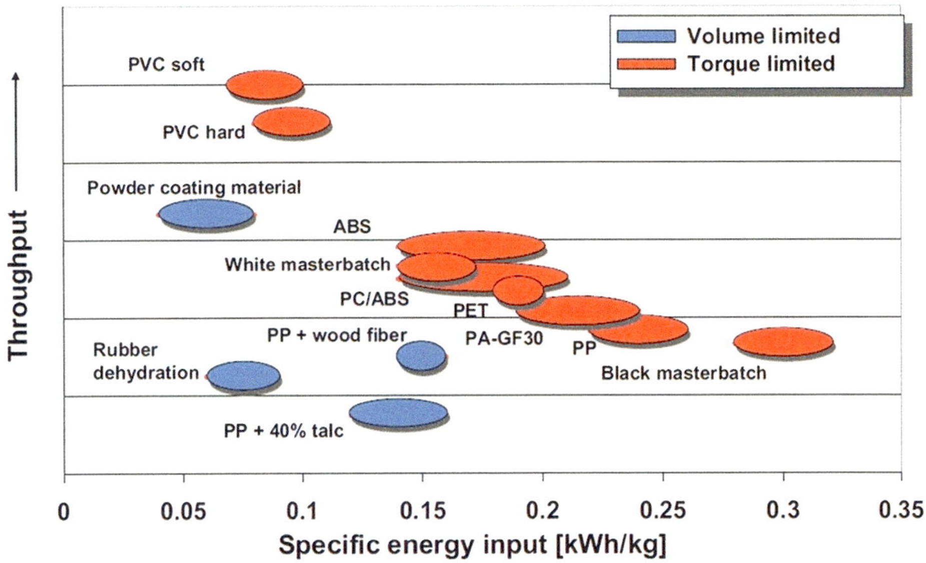

Figure 1.68 Specific mechanical energy input for different volume- and torque-limited processes (according to Coperion Werner & Pfleiderer)

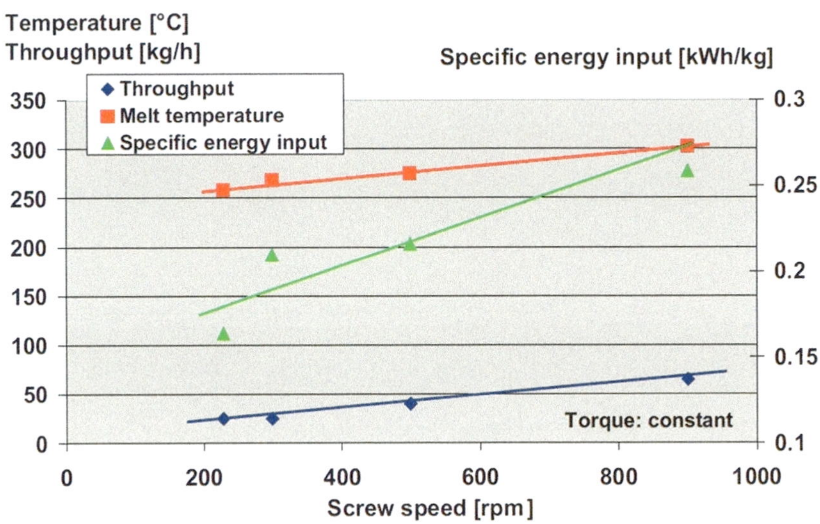

Figure 1.69 Influence of screw speed on melt temperature, throughput, and the specific mechanical energy input

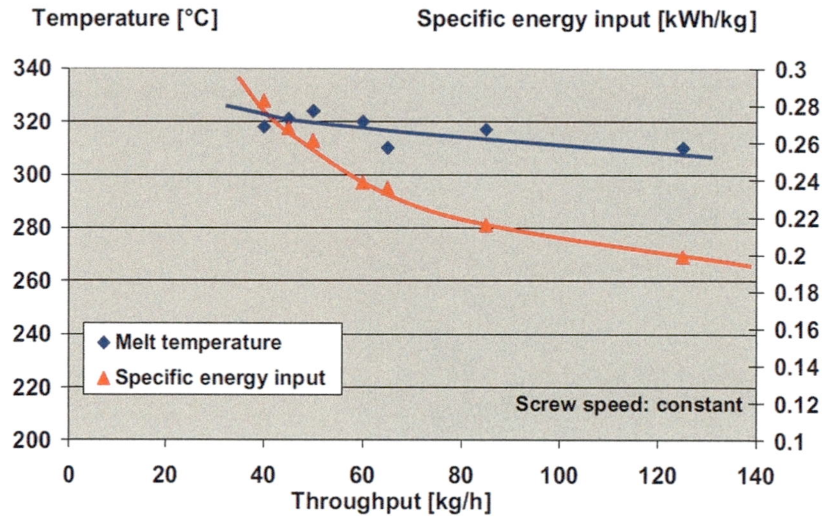

Figure 1.70 Influence of throughput on melt temperature and the specific mechanical energy input

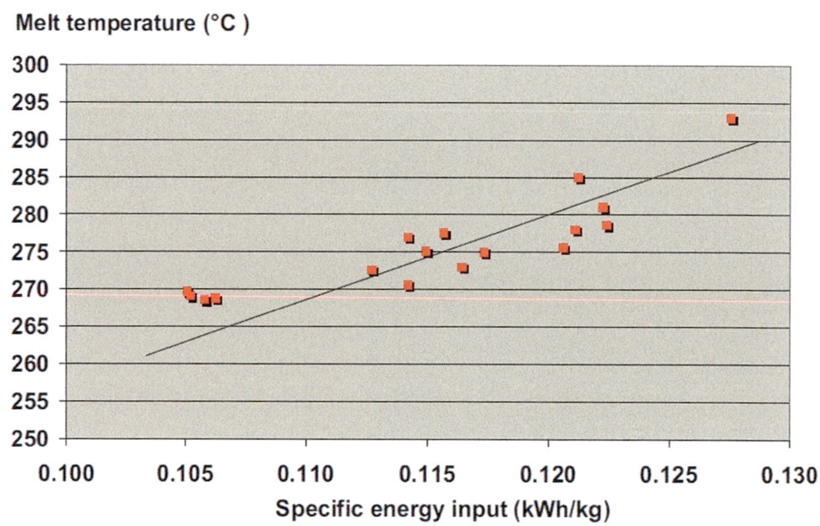

Figure 1.71 Influence of the specific mechanical energy input on the melt temperature

1.3.3.2 Residence Time Characteristics

The residence time characteristics in an extruder have a great influence on product quality, for instance when processing heat-sensitive products, in reactive processes, and in the case of dispersion and melting. As with all continuous processes, there is no precisely defined residence time in co-rotating twin-screw extruders, but a residence time distribution.

While the minimum residence time is significant for dispersive mixing, the residence time distribution in the extruder influences distributive mixing as a measure for longitudinal mixing performance [17]. In partially-filled screw sections the average residence time depends only on screw pitch and screw speed, while in fully-filled screw sections it only depends on throughput.

Along with the process parameters, the residence time distribution in a co-rotating twin-screw extruder depends substantially on the screw configuration (Figure 1.72). A simple conveying screw (1), for instance, has a very narrow residence time distribution, whereas the screw with toothed mixing elements (5) has the widest residence time distribution and the largest minimum and average residence times. Kneading block zones (2–4) also have a wider residence time distribution than simple conveying screws. A kneading block zone with narrow kneading discs (3) has a wider residence time distribution than one with broad discs (2), but has lower minimum and average residence times. Kneading block zones with wide conveying and re-conveying kneading blocks (4) have the widest residence time distribution and the highest minimum and average residence times of all kneading block configurations.

Along with the screw configuration, the residence time distribution in twin-screw extruders is also influenced by the process parameters. As shown in Figure 1.73, throughput has a greater influence on the average residence time than screw speed and pitch [4].

The minimum residence times in compounding twin-screw extruders for an average production run range from 5–15 seconds.

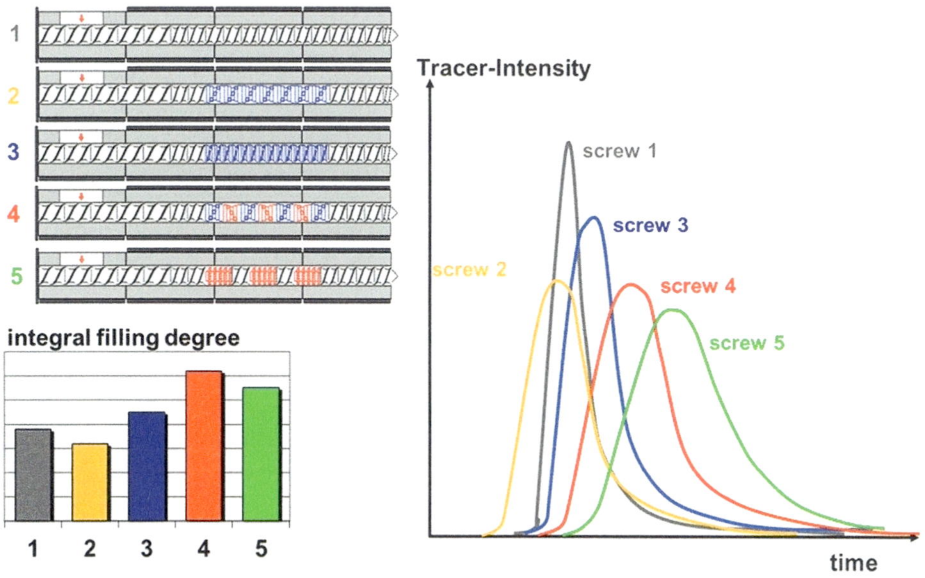

Figure 1.72 Residence time distribution for different screw configurations in a co-rotating twin-screw extruder (constant screw speed and throughput)

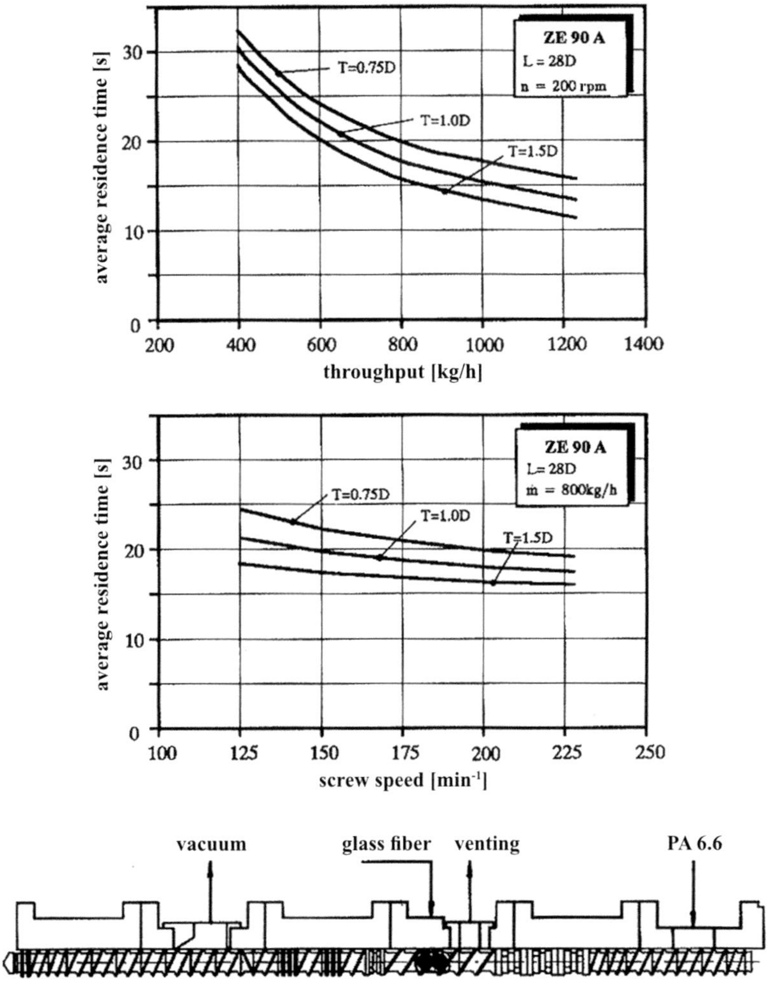

Figure 1.73 Average residence time in relation to throughput and/or screw speed for a co-rotating twin-screw extruder [4]

1.3.4 Process Examples

1.3.4.1 Incorporation of Glass Fibers

Polymers are reinforced with glass fibers to improve their mechanical properties (strength, Young's modulus, and impact strength) and their heat deflection temperature.

Predominantly chopped strands or milled fibers with initial fiber lengths of up to approx. 3 mm are used to reinforce thermoplastics. Recent developments are increasingly including long glass fibers (fiber length: 25–30 mm). The objective when incorporating glass fibers in twin-screw extruders is to distribute the fibers homogeneously in the polymer matrix and to achieve optimum length distribution in the end product. Although excessively long fibers generally improve the mechanical properties of a product, they diminish the flow rate of the melt during subsequent processing (e.g., injection molding). However, if fiber lengths are too short, the mold filling characteristics is improved for injection molding but mechanical properties deteriorate. In addition, it is necessary to ensure that the fiber sizing is not damaged or even evaporated through high temperatures, which would impede the bonding mechanism between fiber and polymer necessary to ensure good mechanical properties.

Therefore, in the compounding process the fibers are generally added to the polymer melt after melting. Feeding fibers at the first intake generally results in the fibers being broken down too much in the plastification zone for most applications (Figure 1.74).

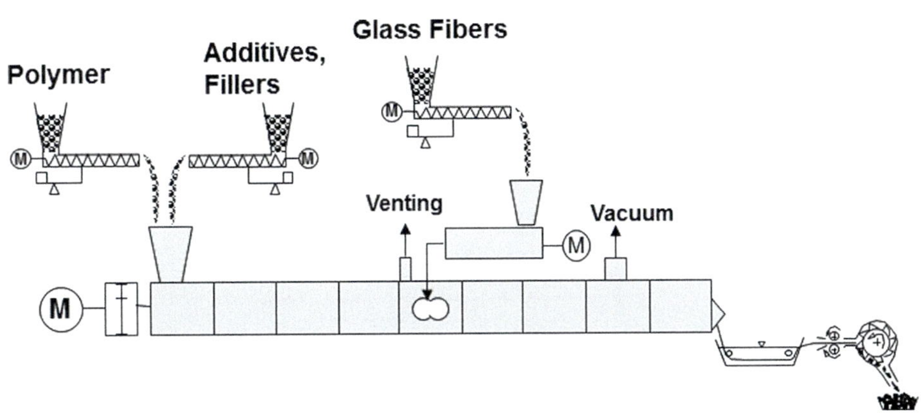

Figure 1.74 Schematic representation of the process of reinforcing polymers

The melt cools down when glass fibers are added. It is therefore necessary that the melt temperature is sufficiently high before adding the fibers to ensure that the polymers do not cool down to their solidification temperature. Too great an increase in viscosity caused by cooling will prevent the melt from wetting the fibers, in which case the melt will cause mechanical stress and damage to the fibers. The size reduction of the fibers depends on the dimensionless kinematic parameter of flow Λ (further details in Section 4.1.1.3.3). The fiber length will reduce as the screw speed increases and/or the throughput decreases.

Figure 1.75 shows that the pressure build-up ability of a melt with long glass fibers differs significantly from that of a pure melt. This is essentially caused by the change of the glass fiber length with Λ. For $\Lambda = 1$, the pre-distributed fibers form an internal fiber structure which is conveyed solely by drag flow (Figure 1.76 left). The pressure build-up ability reaches a maximum when $\Lambda \approx 1/3$, at which point this structure is almost completely destroyed (Figure 1.76 right). When $\Lambda \approx 0$, the individual broken fiber pieces can move freely in the melt. The increased destruction can be explained by the increase in pressure flow at decreasing Λ values, as these apply greater stress to the fiber structure, eventually resulting in its destruction [3].

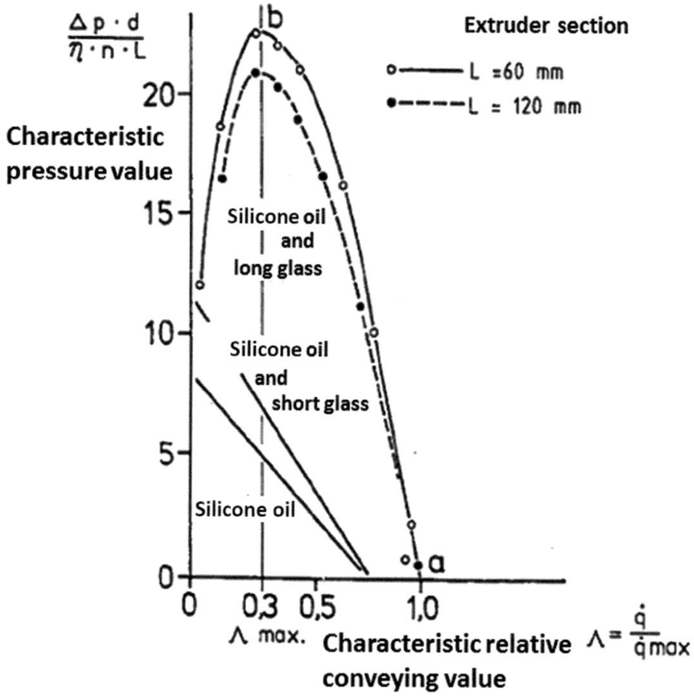

Figure 1.75 Pressure build-up ability of silicone oil with and without long and/or short glass fibers (lengths 6 mm and 12 mm) [3]

To prevent significant reduction of the fiber length, the lowest possible screw speed and/or the highest possible throughput and degree of filling should be selected. However, the opposite process parameters are required to ensure effective distribution of the fibers (see Section 1.3.2.4). Therefore, glass fiber incorporation is an optimization problem.

After wetting, dispersion, and pre-comminution of the fibers, the melt is devolatilized to achieve a perfect pumping and constant material flow within the die. Comminution must be completed prior to the devolatilization zone to prevent fibers from being separated and deposited in the devolatilization area.

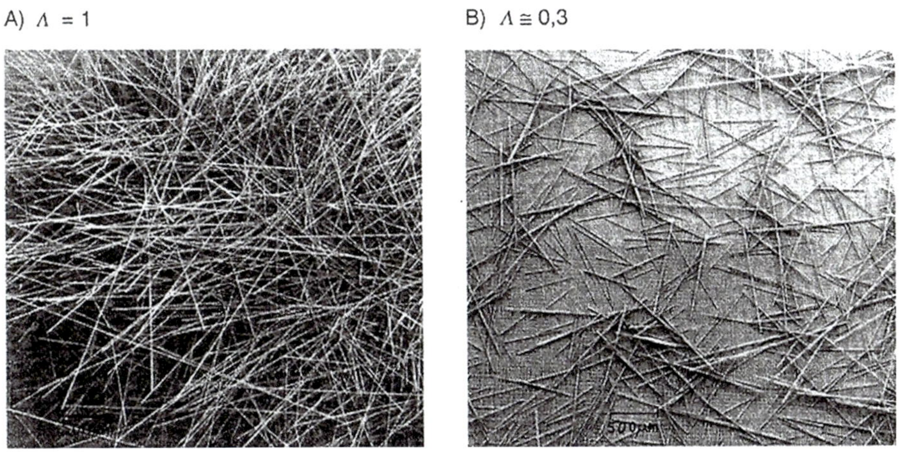

Figure 1.76 Glass fiber structure after stress in the screw channel (initial fiber length: 3 mm) [3]

Fiber comminution is heavily influenced by the viscosity of the pure melt. The fibers in high-viscosity melts are already comminuted to a greater extent in the screw channel during the feed process, but in low-viscosity melts it is necessary to use kneading and mixing elements for dispersion [3]. Together with the viscosity, the fiber content has a significant influence on the fiber comminution (Figure 1.77) [18]. The higher the fiber content, the higher the shear stress and the more the glass fibers break down.

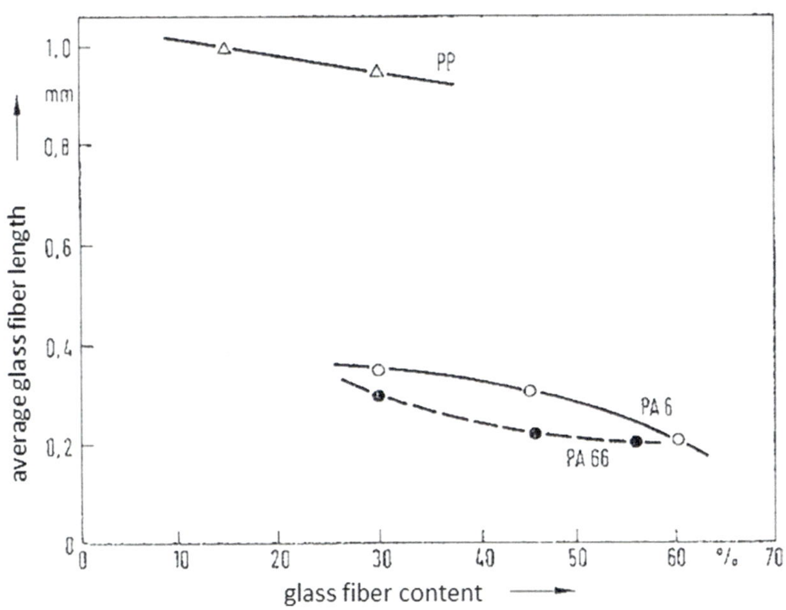

Figure 1.77 Influence of the glass fiber content on the average fiber length for GF-PA6, GF-PA66, and GF-PP [18]

1.3.4.2 Incorporation of Fillers

Fillers are added to polymers either to improve the polymer properties or to reduce the price of the compound. The properties of filled polymers are heavily influenced by the interaction between particles and polymers as well as by particle size, particle size distribution, and the homogeneity of the particle distribution. The smaller the particles and the more homogeneous their distribution is in the polymer matrix, the better are in general the properties of the compound. Therefore, dispersion plays a key role.

Fillers are primarily incorporated in powder form with filler contents reaching 50% by weight and even up to 80% by weight for highly-filled systems. Problems in the processing of low-density powders are generally caused by the high volume of air and in some cases humidity which is drawn in with the powder and which hinders plastification. Therefore, and because most of the mineral fillers are abrasive, fillers are generally incorporated into the melt. The aim is to ensure that the solid is fully wetted by the melt. The mixing zone is ideally operated partially filled to enable the drawn in air to escape via the partially-filled screw channel and a vent opening without carrying unwetted solids with it. Then, the final dispersion, homogenization, and devolatilization of the melt take place.

For filler contents > 40% by weight, it is advisable to add the filler in two or more stages to enable wetting, to manage the gas quantities, and to prevent the melt from cooling too much [3]. A portion of the filler (up to approximately 10% by weight) can generally also be added at the first intake stage (Figure 1.78).

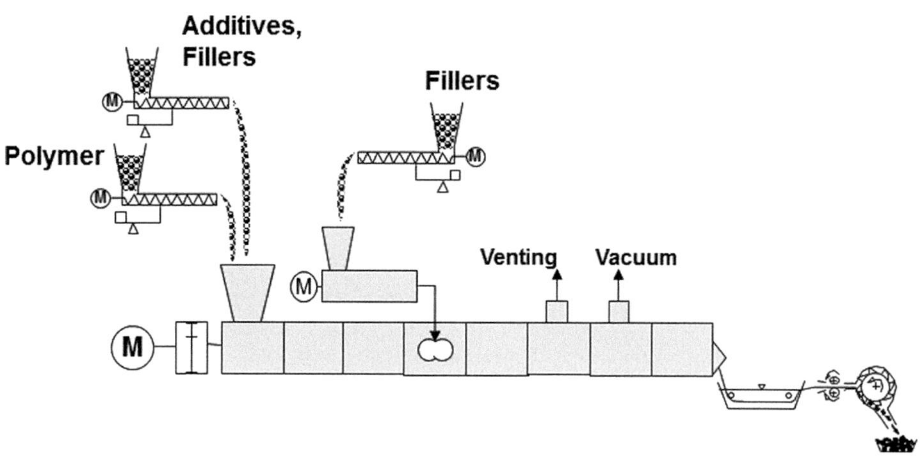

Figure 1.78 Schematic representation of the process of incorporating fillers into a polymer

The incorporation of powders usually is a volume-limited process, in which the feed capacity of the screw reaches its limit.

Fillers can also be incorporated in the form of compacted powder pellets and pelletized masterbatch concentrates. Both allow greater throughput rates in comparison to powders, because the greater bulk density means a higher feed capacity on the one hand and less gas to be removed on the other hand. Consequently, mixing zones can generally be operated completely filled. However, the increased throughput is accompanied by higher costs for the pellets or masterbatches and the economic feasibility of any application should be determined in advance.

Frequently used fillers include talcum, calcium carbonate, and kaolin, which have the following processing properties:

- Talcum is generally platelet-formed and therefore abrasive
  - Preferably added to the melt
- Calcium carbonate is bead-formed or cubic and is less abrasive
  - Can also be added at the first intake stage
- Kaolin contains approx. 14% chemically bonded water and is relatively soft and crumbly. Its lack of rigidity and its swellability in water means that it is generally only used as a filler in relatively small levels and only in its calcinated form (= fired, i.e., de-watered).
  - Can be added at the first intake stage as well as to the melt.

1.3.4.3 Production of Masterbatches

The term masterbatch generally refers to pelletized concentrates of pigments or additives that are embedded in a polymer matrix. They serve to ensure dust-free, accurate, and simple addition of pigments and additives to polymers and compounds.

Preparing a masterbatch is a demanding compounding task, because high quantities of pigments/additives (up to 80% by weight) have to be incorporated into a base polymer. Base polymers can be either universal carriers such as PE, EVA, or waxes, which are compatible with many polymers, or material-specific carriers if there is a risk, for example, of affecting the compound properties, particularly with high quantities of additive. Base polymers should be of equal or lower viscosity than the polymers that are to be filled or colored, because they disperse better in a high-viscosity polymer on the one hand and on the other hand because the viscosity is increased by the filler. In masterbatch production, we differentiate between premix, split-feed, and color matching processes. The three production processes result in differing throughputs, as well as in various advantages and disadvantages that are described in the following (Figure 1.79).

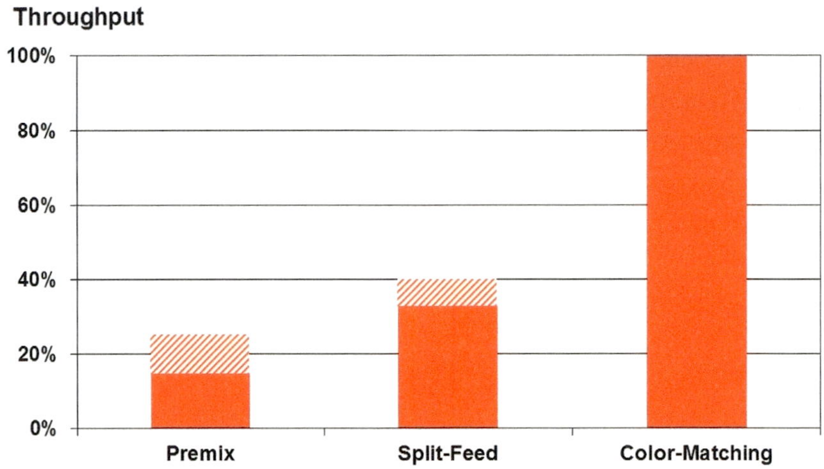

Figure 1.79 Achievable throughputs with different masterbatch production processes

1.3.4.3.1 Premix Process

In the premix process, all additives are premixed in a mixer and then fed to the twin-screw extruder at the first intake stage via a volumetric feeder. A portion of the polymer is generally mixed with the pigments in powder form to achieve a better pre-distribution of the pigments (Figure 1.80).

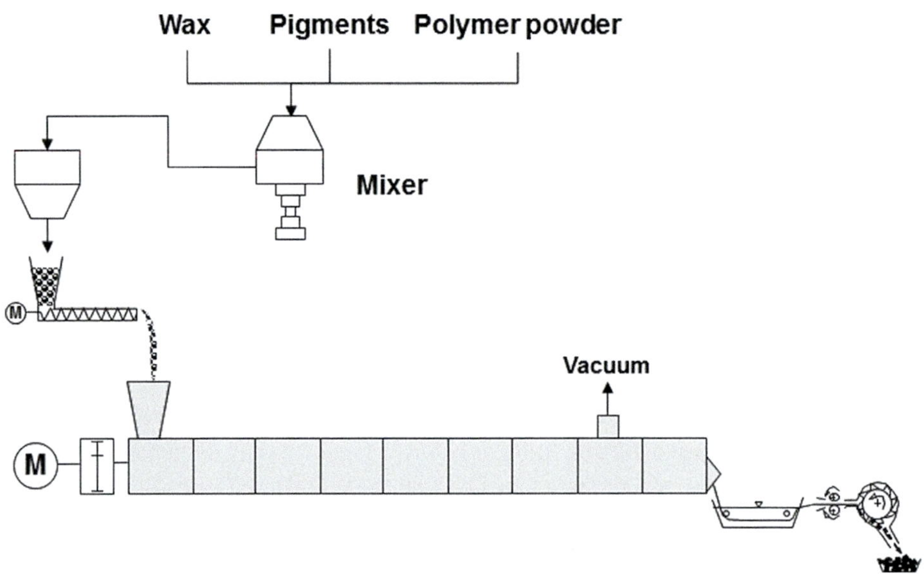

Figure 1.80 Schematic representation of the production of a masterbatch using the premix process

The following advantages make this process extremely popular [19]:

- No product loss at start-up
- Simple operation of all system components
- Extremely good shear effect for the breakdown of pigment agglomerates
- Easy to clean when changing colors
- Relatively low investment costs

1.3.4.3.2 Split-Feed Process

In the split-feed process, the additives are fed into the polymer melt via a side feeder downstream of the plastification zone. Here, gravimetric feeding is always used (Figure 1.81).

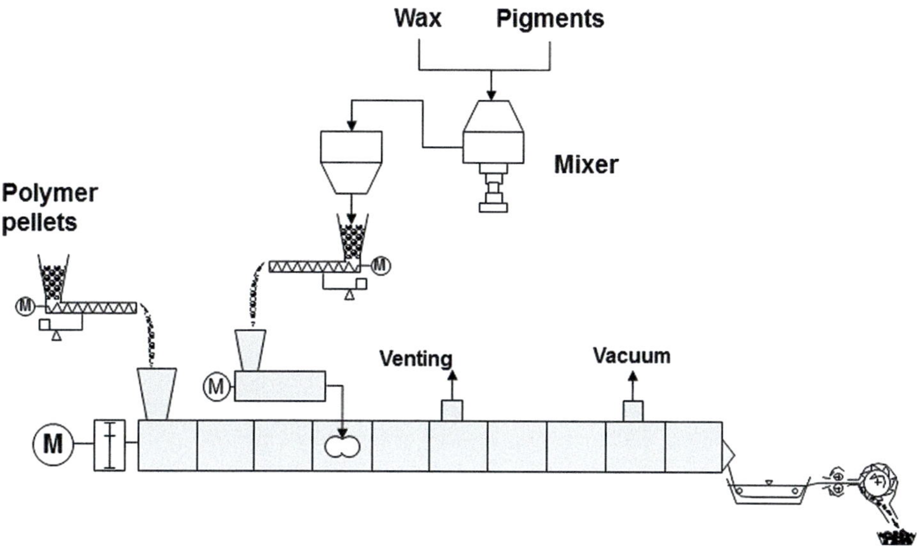

Figure 1.81 Schematic representation of the production of a masterbatch using the split-feed process

This process is particularly suitable for large batch sizes and has the following advantages [19]:

- Good product quality due to gentle product handling
- Minimal wear of screw elements and barrels
- High volumes of pigments and/or additives possible

The main disadvantages are:

- Increased cleaning time
- Higher investment costs due to larger number of system components and gravimetric rather than volumetric feeding.

1.3.4.3.3 Color Matching

In the color matching process, various masterbatches containing a single pigment type, what is referred to as a monobatch, are fed either premixed or separately into the extruder via the first intake (Figure 1.82). The two compounding processes (1. monobatch production, 2. masterbatch production) give color matching the advantage of extremely good dispersion. However, it comes at higher costs.

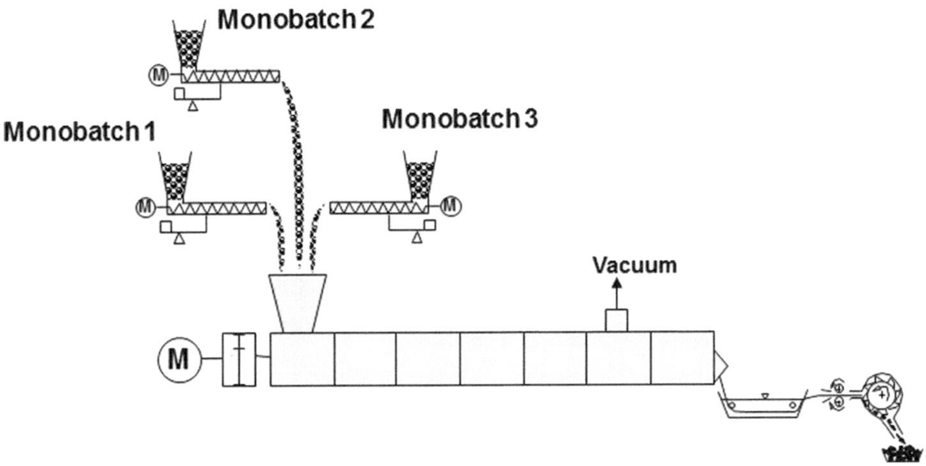

Figure 1.82 Production of a masterbatch using the color matching process

In general, during masterbatch production pigments should not be subjected to cold compression, which can produce secondary agglomerates that are extremely difficult to break down. This can be avoided in two ways: firstly, by ensuring that pigments are not heavily compressed in the solids feeding zone during the premix process (or in the side feeder in the split-feed process). Secondly, in the split-feed process, by ensuring that, before the pressure builds up, the pigments are wetted as quickly as possible by the melt as they are fed into the extruder.

1.3.4.4 Coloring

The objective of coloring polymers is to create the desired color both consistently and reproducibly and to meet the requirements of brilliance, depth of color, and covering power.

In general, coloring of compounds is not carried out in a separate process; it effectively takes place alongside the compounding process. A standard screw configuration for coloring includes the plastification zone plus one dispersion, ventilation, homogenization, devolatilization, and pressure build-up zone each.

Colored pigments can be incorporated in the form of

- Loose powdery pigments

- Masterbatch pellets
- Color compounds
- Liquid dyes

Coloring by means of masterbatches predominates by a significant degree according to an AMI study [20] (Figure 1.83).

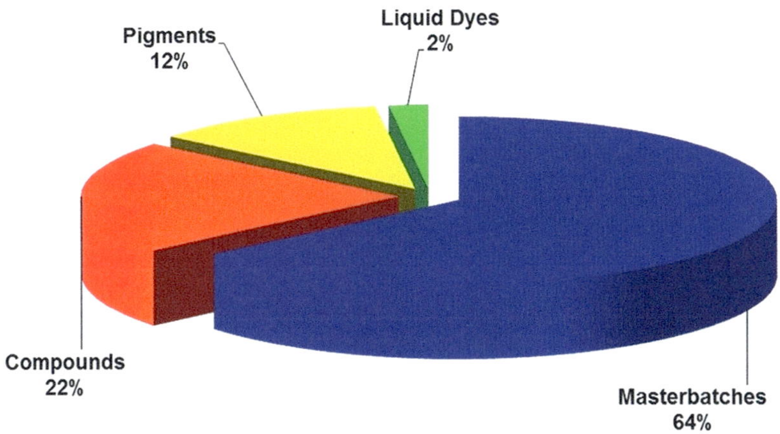

Figure 1.83 Distribution of coloring methods [20]

The type of colorant used depends, among other things, on the handling of the colorant, costs, cleaning time, quality requirements, available equipment, and the compound formulation.

Pigment properties, such as particle size distribution, chemical structure, and thermal stability, as well as the process have a significant impact on the dispersion process. The primary particle diameter of the pigments is generally less than 1 µm.

The following guidelines, among others, should be followed during the coloring process.

To prevent damage to the sometimes temperature-sensitive pigments, incorporation must be carried out with as little energy dissipation as possible. It is therefore beneficial to carry out incorporation at high degrees of filling. The shear stress of the pigments can be further reduced by adding the pigments after the plastification zone in the split-feed process.

Reduced energy dissipation is also beneficial to avoid changing the inherent color of the polymer due to excessive temperature stresses. This will have a negative effect on the compound color and/or on the content of pigment required to achieve the desired color.

Due to the generally unavoidable polymer decomposition during the compounding process, particularly with high pigment contents of up to 15% by weight, the feedstock polymer should have a higher viscosity than the target viscosity of the end product.

When feeding powdered pigments in small quantities, premixing of the pigments with other formulation ingredients or flaking of the pigments on the polymer pellets is recommended, because the slightest feeding fluctuations will result in an immediate color discrepancy. A wide range of material-specific dispersion additives also facilitate dispersion.

Incorporation is easiest when using masterbatches, because it requires only distributive mixing of the pre-dispersed pigments. A further advantage of masterbatch use is the lower effort for cleaning of the dosing units and conveying lines as well as the omission of aspiration to avoid contamination of the environment with dust. However, costs of masterbatches and the associated raw material logistics are higher than for pigments in powder form.

Another option for adding pigments is to use liquid dyes in which the pigments are suspended in. The liquid colors are dosed into the melt after the melting zone. The advantages of coloring with liquid dyes are a more even coloring, lower consumption of colorants, as well as easier cleaning and shorter product changeover times. It must be ensured that the liquid carrier does not have any negative effects on the properties of the compound (e.g., by polymer degradation). However, larger pigment portions are difficult to incorporate into polymers with liquid dyes because large amounts of liquid have to be evaporated or remain in the compound. The liquid colors must be stirred regularly to prevent segregation. The storage stability of liquid colors is generally lower than that of pure pigments.

References for Section 1.3

[1] Brandstetter, F., Kunststoffe – ein reifes Arbeitsgebiet, VDI conference: Aufbereitungstechnik 2003, Baden-Baden

[2] Hohenberger, W., Füllstoffe und Verstärkungsmittel, *Kunststoffe* (1999), 89(7), pp. 101–104

[3] Löhr, D., Herstellen von Compounds, in: *Kunststoff-Handbuch*, Becker, G.W., Braun, D., Eds, Vol. 3/4 – Polyamide, Hanser, Munich (1998), pp. 299–314

[4] Schuler, W., Auslegung und Ausführung der Verfahrenszonen, in: *Der Doppelschneckenextruder*, VDI-Verlag, Düsseldorf (1998)

[5] Hofmann, J. et al., Energieeintrag, Scherung und Temperaturspitzen in Compoundern, VDI conference: Polymeraufbereitung 2002, Baden-Baden

[6] Bastian, M., Gabor, T., Plastifizierung von Polymeren in Gleichdrall-Doppelschneckenextrudern. VDI conference: Polymeraufbereitung 2002, Baden-Baden

[7] Melisch, U., Grundlagen zur Simulation des Plastifizier- und Förderprozesses eines dichtkämmenden Gleichdrall-Doppelschneckenextruders. Dissertation Uni-GH Paderborn (1998)

[8] Black, T., in: *Plastics Compounding – Equipment and Processing*, Todd, D.B., Ed., Hanser, Munich (1998)

[9] Hensen, F., Knappe, W., Potente, H., *Kunststoff-Extrusionstechnik*, Vol. I, Hanser, Munich (1989)

[10] Scheel, G., Einfluss der einzelnen Verfahrensschritte von gleichlaufenden Zweischneckenextrudern auf die Produkteigenschaften bei der Polymeraufbereitung VDI conference: Aufbereitungstechnik 2003, Baden-Baden

[11] Pawlowski, J., *Die Ähnlichkeitstheorie in der physikalisch-technischen Forschung*, Springer-Verlag, Berlin, Heidelberg, New York (1971)

[12] Damsky, W., Löhr, D., Mischen in der viskosen Phase, in: *Handbuch Mischen von Kunststoffen*, VDI-Verlag, Düsseldorf (1983)

[13] Training documents, Coperion Werner & Pfleiderer

[14] Martin, G., Untersuchung der Homogenisierungsfunktion von Einschneckenextrudern für die Kunststoffverarbeitung. Dissertation, Karlsruhe University (1972)

[15] Potente, H., Rechnerunterstütze Extruderauslegung, Kunststofftechnisches Seminar, Paderborn (1992)

[16] Weimann, M., Dispergierhilfsmittel zur Herstellung von Farbkonzentraten und Compounds, VDI conference: Einfärben von Kunststoffen, Baden-Baden (2001)

[17] Kessler, H., Modell zum stationären und instationären Mischen in konventionellen Einschneckenextrudern, Dissertation, Paderborn University, (1991)

[18] Hess, K. M., Verstärken von Thermoplasten mit Fasern, in: *Handbuch Mischen von Kunststoffen*, VDI-Verlag, Düsseldorf (1983)

[19] Homepage Coperion Werner & Pfleiderer, *https://www.coperion.com/en*

[20] AMI Study, in: *Kunststoffe* (2004), 94(9), p. 155

■ 1.4 Process Understanding – Overview and Evaluation of Experiments and Models

Klemens Kohlgrüber

1.4.1 Introduction

To introduce the topic experiments and models, two process technology tasks shall be regarded:

1. Degassing of a polymer with 10% of volatile materials (monomers, solvents, etc.) to few ppm, mixing in of additives, and a polymer filtration. This may be a typical task of the polymer processing of a primary production of a raw material producer (Figure 1.2).
2. Additives and fillers will be mixed in and dispersed into a polymer that is available as granulate. This task is also illustrated in Figure 1.2.

Due to pre-considerations, co-rotating twin-screw extruders should be used for both tasks. At predetermined throughput and predetermined degassing capacity the question arises: how should a corresponding extruder be configured? The extruders for both tasks will differ greatly. The following variances are on hand for the extruder configuration:

- *Size of the machine:* diameter, length, etc.
- *Housing selection:* closed, open, length, bearing positions, etc.
- *Screw element selection:* type, pitch, number of threads, clearances, etc.
- *Arrangement (positions)* of the housings and elements
- *Dosing Technique* (liquid, solid, gravimetric) and positions

Further variations are the process values speed, heating, cooling, entrainers, etc.

The ideal scenario would be for a model to predict the precise screw configuration for a given operation, including all material and dimensioning issues and to allow expected operating conditions to be calculated using the process values. So far, no electronic program exists that can propose in detail the extruder configuration with the variances of geometry and process parameters in dependence of the task. Experts have to determine the extruder configuration of the respective task.

In the case of a product for which little experience exists, usually tests become necessary. For financial reasons, the tests will normally be executed on a small scale. Apart from production scale, the tests must be planned carefully by a scale-down, so that the results will be comparable with a production line.

On a small scale, the whole process or parts can be carried out on a small extruder but it has to be taken into account that screw dimensions of different manufacturers can be different. An example is shown in Table 1.3 [1].

Table 1.3 Data of laboratory extruders [1]

	Diameter In mm	D_a/D_i [1]	M_d/A^3 in Nm/cm³	Flight Depth in mm
Berstorff	33/35	1.46(1.65)/1.74	12.2/(16.0)/10.5	5.3/7.6
Coperion	32/34	1.55/1.80	11.6 to 18	5.6/7.4
Leistritz	28	1.66	12.5	5.6

The clearances, which are not listed in Table 1.3, can also be different. The test results cannot be compared directly due to the different dimensions. Depending on the product and process being on hand, it will be more or less relevant.

If individual functional zones of an extruder should be studied in detail, then these will be advantageously considered individually. If this is executed consecutively for several functional zones, then, for example, a compounding process will be optimized successively. One example procedure is described in [2]: “In order to be able to consider in isolation the influence of process parameters for individual extruder zones, the layout design of twin-screw extruders should be carried out piecewise.” This corresponds approximately to the procedure with batch kneaders for determination of the capacity input. The temporal energy flow in the batch kneader is indicated along the extruder.

The procedure by means of tests to characterize individual functional zones or the whole process is especially useful when complex materials or material blends are on hand. To acquire basic process technology knowledge with extruders and screw geometries, model screws with model liquids are used. This is described in several sections of this book (e.g., Section 1.2.1).

A disadvantage of these experiments is that only some process values can be measured. For example, by means of usual melt temperature sensors the product temperatures along the extruder can only be measured to a limited extent. In total you will receive essentially integral values (e.g., the specific drive capacity). The advantage that all real product characteristics will be considered in experiments can also be a disadvantage if single effects should be studied. In a model, for example the product viscosity and the screw clearance can be simply modified. On the other hand, it is often difficult in models to receive product data under product conditions. The process behavior can change considerably for example when adding lubricants; waxes change the friction between melt and wall.

Finally, the "ultimate" method to design extruders and to understand the process does not exist. This is why it is useful to learn as economically as possible from the experiment and from the model, i.e. to use the strengths of the respective "methodology". Thereby, the experiments should be projected model-based and models should be supported by experiments. Possible modules of a trial-based modeling are depicted in more detail in Figure 1.84.

Example: By means of 3D modeling and a laboratory screw with a model liquid (e.g., silicone oil) the conveying parameters (= profile parameters) of a particular screw geometry will be determined. To consider the real material data, tests in the technical center will be additionally executed. With these three methods characteristic diagrams for the conveying parameters and the product will be generated. Then these can be incorporated into a one-dimensional modeling.

A production-based model is depicted in Section 1.4.6.3. A model calculating for example non-measurable process factors is named an observer model. The model can be incorrect or the screw is subject to alterations, e.g. by wear. Then the observer model must be corrected. This correction by means of a so-called corrector. The corrector uses isolated measured data which will also be calculated by the observer. For example, the product outlet temperature is measured occasionally with a probe thermometer. The measured temperature will be compared with the calculated temperature. In there are deviations the corrector can use the information for adapting the model reasonably. In the prospect it is described which further possibilities are existent for future automations with a predictor model.

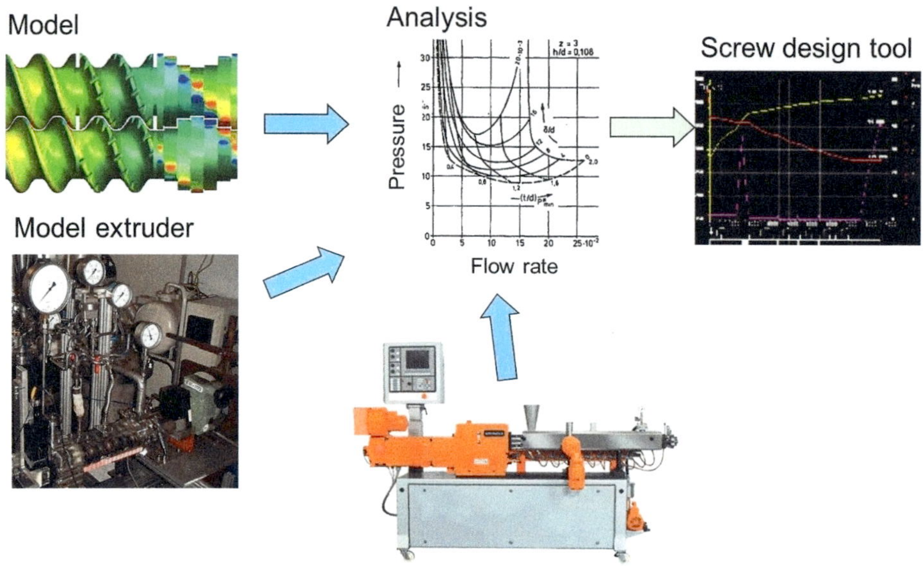

Figure 1.84 Example for modules of a trial-based modeling

1.4.2 Classification of Models and Experiments

With regard to a modeling also the following terms are used:

- So-called "rigorous modeling"; this goes without measurement data
- Part-empirical; combination of experiment and model
- Key-figure-oriented description, with and without measurement data

The key figures may be dimensioned (e.g., the specific energy input) or dimensionless (e.g., the most important extruder key figure throughput).

A further method of approach and classification of measurements and evaluations can be made according to spatial dimensions (Figure 1.85).

Regarding a screw "through a telescope", the basic question is, what goes in and what comes out. The balance of the complete screw or individual sections can be described as zero-dimensional. Even zero-dimensional models can supply interesting predictions as shown in Section 4.3 at the energy balance.

Although the real world is three-dimensional, zero-dimensional measurements can be performed (e.g., a torque measurement generally relates to the entire screw). Measurements have the advantage that they reflect real product behavior. On the other hand, individual effects are generally impossible or difficult to detect due to the number of factors involved. Furthermore, some "fields" are impossible or extremely difficult to measure (e.g., over the screw cross-section). Temperature measurement is an example. It is difficult to obtain an estimated product temperature measurement along the screw. The housing usually has a great influence on temperature measurement. There is no method available for measuring the temperature distribution over a screw cross-section.

As stated before, models are a valuable aid for the design and scale-up/down of extruders. Depending on the model depth (zero- to three-dimensional) they make calculation of process and key values possible which cannot be determined by technical measurements. From the author's point of view, one- and three-dimensional models are especially important.

In addition to the spatial model dimensions, the time during which non-steady processes (e.g., when starting a screw) occur can also be of importance. Normally, co-rotators are, however, operated continuously so that the gravity center is based on steady-state processes.

Besides the already mentioned points for modeling, support for designing extruders, increase of process understanding, and calculation of missing measured values, the models are useful for further aspects which can be listed under the term "Motivation for Modeling".

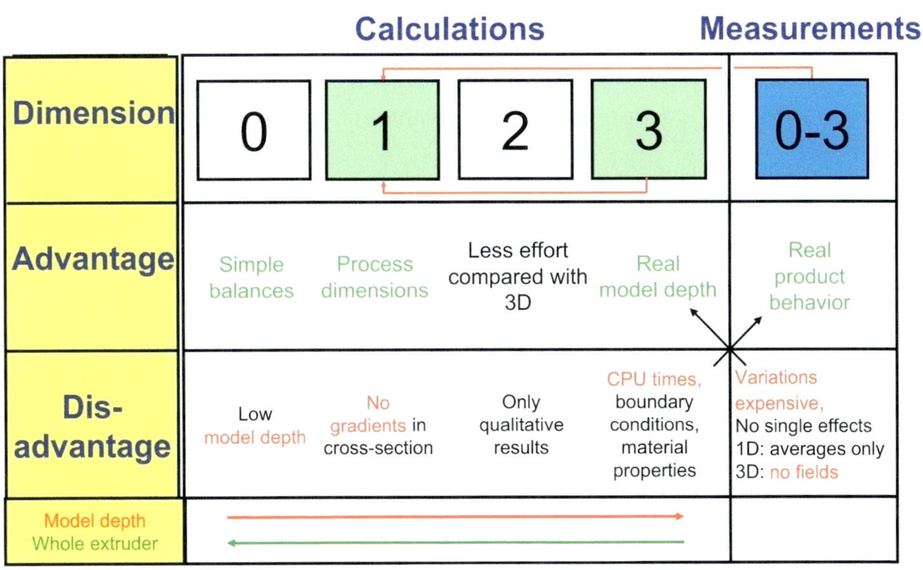

Figure 1.85 Classification of measurements and evaluation according to spatial dimension

Motivation for Modeling

- Screw design
- Increase of the process understanding
  - Simulation of individual effects, qualitative studies
  - Simulation of system-critical process conditions
- Calculation of inaccessible or difficult-to-access process values
- Scale-up/scale-down
- Saving of testing costs, support in test planning
- Training purposes (offline)
- Process monitoring and control (online)

The ultimate aim of modeling is the "glass screw". The actual situation regarding this aim is detailed below.

1.4.3 Solid Materials

In the first example of Section 1.4.1, the product is dosed into the degassing extruder most of the time as a liquid, whereas in the second example the plastic is supplied as granulate into the extruder, most of the time by a gravimetric dosing unit. Thus, normally an "auxiliary dosing" happens at the co-rotating twin screw, contrary to the self-dosing of many single-shaft screws. Auxiliary dosed means that the throughput of the extruder is pre-set by the dosing unit and is thus independent of the screw speed. The screw speed is a "freedom parameter" and can, within certain limits, be adjusted independently of the throughput.

In the plastics industry, the plastic is normally available as an intermediate product, i.e. granulate or powder. In the food industry, intermediate or final products are available in different product forms, and hence equally diverse are the influencing factors in conveying of these solid materials:

- Density of the material
- Density of the bulk material
- Humidity
- Size of the particles
- Size distribution of the particles
- Shape of the particles
- Strength of the particles
- Flowability
- Air retention capacity (important in pneumatic conveying)

These product parameters are very difficult to describe in a model. Depending on the conveying, in the hopper or in the extruder, not all parameters are, of course, of the same relevance. Besides physical models, neural networks can be used for a description of the feed behavior in an extruder. However, if a neural network model is not tuned in to the relevant parameters, then the transfer of trials with product A to product B, for example, will not succeed.

The above listing shows the parameters of a solid material, which are more or less relevant in conveying. Regarding the dosing into an extruder, the geometry of the extruder, the housing, and the screws is still relevant. The formation of "solid material bridges" would be the worst case. For powders with low bulk density, the use of an extruder can also be uneconomic, especially when using bigger production extruders (see Section 4.1). Therefore, the feed limits at extruders comprise an important topic.

The transport of solid material in form of granulates in the extruder can be described by means of DEM simulation. A comparison with experiments shows a good agreement. Seen in the product flow direction, the melting takes place in the extruder after the granulate feeding. The melting is a complex process and the corresponding model approaches are not yet able to describe this process satisfactorily.

Further solid materials like fillers and reinforcements can be added to the plastic melt during the compounding process. The material data change during the process and are often unknown. Appropriate mixing rules are given in Section 3.3. The descriptiveness of the material data of blends for application in practice is still restricted. A new viscosity approach for the melting with fillers is given in Section 3.5.2.

If the material data of a melt can be described well, then the process values of an extruder can be calculated easily. The possibilities as well as the limits of the corresponding models are described in the following section.

1.4.4 Highly Viscous Liquids

1.4.4.1 One-Dimensional Models

From the perspective of process technology, a screw is often divided into different functional zones, e. g.:

- Feed (twin-screw extruders are generally run "starved", i. e. fed from an external feeder)
- Melting
- Filler intake
- Mixing
- And others

The process in the extruder is thus perceived as a one-dimensional process. The individual process happens along its axis in the functional zones. The corresponding models are designated as one-dimensional models. In this book, the author favors the approach introduced by Pawlowski [3]; explanations on this in Section 1.2.1.

Pawlowski shows in his book the following dependence on pressure p and capacity P for non-Newtonian materials "without memory", Equation (1.3) (Equation 6.3.1 in [3]):

$$\frac{\Delta p \cdot D}{H \cdot n \cdot L} = f_{\Delta p}\,, \quad \frac{P}{H \cdot n^2 \cdot L \cdot D^2} = f_P \tag{1.3}$$

The two functions $f_{\Delta pP}$ depend on Equation (1.4):

$$f_{\Delta p,P} = f_{\Delta p,P}\left(\frac{\dot{V}}{n \cdot D^3},\, Re,\, \frac{L}{D},\, n \cdot \theta,\, \pi_{rheological,i}\right) \tag{1.4}$$

and differ for p and P. Two dimensional material properties assume the dimensionless process values for pressure and capacity: H, for example, as zero shear viscosity and θ as time constant in the rheological description. $\pi_{rheological,I}$ are dimensionless material constants "needing" no process values. The individual dimensionless parameters and key figures, respectively, in Equation (1.4) are explicated as follows:

Reynolds Number

The second key figure is the Reynolds number, Equation (1.5):

$$Re = \frac{\rho \cdot n \cdot D^2}{H} \tag{1.5}$$

Written description (Equation (1.6)):

$$\text{Reynolds number} \quad Re = \frac{\text{"measurement for motion"}}{\text{"measurement for friction"}} \tag{1.6}$$

For extruders also two Reynolds numbers can be considered: with $H = \eta$ for Newtonian materials, a Reynolds number of the screw speed, Equation (1.7):

$$Re_n = \frac{n \cdot D^2 \cdot \rho}{\eta} \tag{1.7}$$

and a Reynolds number of the throughput, Equation (1.8):

$$Re_{\dot{V}} = \frac{\dot{V} \cdot \rho}{\eta \cdot D} \qquad (1.8)$$

The quotient of the two Reynolds numbers is introduced as a new key figure and is the dimensionless throughput, Equation (1.9):

$$\dot{V}^* = \frac{Re_{\dot{V}}}{Re_n} = \frac{\dot{V}}{n \cdot D^3} \qquad (1.9)$$

This is the first key figure $\frac{\dot{V}}{n \cdot D^3}$ in Equation (1.4), which is described in detail in Section 4.1.1.3 and which is to be regarded as the most important extruder key figure. This is the throughput key figure.

The Reynolds number as per Equation (1.7) characterizes the flow condition in the extruder. Normally, the Reynolds number is used for classification of the flows into turbulent and laminar flows. In the here-considered highly viscous liquids even a so-called creeping flow exists. At a creeping flow the friction forces dominate the inert forces and the Reynolds number has smaller values. Mathematically, this occurs for $Re \rightarrow 0$; practically, there is a transition zone which is depicted in Figure 1.86.

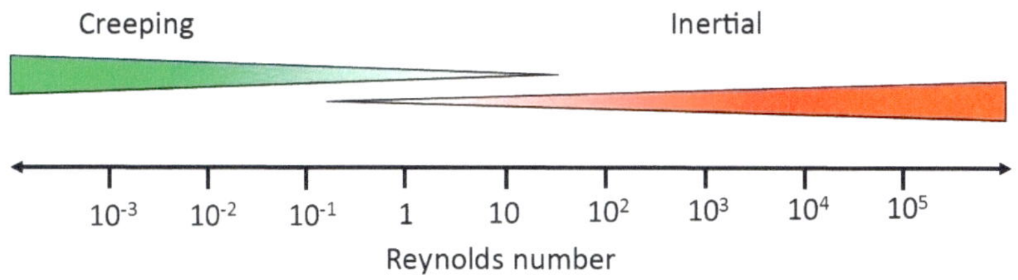

Figure 1.86 Classification of the flow types subject to the Reynolds number

Pawlowski's results of measurement for the pressure characteristic of single-shaft screws show that up to approx. $Re < 100$ a creeping flow is existent [3]. Today, it is possible to calculate flow fields three-dimensionally and the qualitative characterization in Figure 1.86 can be determined quantitatively. Thus, for small Reynolds numbers the dependence in Equation (1.4) on the Reynolds number can be omitted.

Hydrodynamic Inlet Flow

In Equation (1.4) the length L (dimensionless L/D) describes the dependence of the hydrodynamic inlet flow. This represents the length L, at which the velocity distribution after a geometrical change to "x percent" (often 90%) aligns with the definitive velocity distribution. This is very low for creeping flows. Without going into further details, a relation for tubes stated in [4] is mentioned here, Equation (1.10):

$$\frac{L_{inlet\ length}}{L_{characteristic}} = \left(a^{1.6} + \left(b \cdot Re\right)^{1.6}\right)^{1/1.6}, \quad a \approx 0.6;\ b \approx 0.05 \tag{1.10}$$

Very small Reynolds numbers $Re \rightarrow 0$ result in $L_{inlet\ length}/L_{characteristic} = a < 1$. If $L_{characteristic}$ is interpreted in rough approximation as screw diameter D, then the inlet length is $L/D = 1$.

Thus, the dependence of L/D in Equation (1.4) can be disregarded for many applications. This means that at a geometrical change of the screw in the extruder a flow condition appears on short length, the inlet length, which will remain the same as from this length. The conveying parameters (will be introduced in the following) of a screw geometry are then constant as well as from this length and for not too short screw elements independent from the length. However, the L/D dependence can be greater at some screw geometries, for example with conventional single-screw elements. Here the broad tips close the intermeshing area and an L/D independence only occurs as from L/T approx. 2, wherein T represents the screw pitch.

Newtonian Liquids

Here the dependence on $n{\cdot}\theta$ and $\pi_{\text{rheological},i}$ does not arise and the equation simplifies with the above preconditions to Equation (1.11):

$$\frac{\Delta p \cdot D}{\eta \cdot n \cdot L} = \Delta p^*, \quad \frac{P}{\eta \cdot n^2 \cdot L \cdot D^2} = P^* \quad = f_{\Delta p,P}\left(\dot{V}^*\right) \tag{1.11}$$

Pawlowski [3] and Böhme [5] have shown that at creeping flows the functions $f_{\Delta pP}$ for pressure and capacity are linear.

This linear dependence is derived clearly for the pressure in Section 4.1 and for the capacity in Section 4.2. This results in the basic equations for extruders, Equation (1.12):

$$\Delta p^* = A_2 \cdot (1 - \Lambda); \quad P^* = B_2 \cdot \left(1 - \frac{\Lambda}{\Lambda_B}\right); \quad \Lambda = \frac{\dot{V}^*}{A_1}; \quad \dot{V}^* = \frac{\dot{V}}{n \cdot D^3} \tag{1.12}$$

As shown in Sections 4.1 and 4.2, the basic equations for extruders can be depicted in the form of the axis intercepts, Equation (1.13):

$$x + y = 1 \tag{1.13}$$

in which here x and y are representing symbolically dimensionless process values.

In filled screw sections the conveying parameters for extruders (Equation (1.14)) arise in these basic equations:

$$A_1,\ A_2,\ B_1,\ B_2 = \ldots\left(geometry\right) \tag{1.14}$$

As these only depend on the geometry of the screw, they are also referred to as profile parameters.

Still further relations can be derived which are not discussed here. So far, the key figures have received little attention in the literature. One reason is that Pawlowski in [3] and further publications addresses rather theoretically interested readers. A clear elementary derivation is therefore given in Chapter 4. A further reason is that the conveying parameters for the different screw geometries must be measured in a demanding process or determined by an elaborate three-dimensional calculation. For usual conveying elements, the conveying parameters are therefore stated in this book in the following Section 1.5. The already heard argument that Pawlowski's reflections are relating only to simple thread geometries for single-shaft screws and are outdated must be commented on. In fact, Pawlowski's measurements have been realized with simple geometries, but the depicted basics are also valid for complex geometries and also for multi-shaft extruders, co- and counter-rotating. As an example, measurement results from Ullrich for a co-rotating screw and a counter-rotating screw [10] are shown in Figure 1.87. The pressure build-up of the counter-rotating screw is considerably better. The maximal pump efficiency with approx. 30% in the shown example is approx. three times higher.

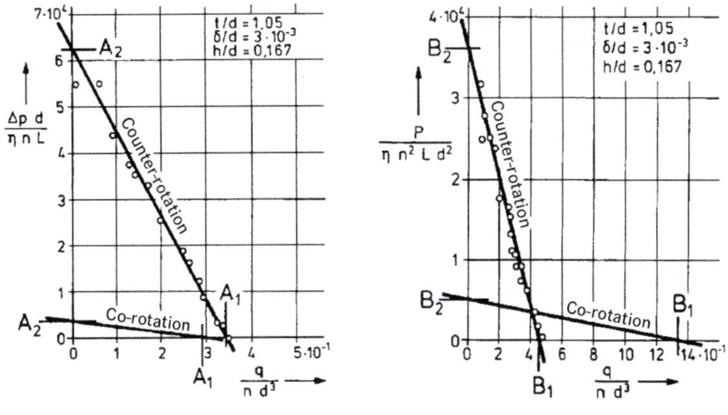

Figure 1.87 Measured conveying characteristic and conveying parameters for a co-rotating and counter-rotating screw

In the outlook it is proposed to enlarge the characterization of screw geometries to further key figures to better judge conventional and new screw elements for mixing and dispersing.

The conveying parameters, independent of the complexity of the screw geometry, are just as well as the number π (also a geometry value: circumference divided by diameter of the circle), to be considered as "fixed numbers". It would be desirable that the conveying parameters for every screw element would be available and indicated. This proposal will be explained in detail and commented in the prospect. By means of the key figures, a screw element can be classified regarding conveying and performance behavior without additional experiments. This shows that, even if a reader does not use models, by considering the basics and the results of the models, the process understanding will be increased and new questions regarding the practice will arise. Thereto, two simple examples regarding conveying parameter A_1 are presented.

Examples Regarding Conveying Parameter A_1 of a Screw Element "*S*"

Question 1: Which throughput can be maximally conveyed by a conveying active screw element "*S*"?

Answer: Calculate with: $\dot{V} = A_1 \cdot n \cdot D^3$

Question 2: With the element "*S*" a pressure is built-up in the extruder, e.g. a pressure in front of a die. Is the extruder operated at the optimal working point so that the unavoidable increase of the product temperature, being required for the pressure build-up, will be as low as possible?

Answer: Calculation of the throughput key figure $\dot{V}^* = \frac{\dot{V}}{nD^3}$. If the value is approx. $A_1/2$, then the answer is "yes".

The presented one-dimensional model for pressure and capacity is fundamental for the calculation of zones filled or partly filled with melt. For further chemical engineering processes like mixing, degassing, etc., the basic equations are also required and extended with other equations. Regarding a formulation of the basic equation as a differential equation system for consideration of the product temperatures and of variable material data, please refer to Section 4.3.

1.4.4.2 Three-Dimensional Models

Three-dimensional models are directed toward the "glass screw". This is a big advantage because also fields of process values and variable material data can be depicted. The following Figure 1.88 shows as an example a temperature distribution on two screw elements.

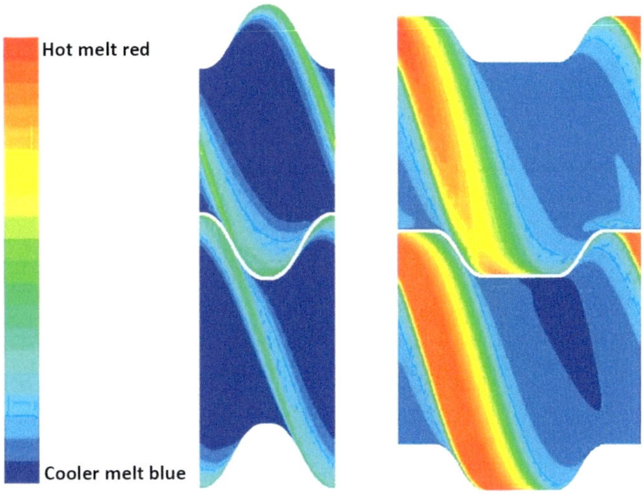

Figure 1.88 Temperature on a double-flight (left) and a single-flight (right) screw element [6]

The accuracy of the modeling, however, depends on the used material data and the marginal conditions. The marginal conditions must be specified and given for the spatial limits of the modeling (is the limit the product, the screw housing, …?). Because calculation times of three-dimensional models are significant they are still today not usable “online”, but also “offline” there are restrictions for three-dimensional modeling. The calculation of partly filled sections is at the beginning. Results thereto are shown in Section 4.7.3. Regarding melting no model of practical relevance is yet available.

With one-dimensional models, a total screw can be calculated, except for the melting zone. Of course, the pre-condition is that the material data are available and that suitable average values over the screw cross-section can be determined. As up to now only screw sections filled with viscous liquids can be exactly modeled with 3D models, a 3D modeling of the whole screw is not possible for many applications. For the calculable sections, 3D modeling offers a wealth of results (see Figure 1.89).

A three-dimensional calculation offers especially fields (e. g., viscosity distribution and temperature fields), which are impossible or difficult to obtain using measurements. From these field quantities scalar (integral) values such as the drive power can be calculated. Examples can be found in Section 4.1 in which measurements and calculations for the dimensionless throughput A_1 and the dimensionless pressure A_2 are compared.

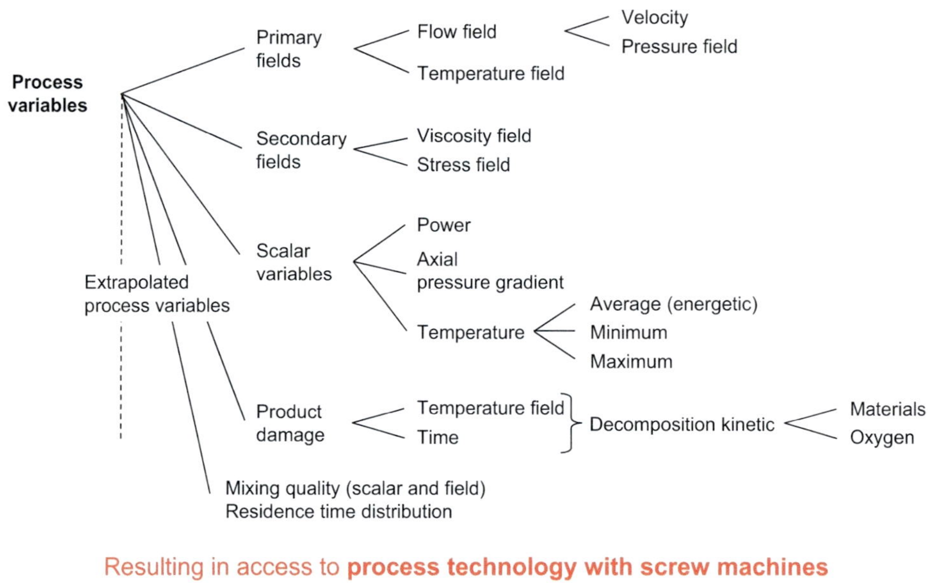

Figure 1.89 Results of a three-dimensional calculation

Three-dimensional models can provide amongst others the following results:

- Calculation of the conveying parameters A_1, A_2, etc. for the one-dimensional models which could only be measured in the past.
- Calculation of representative average values for one-dimensional models.
- Studies of individual influences, e.g. separate influence of the structural viscosity and the temperature dependence of the viscosity.
- Identification of peak loads and their interpretation, e.g. how are areas with high temperatures "mixed in" by the flow field.
- Optimization of existing screw elements (clearance, pitch, etc.) and support/evaluation at new screw elements.

These results enable calculations of chemical engineering processes like mixing, reaction, etc. being really of interest. Process tasks executed in screw machines are shown in Table 1.1.

1.4.5 Summary

The aim of the description of the screw models in this book is to develop the connections from simple to difficult. The focus was placed on the description of models with mostly secured basics. Some less practically relevant models still needing considerable further development have not been specified. An example is the different melting models for plastic granulate.

The modeling of solid materials and the three-dimensional calculation of screw sections partly filled with melt is still in the early stages. Also, the description of complex material blends, as they can occur during compounding, is less practicable. Nevertheless, the one- and three-dimensional models are giving a considerable improvement of knowledge.

With this book, the introduced one-dimensional models are available to the reader. The three-dimensional models are very time-consuming and expensive. For a CFD program normally royalties must be paid and – as not merchantable – an efficient net generator must be developed for the screw geometry. Three-dimensional flow calculations are, therefore, normally not available to smaller companies.

Also, the evaluation of the modeling can therefore be different (see the introductory examples mentioned in Section 1.4.1):

- Task 1: The extruder is fed with the melt; the extruder contains a base polymer with solvent of which all material data are known, main process step is the degassing.
- Task 2: Extruder is fed with granulate; due to the dosing occurs in the extruder a complex blend of which the material data are not existent, there are several process steps (functional zones).

Obviously, the first example can be better described with the existent models that are introduced in this book. In addition, Task 1 occurs normally at a producer of raw materials. This producer normally has a team which can take care of the solutions of the modeling.

The state of knowledge for the modeling of Task 2 is lower. The melting cannot be calculated. The material data are not known. If the compounding is done by a smaller company, most of the time there will be no team of process engineers available.

Conclusion of the example: A smaller company having a relatively more difficult task will rather be inclined to conclude that the modeling is not very helpful.

Conclusion of the author: Even if the process values, due to lacking information with models, cannot always be described in detail, the improvement of knowledge will be increased considerably by the models. This also applies in the case when the models are not used at all, but only the basics of modeling and results on the models are considered.

1.4.6 Prospects and Proposals

1.4.6.1 Program for Extruder Configuration

It was noted in the introduction that no electronic program can propose in detail the extruder configuration with the variances of geometry and process parameters in dependence on the definition of the project. The extruder configuration of the particular task must be executed by experts. The question is whether and when a computer program will be able to perform this.

1.4.6.2 Further Development of Models

The prospect concerning the further development of one-dimensional computer programs is stated in Section 4.4. In the future, three-dimensional models will be able to calculate the screw sections partly filled with melt better, and consequentially new knowledge can be obtained. This also applies to the results of the granulate transport in the extruders. The description of powders in extruders is not yet practicable and tests with the products are required.

The melting of granulate is one of the economic core functions of an extruder (see Section 1.1). This is emphasized in the introduction and the success of extruders is associated with the success of the plastics. Practicable models are not known. Corresponding to the significance of the plastics industry, a research project should be established to push on modeling and understanding of the melting. New screw elements can be developed due to this understanding. The aim should be to increase efficiency and product quality.

1.4.6.3 New Model Applications – Online

Normally, the models for extruders are used "offline", i.e. they are not linked with the process. For example, tests or operating conditions are "recalculated" or computations for the design of an extruder are carried out.

The screw modeling can also be used "online" for process observation, controls, and regulation [7]:

One part of the model is the "observer". It calculates non-measurable or process values being difficult to measure, parallel to the running operation. Figure 1.90 shows an example for the input and output values of an observer relating to a degassing screw.

If the observer also calculates one or several values that can be measured, then deviations between calculated and measured process values will be possible, i.e. the model is faulty (or the measured values are faulty). A further model part, the "corrector", can now use these deviations to adaptively improve the observer model or, at least, to "adjust" model parameters.

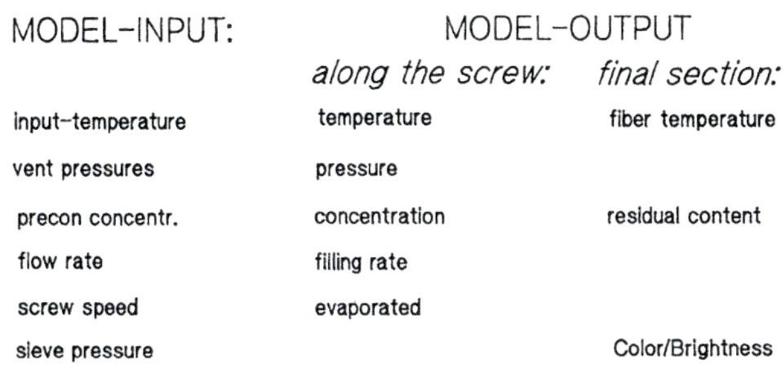

Figure 1.90 Example for the process values of an observer

Whilst the observer model is running parallel to the process and follows quasi the process and calculates missing measuring values, the modeling can also be applied for the "forecast". This model part is named "predictor". The predictor calculates the control factor for a process in that way that a required running condition occurs. A guiding system is placed over the normal control system (Figure 1.91).

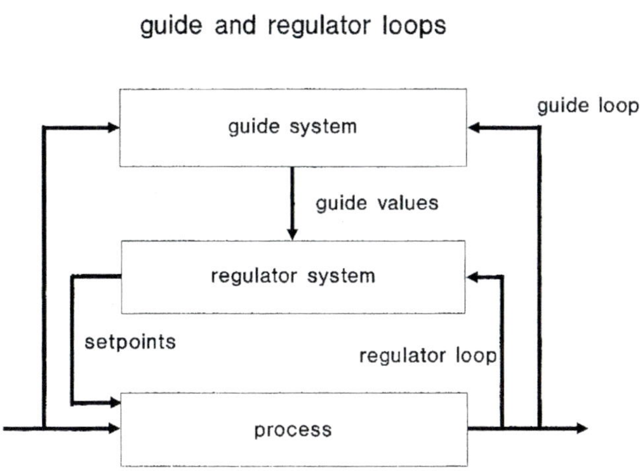

Figure 1.91 Normal control system with super-ordinated guiding system

The pre-settings of the predictor are of course subject to restrictions (e.g., physical limits). Furthermore, limits will arise if not all influences can be indicated to the necessary exactness during the process.

As the cited original literature source is not generally available, the reader is referred to reference [8]. It remains to be seen whether such online models will gain importance for the extruders. "Mini observers" are already installed by calculating for example a specific energy input from driving power and throughput.

1.4.6.4 Process Characterization of Screw Elements by Key Figures

Essentially all elements of the co-rotating twin-screw extruder are a modification or variation, respectively, of the basic geometry as per "Erdmenger", citation from [9]. The modifications of the screw geometries, kinematically not self-cleaning, are manifold. A corresponding number of geometrical different screw elements are offered by extruder manufacturers and producers of screw elements. As per manufacturer's data each screw element has its special advantages. For the user it is difficult to judge the mostly qualitative information. Then the user often executes tests to check whether the advantages will also be reached with his products. This is costly and time-consuming. Therefore, the author would like to propose the characterization of the screw elements by process key figures in the future. These key figures could be provided by producers of extruder and screw elements, universities, and specialized companies. The problem, however, is that companies probably do not want to publish their know-how. Therefore, this would be a research project for the universities. In addition, these key figures will be calculated expediently with three-dimensional models. The efforts are considerable and cannot be executed "on the side". Of course, there will be counter-arguments: key figures are also part of an "information destruction", but this is like the indication of a mix quality when mixing: a three-dimensional piece of information, a concentration distribution, is zero-dimensionally described by a number, a mix quality. If it is specified how the mix quality is defined and on which size of the control volume the mix quality relates, then it is a useful indication for the user and can be used as a quality indicator for products. In the same way, process parameters for screw elements should be developed.

Which process key figures are possible? For the conveying, pressure and performance characteristic of a screw element are these at least the conveying parameters A_1, A_2, B_1, and B_2. It is referred to the above statements as from Equation (1.14) and to the simple example there of A_1 of a screw element "*S*".

As one of the core functions of a co-rotating twin-screw extruder is the mixing and dispersing (Section 1.1), new key figures should be developed and indicated. There are initial, more qualitative approaches, for example at presentations of the company Extricom.

In the future, standardized, representative key figures of a screw element, corresponding to the chemical engineering process such as mixing and dispersing, should be developed. This is initially a demanding theoretical task. On the other hand, in practice, screw elements are constantly being further developed, which must be evaluated for the process. With key figures, the expert and the user can quantify and compare currently available and new screw elements more quickly and ultimately save time and costs.

References for Section 1.4

[1] Bierdel, M., Vergleich der Extrudersysteme, in: Kohlgrüber: VDI-Forum "Zwei- und Mehrwellenextruder", Fall 2015 (annual forum)

[2] Michaeli, W., Schneckenauslegung bei der Herstellung von Polymer-Schichtsilikat-Masterbatches, *Chemie-Technik*, October 9, 2010: *http//www.chemietechnik.de/schneckenauslegung-bei-der-herstellung-von-polymer-schichtsilikat-masterbatches/*, accessed May 2016

[3] Pawlowski, J., *Die Ähnlichkeitstheorie in der physikalisch-technischen Forschung*, Springer-Verlag, Berlin, Heidelberg, New York (1971)

[4] Kuhlmann, H., *Strömungsmechanik*, Pearson Studium (2007), p. 226

[5] Böhme, G., Theoretische Betrachtungen über schleichende Strömungen. In: Festschrift, Universität-GH Essen (1995), pp. 27–40

[6] Conzen, C., Historische Entwicklung und energetische Beurteilung der Gleichdralldoppelschnecke, oral presentation at the VDI conference: Aufbereitungstechnik (2009)

[7] Kohlgrüber, K., Model concept for ... for process control (observer, predictor, corrector) and offline use, internal report, Bayer, December 2, 1988

[8] Kohlgrüber, K., Woelk, G., Energetische Optimierung von Industrieöfen, 9. Folge: Rechnergestützte Führung von Industrieöfen, gas wärme international (1987), Vol. 36, pp. 438–441

[9] Extricom GmbH, Web page: *http//www.extricom.de/index.php?option=com_content&view=article&id=8&Itemid=35&lang=de*, accessed May 2016

[10] Ullrich, M., Das Förderverhalten von Doppelschnecken, internal report, Bayer, February 9, 1989

1.5 Conveying and Power Parameters of Standard Conveying Elements

Thomas König

The parameters A_1, A_2, B_1, and B_2 are required to calculate pressure and power input. In general, these parameters depend on all geometry parameters of the screws.

In addition to the large influence of the ratio of centerline distance to diameter and pitch, the influence of the clearances is also important. This applies to the pressure characteristics, but even more so to the power input.

Different extruder manufacturers have different clearance strategies; sometimes, even screw elements within a series are not geometrically similar. Therefore, even for one manufacturer, several diagrams would be required to display the parameters A_1, A_2, B_1, and B_2. It would be beyond the scope of this book to give the dimensionless parameters for all screw elements on the market. Nevertheless, to provide a possibility for practical calculations, "typical" games of conveying elements were selected and the parameters were determined for different ratios of A/D and DA/DK using CFD calculations to provide the reader with an indication (Figure 1.92 and Figure 1.93).

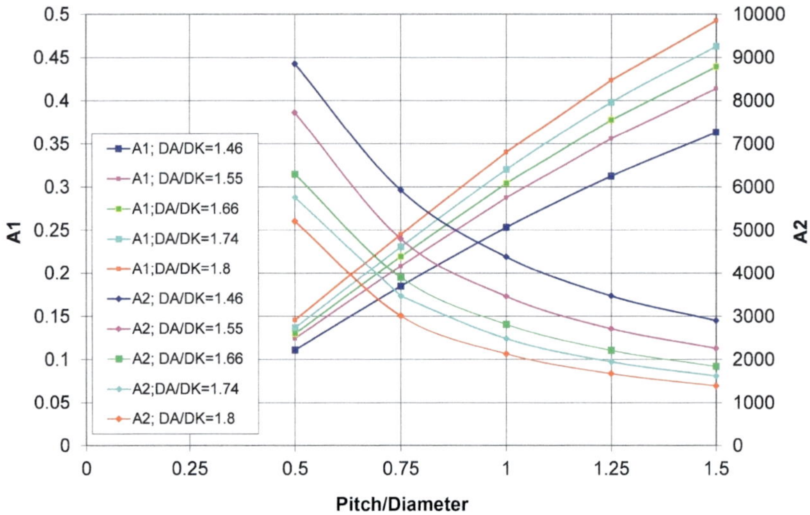

Figure 1.92 Conveying parameters for typical two-lobe extruder elements of different *DA/DK*

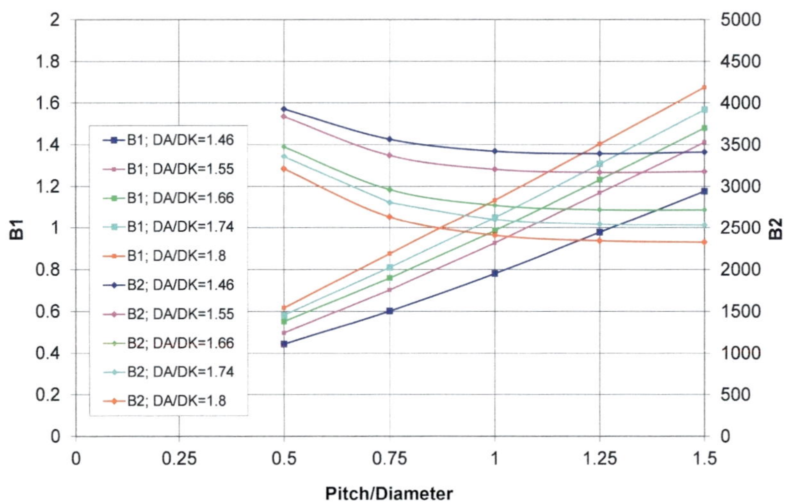

Figure 1.93 Power parameters for typical two-lobe extruder elements of different *DA/DK*

1.6 Frequently Used Symbols

Klemens Kohlgrüber

Process Variables/Parameters	
D_a	Screw outer diameter
D_i	Screw inner diameter
$\dot{q}$	Heat flux density

Process Variables/Parameters	
r	Ratio of melt viscosities $\eta_{dispers} / \eta_{contin}$
$\dot{M}$	Mass flow
$\dot{V}$	Volumetric flow
q	Former symbol for $\dot{V}$
$\dot{V}^*$	Throughput (dimensionless) $\dot{V}^* = \frac{\dot{V}}{nD^3}$, most important extruder parameter
Q	Former symbol for $\dot{V}^*$
Λ	Flow parameter (dimensionless)
n	Speed
$\dot{\gamma}$	Shear rate $\dot{\gamma} = \frac{u}{H}$
Δp	Pressure difference, for example in a certain extruder part with length L
Δp^*	Dimensionless pressure difference $\Delta p^* = \frac{\Delta p D}{\eta n L}$
$\Delta\tilde{p}$	Basic equation for pressure difference $\Delta\tilde{p} = \frac{\Delta p^*}{A_2} = \frac{\Delta p}{\Delta p_{max}}$
P	Power, for example in a certain extruder part with length L
P^*	Dimensionless power number $P^* = \frac{P}{\eta n^2 L D^2}$
$\tilde{P}$	Basic equation for pressure difference (dimensionless) $\tilde{P} = \frac{P^*}{B_2} = \frac{P}{P_{max}}$
M_d	Torque
η_p	Pumping efficiency (dimensionless) $\eta_p = \frac{\dot{V}\Delta p}{P}$
ε	Degree of fill, e.g. definition via Λ
ϑ, T	Temperature
τ	Shear stress $\tau = \eta\dot{\gamma}$
t	Residence time
e	Specific energy input $e = P / \dot{M}$

Geometry Parameters: Six Parameters of the Classic Erdmenger Profile	
D	Barrel diameter
T	Pitch
A	Centerline distance
Z	Number of flights
S	Screw-screw clearance
δ	Screw-barrel wall clearance

Further Symbols in Section 2.1, Table 2.1.

Dimensionless Process Factors, also Called Profile Parameters, A_i und B_i of an Extruder Element	
A_1	Dimensionless intrinsic throughput ($\Delta p = 0$)
A_2	Maximum dimensionless at zero throughput
B_1	Dimensionless throughput at the "turbine point" ($P = 0$)
B_2	Dimensionless power at zero throughput ($\Lambda = 0$)
B_4	Dimensionless power for a partially-filled screw
B_5	Dimensionless power for a screw operating at intrinsic throughput ($\Delta p = 0$)

$$A_1 = \dot{V}^*_{max} \qquad A_2 = \Delta p^*_{max}$$

$$\Lambda = \frac{\dot{V}^*}{A_1} = \frac{\dot{V}^*}{\dot{V}^*_{max}} = \frac{\dot{V}}{\dot{V}_{max}} \qquad \Delta\tilde{p} = \frac{\Delta p^*}{A_2} = \frac{\Delta p^*}{\Delta p^*_{max}} = \frac{\Delta p}{\Delta p_{max}}$$

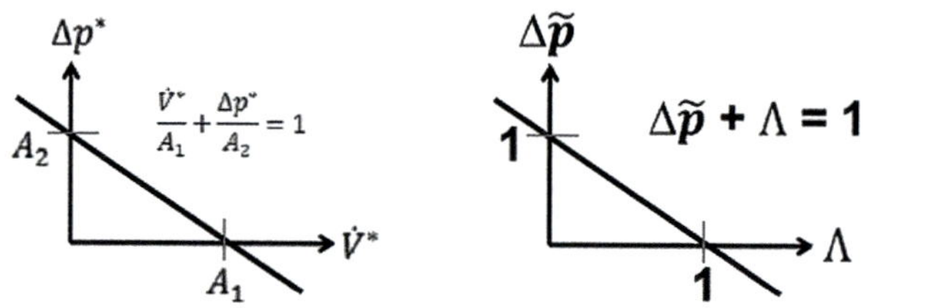

$\dot{V}^*$ = Throughput number

Λ = Flow parameter = "Reynolds number" for axial transport

Figure 1.94 Extruder characteristic curves for the pressure-throughput characteristic for Newtonian liquids

Material Parameter	
η	Dynamic viscosity, for example described by the Carreau approximation $\eta(\dot{\gamma}) = \frac{\eta_0}{\left(1 + \frac{\dot{\gamma}}{\dot{\gamma}_c}\right)^m}$

Explanation of the equation and the notation in Section 3.1, Equation (3.7).

c	specific heat capacity
ρ	density $\rho = \frac{\dot{M}}{\dot{V}}$

2 Basics – Screw Elements

2.1 Geometry of Co-Rotating Extruders: Conveying and Kneading Elements, Including Clearance Strategies

Thomas König

2.1.1 Introduction

The geometry of closely intermeshing co-rotating twin screw extruders is distinguished in practice by the fact that adjacent screw elements are identical in geometry, are symmetrical, and rotate at the same speed [1]. Geometries in which the two shafts rotate at different speeds and which have different numbers of threads [2] have not achieved practical significance and are therefore not considered here.

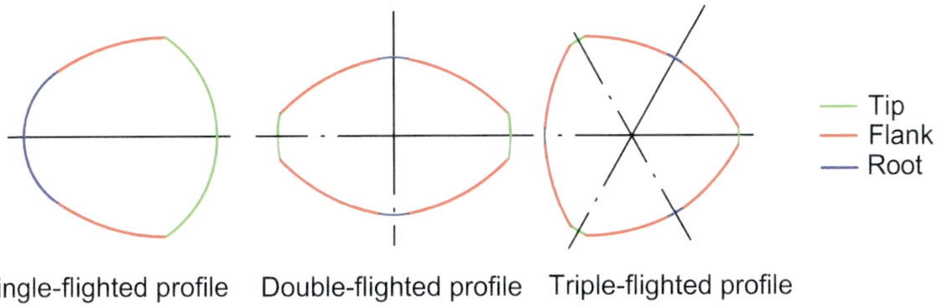

Figure 2.1 Screw profiles with lines of symmetry

A screw profile with Z threads can be divided into $2Z$ symmetrical parts, as illustrated in Figure 2.1. A screw profile is composed of three parts within this symmetrical area: tip, flank, and root. The tip comprises an arc whose diameter equals the external diameter of the screw profile and whose center point is the center of the circle. One edge of each tip borders the adjacent flank. For screws that wipe the inside of the barrel, the flanks each comprise an arc whose radius equals the centerline distance. The flanks merge tangentially into the root area. The diameter of the root area corresponds to the diameter of the screw core, and its center is the center of the profile. The tip cleans the root of the opposite screw and vice versa. The corner of the profile between the tip and the flank cleans the opposite flank. This is the basic geometry first described in [3].

2.1.2 The Fully Wiped Profile from Arcs

The basis of closely intermeshing profiles is the fully wiped screw. The individual sections of the profile of the fully wiped screw are arcs. This can be explained on the basis of the kinematic equivalence in which the rotation of the screws is replaced by holding one screw in a fixed position and rotating the other in a circle at a radius equal to the centerline distance.

The first screw (the "generated" screw) is held still in this consideration, and the second screw (the "generating" screw) is moving. After prescribing a part on the generating screw, it is investigated what sort of profile will be generated on the generated screw. In a sense, the generated screw is "cut out" by the generating screw.

First, a point on the generating screw is considered. This point is situated where the tip and the flank merge. This point, together with every other point on the second screw, moves on a circular trajectory with the radius equal to the centerline distance around the first. Examining this line, one obtains the flank; see Figure 2.2.

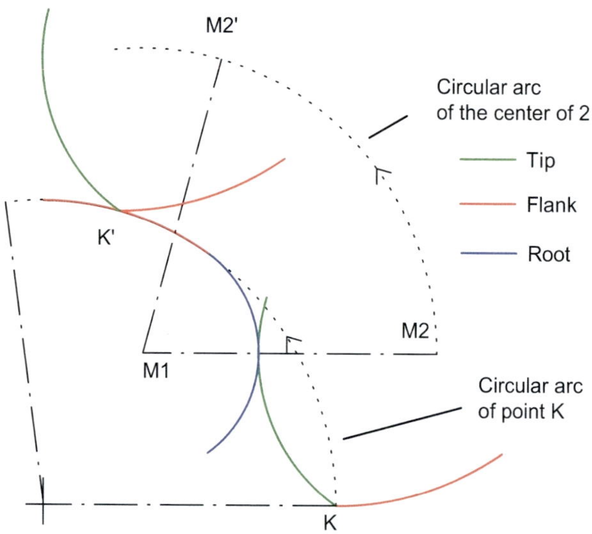

Figure 2.2 Generation of the flank from the transition point between tip and flank of the generating screw

Let us now consider the tip of the generating profile as an arc of radius *R*. The center of this arc rotates at distance A from the center of the first profile. The contact point between these two profiles, which lies on the root of the generated profile, is always on a line connecting the two centers and is at a distance $RI = A - R$ from the center of the first profile; see Figure 2.3.

Just three variables are required to describe a fully wiped profile (Figure 2.4):

- external diameter *DE*
- centerline distance *A*
- number of threads *Z*

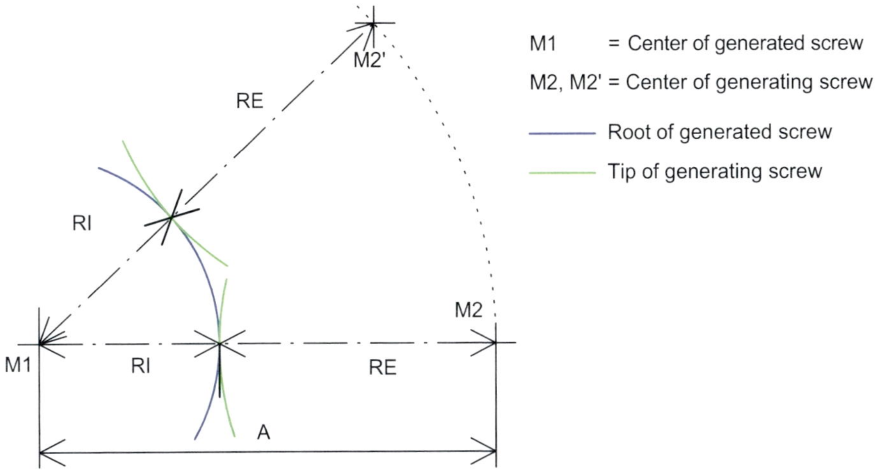

Figure 2.3 Generation of the root as an arc by the tip of the generating screw

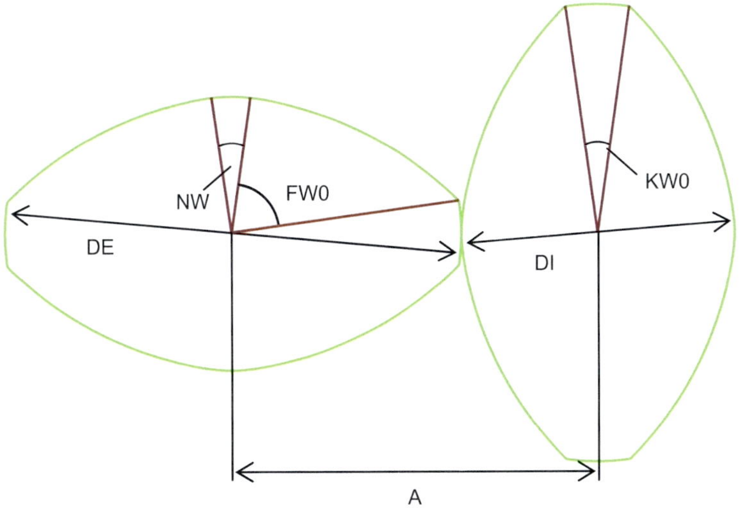

Figure 2.4 Geometric variables of a fully wiped profile

The observations above result in the following equation

$$\frac{DE}{2} + \frac{DI}{2} = A \qquad (2.1)$$

and thus

$$DI = 2A - DE \qquad (2.2)$$

We also see that the root angle is identical to the tip angle of the fully wiped profile.

2.1.3 Geometric Design of Fully Wiped Profiles

What follows is an outline of a geometrical design that enables fully wiped profiles to be produced using elementary geometry. As a first step, we determine the center of rotation M of the profile and draw the symmetry lines 1 and 2.

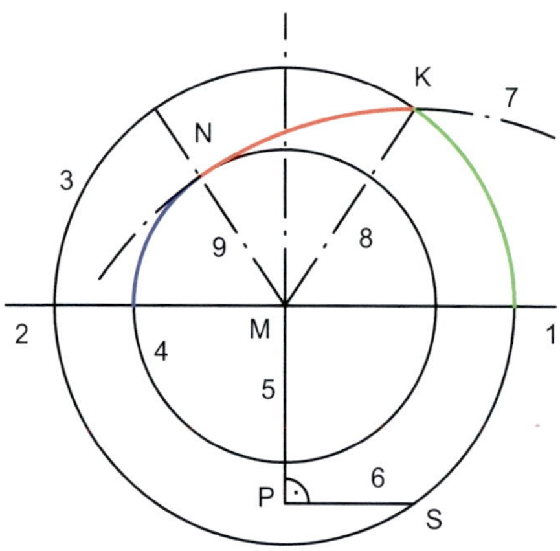

Figure 2.5
Geometric design of a single-flighted profile

If $Z = 1$ (single-flighted profile, Figure 2.5), the lines of symmetry merge at the center of the profile. The two lines are the line of symmetry of the profile. The distance of the profile from its center of rotation is greatest at line 1 and smallest at line 2.

Between the lines of symmetry 1 and 2, we draw a line 5 bisecting the angle on the side facing away from the profile to be drawn. Following this line 5 by half the centerline distance, we obtain point P.

We construct a vertical line with the base P on the side facing the line of symmetry 1. The point at which the vertical 6 intersects the circle 3 is marked S. We draw an arc 7 around S with a radius corresponding to the centerline distance and check point K, which is the point at which this circle intersects the external diameter. This point represents the edge of the screw tips. It is now possible to draw the connecting line 8 between the bisecting point K and the center M. The angle between line 1 and the connecting line is equal to the tip angle.

Mirroring point K on the bisector 5, we obtain point N. The angle between line 9, which connects point N and center, and the line of symmetry 2 corresponds to the root angle.

The profile between the lines of symmetry 1 and 2 is made up of the following arcs: Between line 1 and line 7 is an arc with an external diameter, i. e. a portion of circle 3. It bends sharply at point K and merges into the arc that is part of circle 7. This in turn merges tangentially at point N into an arc that is part of circle 4.

The design of the profile can fail if we make the internal diameter to small. This is the case if point K is outside the lines of symmetry, i. e. the screw can no longer achieve the external diameter.

The geometric design of a double-flighted profile is shown in Figure 2.6.

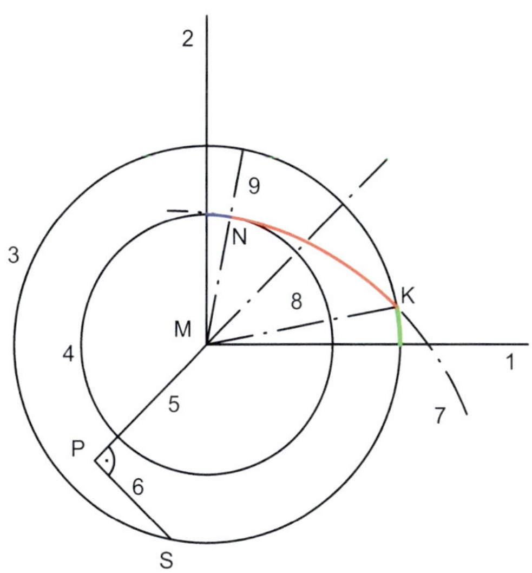

Figure 2.6
Geometric design of a double-flighted profile

2.1.4 Dimensions of Screw Elements with Clearances

A pair of screw elements (see Figure 2.7) gains conveying capacity if the flat profiles are extended perpendicular to its plane by helical rotation. In this case, we speak of a threaded element. The pitch T of the element is another quantity characterizing the element.

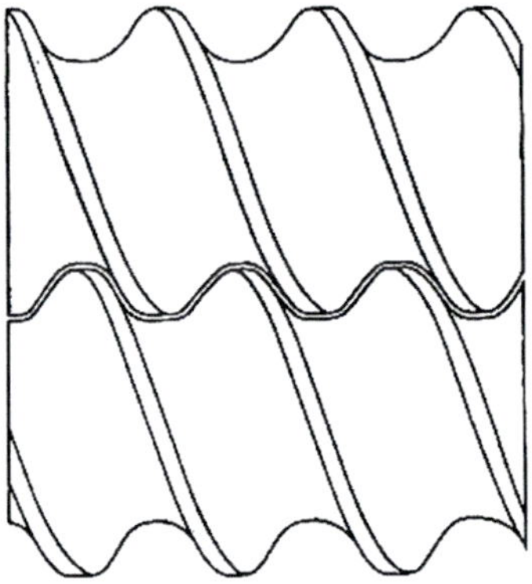

Figure 2.7
Screw element (schematic)

All technical screw elements have clearances between the screws and between the screw tip and barrel wall for the following reasons:

- to prevent metallic erosion
- to compensate for the manufacturing tolerances and unevenness
- to compensate for angle discrepancies
- to compensate for uneven expansion due to different temperatures
- to avoid excessive product stress due to insufficiently wide clearances

Two more dimensions are added to the geometric parameters of the screws: *s* (clearance between the screw screws) and δ (clearance between barrel wall and screw tips).

Figure 2.8 and Figure 2.9 provide an overview of the geometric variables in a screw. In Figures 2.8 and 2.9, the diameter *DE* of the fully wiped screw is larger than the barrel diameter *D* because of the clearances selected for this particular case.

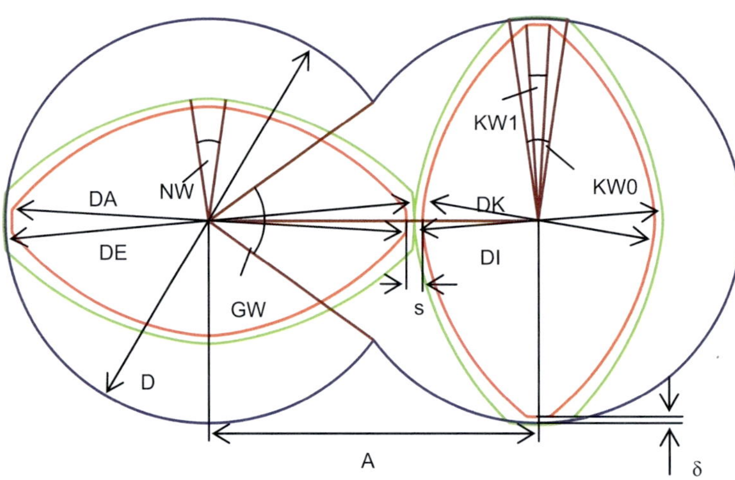

Figure 2.8 Geometric variables of a double-flighted screw profile, cross section; green: fully wiped contour; red: actual contour; blue: barrel wall

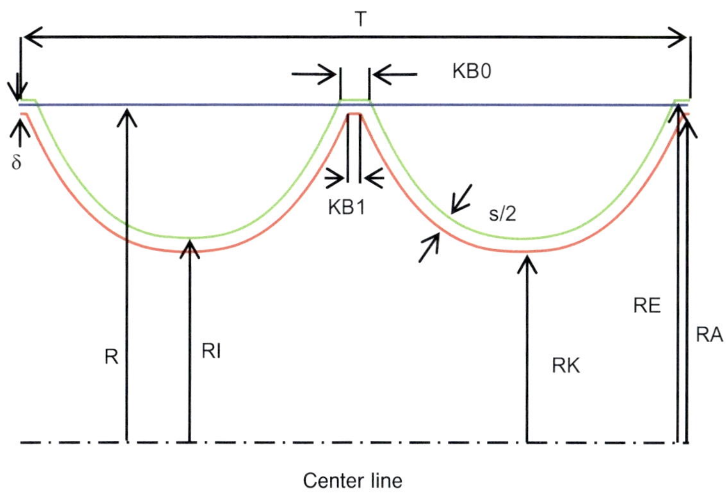

Figure 2.9 Geometric variables of a double-flighted screw profile, longitudinal section; green: fully wiped contour; red: completed contour; blue: barrel wall

Table 2.1 lists the geometric variables of a screw profile.

Table 2.1 Geometric Variables of a Screw Profile, with an Example (Carried out as Planar Offset)

Variable	Symbol	Calculation	Example
Number of threads	Z	Given	2
Centerline distance	A	Given	50 mm
Barrel diameter	D	Given	61 mm
Pitch	T	Given	40 mm
Screw-barrel wall clearance	δ	Given	0.4 mm
Screw-screw clearance	s	Given	0.6 mm
Barrel radius	RG	$RG = D/2$	30.5 mm
Barrel angle	GW	$GW = 2a\cos(A/D)$	69.896° = 1.2199 rad
Screw external diameter, actual	DA	$DA = D - 2\delta$	60.2 mm
Screw external radius, actual	RA	$RA = DA/2$	30.1 mm
Screw core diameter, actual	DK	$DK = 2A - DA - 2s$	38.6 mm
Screw core radius, actual	RK	$RK = DK/2$	19.3 mm
Screw external diameter, fully wiped	DE	$DE = DA + s$	60.8 mm
Screw external radius, fully wiped	RE	$RE = DE/2$	30.4 mm
Screw core diameter, fully wiped	DI	$DI = 2A - DE$	39.2 mm

Table 2.1 Geometric Variables of a Screw Profile, with an Example (Carried out as Planar Offset) *(continued)*

Variable	Symbol	Calculation	Example
Screw core radius, fully wiped	RI	$RI = DI/2$	19.1 mm
Flank angle, fully wiped	$FW0$	$FW0 = 2a \cos(A/DE)$	69.355°
Tip angle, fully wiped	$KW0$	$KW0 = 180°/Z - FW0$	20.645°
Root angle	NW	$NW = KW0$	20.645°
Tip angle, actual	$KW1$	Iterative, depending on clearance strategy	16.645°
Flank angle, actual	$FW1$	$FW1 = 180°/Z - (NW + KW1)/2$	71.355°
Tip width, fully wiped	$KB0$	$KB0 = T\,KW0/(2\pi)$	2.29 mm
Tip width, actual	$KB1$	$KB1 = T\,KW1/(2\pi)$	1.85 mm

If the ratio of the centerline distance to the barrel diameter is too small, the screw cannot properly wipe the external wall. This occurs when the tip angle is calculated to be negative. The fully wiped geometry can be calculated according to Table 2.1 for $KW0 = 0$; this then gives the condition

$$\frac{A}{DE} > \cos\frac{90°}{Z} \tag{2.3}$$

Table 2.2 Centerline Distance Limits for Different Numbers of Threads for Fully Wiped Geometries

Number of threads	A/D
1	No real limit; >0.707 to cover the intermeshing area
2	> 0.707
3	> 0.866
4	> 0.924

The actual tip angle is always smaller than the fully wiped tip angle, so the minimum values for the centerline distance in Table 2.2 may actually not be enough to guarantee a self-wiping profile.

Figure 2.10 shows the tip angle for fully wiped geometries depending on the ratio of centerline distance to diameter. It also shows the limit of zero tip angles for the number of threads larger than one.

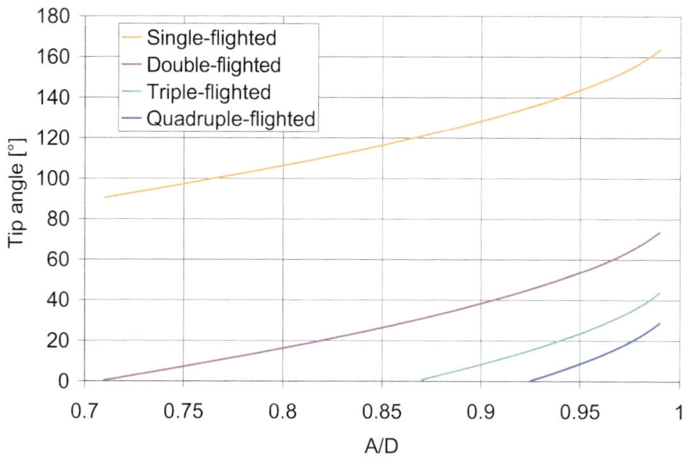

Figure 2.10 Tip angle for fully wiped geometry for thread numbers one to four as a function of the ratio of centerline distance to diameter

2.1.5 Transition between Different Numbers of Threads

Process requirements sometimes mean switching between elements with a different number of threads along an extruder. This always requires transfer elements, spacers in the simplest case, because otherwise a slight axial movement of the screw shafts towards one another will cause them to collide. Figure 2.11 shows this for the transition between triple- and single-flighted screw elements.

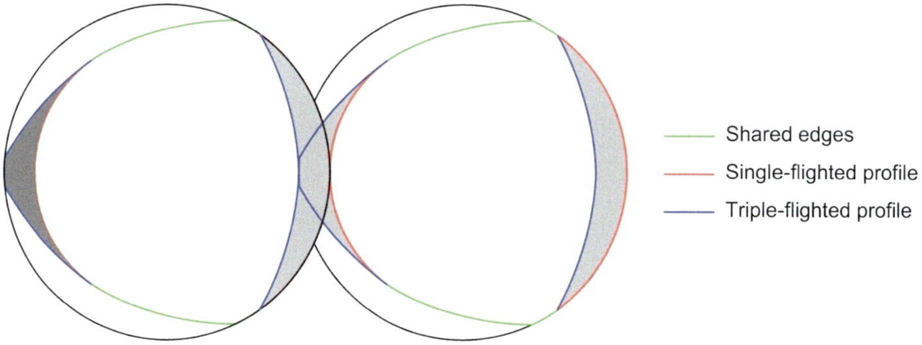

Figure 2.11 Transition between triple- and single-flighted screw elements. The potential collision surfaces are shown in gray

2.1.6 Calculation of a Screw Profile for Production According to Planar Offset

In the case of the planar offset, one starts out with the planar section of the fully wiped profile and moves this inwards by half the screw-screw clearance. To calculate a screw profile according to the planar offset, it is possible to proceed as follows:

From the specified geometry data *DG*, *A*, *Z*, *s*, and δ of the extruder, we obtain the values *DE*, *R*, *DI*, *RI*, *KW*0, and *KB*0 of the fully wiped screw profile according to Table 2.1. Let (x, y) be the coordinates of the fully wiped screw profile and (x_a, y_a) the coordinates of the actual profile. The planar profile is thus

$$y = \begin{cases} RI & 0 < x \leq \frac{KB0}{2} & \text{Root area} \\ f(x) & \frac{KB0}{2} < x \leq \frac{T}{4} - \frac{KB0}{2} & \text{Flank area} \\ RE & \frac{T}{4} - \frac{KB0}{2} < x \leq \frac{T}{4} & \text{Tip area} \end{cases} \tag{2.4}$$

where $f(x)$ describes the profile of the arc of the flank [2]:

$$f(x) = -RE\cos\left(2\pi\frac{x - KB0/2}{T}\right) + \sqrt{A^2 - RE^2\sin\left(2\pi\frac{x - KB0/2}{T}\right)} \tag{2.5}$$

The profile of the screw in the root and tip area is easily determined: We reduce the radius by half the screw-screw clearance and determine the point (x_a, y_a) in the flank that is at a distance $\delta/2$ from the point (x, y) from the curve $f(x)$.

$$x_a = x + \frac{s}{2}\frac{f'(x)}{\sqrt{f'(x)^2 + 1}} \tag{2.6}$$

$$y_a = f(x) - \frac{s}{2}\frac{1}{\sqrt{f'(x)^2 + 1}} \tag{2.7}$$

where $f'(x)$ is the derivative of $f(x)$ with respect to x:

$$f'(x) = \frac{2\pi \mathrm{RE}\sin\left(\frac{2\pi(x - KB0/2)}{T}\right)}{T} - \frac{\pi RE^2\cos\left(\frac{2\pi(x - KB0/2)}{T}\right)}{T\sqrt{A^2 - RE^2\sin\left(2\pi\frac{x - KB0/2}{T}\right)}} \tag{2.8}$$

This approach is illustrated in Figure 2.12.

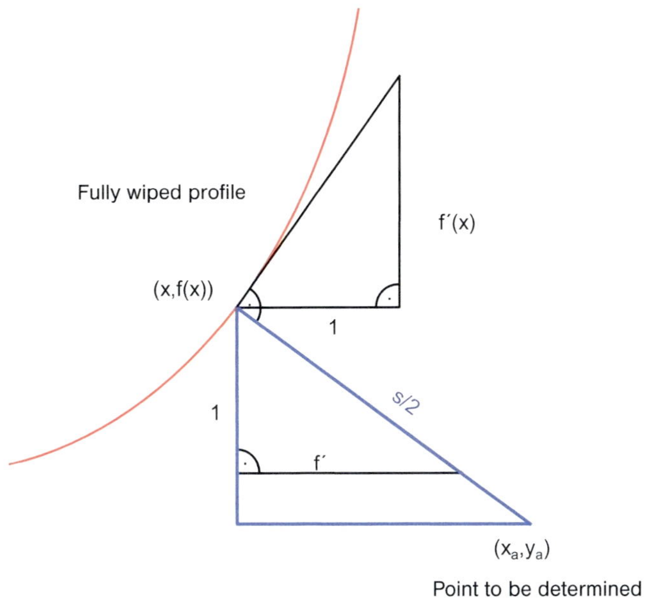

Figure 2.12 Determining a point on the actual screw profile

The tip width of the screw to be produced, *KW*1, is not the same as the tip width of the fully wiped screw. *KW*1 is determined by finding the intersection of the tip with the actual flank curve. The solution is determined mathematically by using the following equation:

$$RA = f(x_K) - \frac{s}{2}\frac{1}{\sqrt{f'(x_K)^2 + 1}} \tag{2.9}$$

which can be determined iteratively. Finally, we also calculate the actual x position of this point:

$$x_{a,K} = x + \frac{s}{2}\frac{f'(x_K)}{\sqrt{f'(x_K)^2 + 1}} \tag{2.10}$$

The tip width in question is thus

$$KB1 = \frac{T}{Z} - 2x_{a,K} \tag{2.11}$$

The coordinates of the actual contour then are

$$\left(x_a, y_a\right)=\begin{cases}\left(x, RK\right) & 0<x\leq\dfrac{KB0}{2}\\ \left(x+\dfrac{s}{2}\dfrac{f'(x)}{\sqrt{f'(x)^2+1}}, f(x)-\dfrac{s}{2}\dfrac{1}{\sqrt{f'(x)^2+1}}\right) & \dfrac{KB0}{2}<x\leq\dfrac{T}{2Z}-\dfrac{KB0}{2}\\ \left(x, RA\right) & \dfrac{T}{2Z}-\dfrac{KB1}{2}<x\leq\dfrac{T}{2Z}\end{cases} \tag{2.12}$$

This procedure is usually implemented in a computer program.

The planar coordinates $\left(x_a, y_a\right)$ can also be used to determine cylindrical coordinates (r, α) according to

$$r=\sqrt{x_a^2+y_a^2} \qquad \alpha=\text{atan2}\left(x_a, y_a\right) \tag{2.13}$$

so that a relationship of the form

$$r=r(\alpha) \tag{2.14}$$

can be obtained. The spatial contour for manufacturing a conveying element then is

$$\vec{x}(\alpha.z)=\begin{pmatrix}\cos\alpha\, r\left(\alpha-\dfrac{2\pi z}{T}\right)\\ \sin\alpha\, r\left(\alpha-\dfrac{2\pi z}{T}\right)\\ z\end{pmatrix} \tag{2.15}$$

Producing profiles according to planar offsets leads to a dependence of the profiles on the pitch, which is especially marked in the tip angle $KW1$. Figure 2.13 illustrates this.

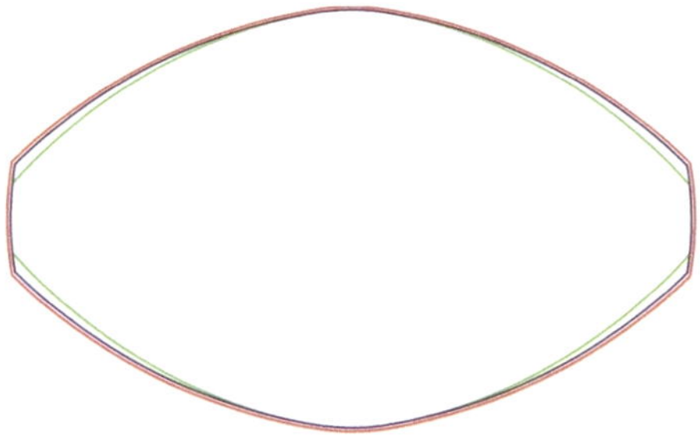

Figure 2.13 Comparison of profiles in which A/D = 0.82, s/D = 0.01, $\dfrac{\delta}{D}=0.01$; red: fully wiped, blue: T/D = 2, green: T/D = 0.4

2.1.7 Free Cross-Sectional Area

The free cross-sectional area A_{free} plays an important role in conveying behavior of threaded elements. The volume of a product with idealized wall sliding could be conveyed with one revolution, and corresponds to the free cross-sectional area A_{free} multiplied by the pitch T (sometimes called "jack conveying"). Since liquids adhering to walls have approximately half of this conveying capacity, this can be estimated to

$$A1 = \frac{1}{2}\frac{A_{free}}{D^2}\frac{T}{D} \tag{2.16}$$

The exact free cross-sectional area depends above all on the number of flights, the barrel diameter, and the centerline distance. A further, less important influence is the clearances, which are not directly included in Equation (2.16), when using a clearance strategy other than centerline distance enlargement.

As an estimate, the free cross-sectional area of the fully wiped contour can be used. This is shown in Figure 2.14.

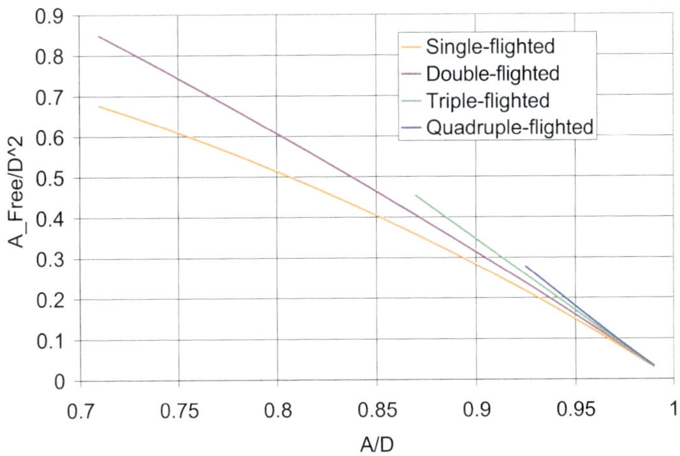

Figure 2.14 Free cross-sectional surface divided by the square of the diameter for different number of threads (fully wiped geometry)

2.1.8 Surface of Barrel and Conveying Elements

The inner surface of a barrel with length L is

$$A_G = (4\pi - 2GW)\, RG\, L \tag{2.17}$$

where the barrel angle GW is given in radians. As a characteristic, the inner surface divided by the length L can be used:

$$A_{G,L} = (4\pi - 2GW)RG \tag{2.18}$$

The surface of a screw element of length L can be calculated when knowing the function $r(\alpha)$ as a surface integral according to

$$A_S = \int_0^L \int_0^{2\pi} \sqrt{\left(\frac{dr}{d\alpha}\right)^2 + r^2 + \frac{4\pi^2}{T^2}\left(\frac{dr}{d\alpha}\right)^2 r^2}\, d\alpha\, dz \tag{2.19}$$

The area of an element pair per length then is

$$A_{S,L} = 2\int_0^{2\pi} \sqrt{\left(\frac{dr}{d\alpha}\right)^2 + r^2 + \frac{4\pi^2}{T^2}\left(\frac{dr}{d\alpha}\right)^2 r^2}\, d\alpha \tag{2.20}$$

so that the surface of the screw element depends on both the contour and the pitch.

It is not possible to give a closed representation of the screw surface, especially with complex clearance strategies. In the following diagram (Figure 2.15), the surface of the screw elements, divided by the barrel diameter, is given for different ratios of centerline distance to barrel diameter, for $\delta/D = 0.007$ and $s/D = 0.007$, respectively.

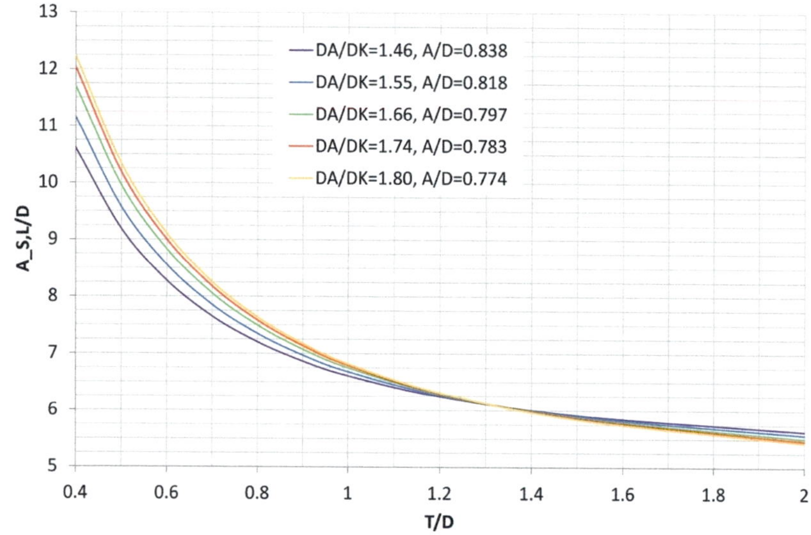

Figure 2.15 Surface area of conveying elements, divided by the barrel diameter D and the length of the element L, as a function of the pitch T per diameter, for screws of different A/D ratio. Clearances $\frac{\delta}{D} = 0.007$ and $\frac{S}{D} = 0.007$

2.1.9 Kneading Elements

Prismatic discs made of self-cleaning profiles, which are offset with respect to one another, are called kneading elements. It is not possible to transport a bulk product along such an element without deformation. If, for example, plastic pellets are conveyed through a section of an extruder with a kneading element, they will be deformed and will be exposed to an energy input that contributes to the melting process. Kneading elements were described for the first time in [4].

If individual discs within kneading elements are arranged in the form of a conveying screw element, they can convey product and also build up pressure. Z-flighted kneading discs, offset at $180°/Z$ with respect to each other, do not convey product and must therefore be overrun, i.e. the product must be forced through them by conveying elements or kneading elements. Backwards-conveying kneading elements are also used predominantly for backward-pumping plastification zones. Figure 2.16 illustrates neutral and conveying kneading elements.

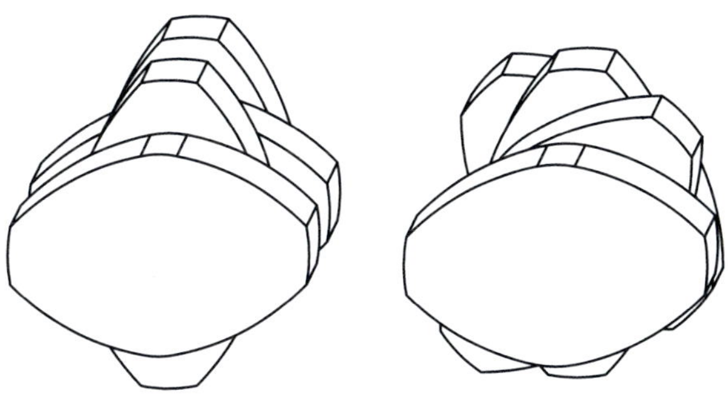

Figure 2.16 Example of double-flighted kneading discs

Special forms of self-cleaning profiles are used in kneading elements. These include circular eccentric elements (see Figure 2.17) [4], and triple-flighted eccentric kneading elements (see Figure 2.18) [5]. The circular eccentric elements are circles with diameter $A/2$. Triple-flighted eccentric kneading elements can also be used in the case of geometries in which a normal triple-flighted geometry is not possible because the centerline distance is too small. With these elements, the wall is cleaned by only one of the three tips.

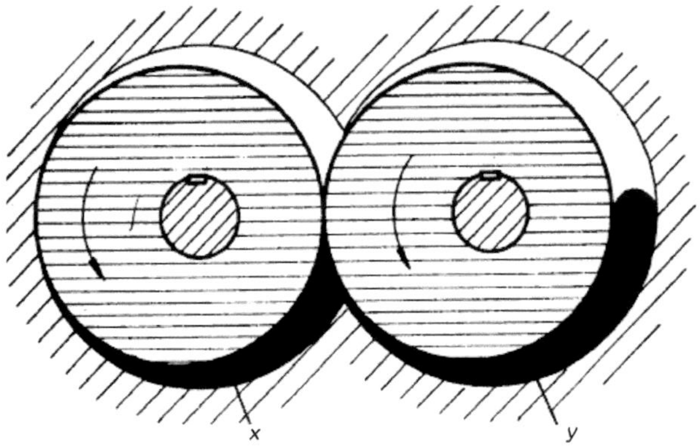

Figure 2.17 Eccentric kneading blocks from [4]

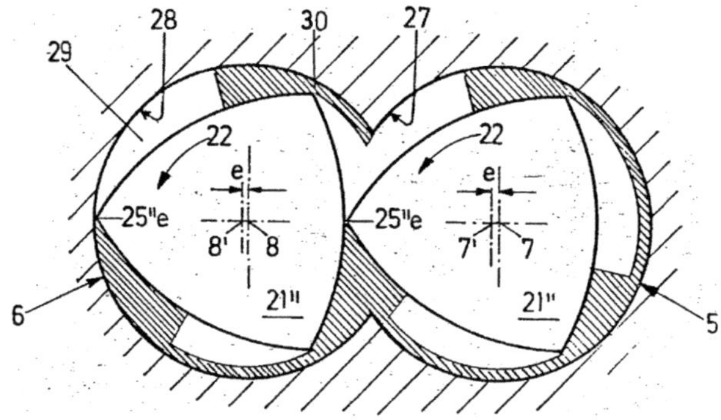

Figure 2.18 Triple-flighted eccentric kneading blocks [5]

The geometry data of kneading discs differ from those of conveying screw elements. There is no pitch. Instead, there is the distance between kneading discs, *s*2, and their widths *B*.

Figure 2.19 gives an overview of their dimensions.

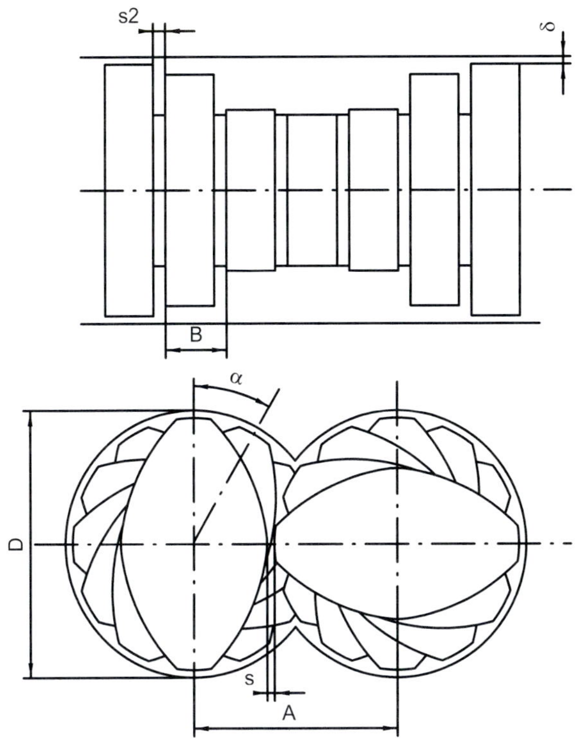

Figure 2.19 Overview of the dimensions for kneading discs

2.1.10 New Developments with Screw Geometries

Beginning in 2009, a different principle for the construction of screw profiles was described in a number of patent applications [6]. These patents are based on the specification of a generating screw element, which must be convex and have a circular arc whose radius r_i may lie between zero and the centerline distance. From this generating screw profile, a generated profile is then derived, which also consists of circular arcs, whose radii then are $A - r_i$.

This principle allows significantly more freedom in the construction than the screw elements described earlier. The patent literature claims that they can be used, for example, to reduce the energy input [7] (see Figure 2.20) or for improved efficiency during pressure buildup [8].

Elements designed according to this principle were not available on the market in 2018.

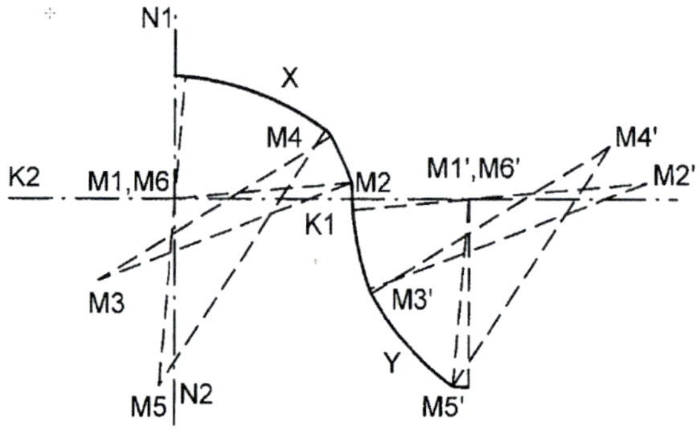

Figure 2.20 Part of a design of a generalized screw element, from [7] with arcs and their center points on the left side (M1 to M6) and the corresponding arcs with their center points on the right side (M1' to M6').

References for Section 2.1

[1] Schuler, W., "Kinematische und geometrische Grundlagen sowie ausgeführte Schneckengeometrien und Fördercharakteristik." In: *Der Doppelwellenextruder, Grundlagen und Anwendungen,* VDI-Verlag (1998), ISBN 3-18-234201-0

[2] Booey, M. L., "Geometry of fully wiped Twin-Screw equipment." *Polym. Eng. Sci.* (1978) 18 (12), pp. 973–984

[3] Meskat, W., Erdmenger, R., *Vorrichtung zum Verkneten, Gelatinieren und Verpressen von plastischen Massen.* German patent application DE862668 (1944)

[4] Erdmenger, R., *Misch- und Knetvorrichtung.* German patent application DE813154 (1949)

[5] Häring, E., Weihrich, G., Wörz, W., *Gleichdrall-Doppelschneckenkneter mit Knetscheiben.* German patent application DE3412258 (1984)

[6] Bierdel, M., König, T., Liesenfelder, U., *Verfahren zur Erzeugung von Schneckenelementen.* German patent application DE102008029304 A1 (2008)

[7] Bierdel, M., König, T., Liesenfelder, U., *Schneckenelemente mit reduziertem Kammwinkel.* German patent application DE102008029305 A1 (2008)

[8] Bierdel, M., König, T., Liesenfelder, U., *Schneckenelemente mit reduziertem Energieeintrag beim Druckaufbau.* German patent application DE102008029306 A1 (2008)

■ 2.2 Screw Elements and Their Use

Ralf Kühn

For continuous processing of highly viscous products, two-flighted, co-rotating twin-screw extruders have proven themselves since their development in the 1950s. They have replaced many discontinuous processes and in future will aid continuous production of further products using new processes.

Certainly, this system can also credit its success to the fact that it is enormously flexible and that its modular components - housing and screw set elements - have been developed continuously. For example, the primary aim of a complete self-cleaning function was abandoned in order to develop new element geometries using different construction principles. Only by continually developing individual components it has been possible to implement a broad variety of processes with a single machine system and to process a variety of products. Today, along with the self-cleaning standard elements - the feeding and kneading elements - there are also a variety of mixing, cutting, stretching, compacting, and homogenizing elements with only partial self-cleaning. Some of them were first developed only for special processes; meanwhile, however, others have established themselves in process engineering and have been universally implemented.

The discovery of new construction methods for element profiles, meanwhile, has enabled the creation of complex elements that move far beyond the Erdmenger patent, and by which the self-cleaning principle can be retained in its entirety (see Section 2.3 Patents: WO 2009/152947, WO 2011/039016). The search for expedient profile forms and their potential areas of use is currently the subject of multiple research projects.

Individual geometries and elements dating back to the origins of development, however, are no longer being implemented. Instead of small-volume extruders in which three-flighted elements are often used, large-volume, two-flighted machines which exhibit a clearly larger D_a/D_i diameter ratio of 1.4 to nearly 2.0 are overwhelmingly in use today.

2.2.1 Construction of Screw Elements

Screw elements are composed of individual geometric building blocks. The most well-known basic module is the Erdmenger profile, consisting of 4, 8, or 12 arcs, respectively, and constituting as such the contour for one-, two-, or three-flighted elements. The flank arcs of such profiles have a radius corresponding to the sum of the exterior and interior radii (i.e., crest and root radius) of the element.

In order to establish the conveying characteristics of the element, the profile can be twisted along the element axis in various ways (see Figure 2.21). According to pitch angle and direction you will get a conveying or reverse-conveying element. Should the profile be carried out only parallel to the element axis, then discs will result which as such would have no conveying effect. The discs are however often combined into one kneading block in order to retain an element which exhibits a defined mixing and conveying effect.

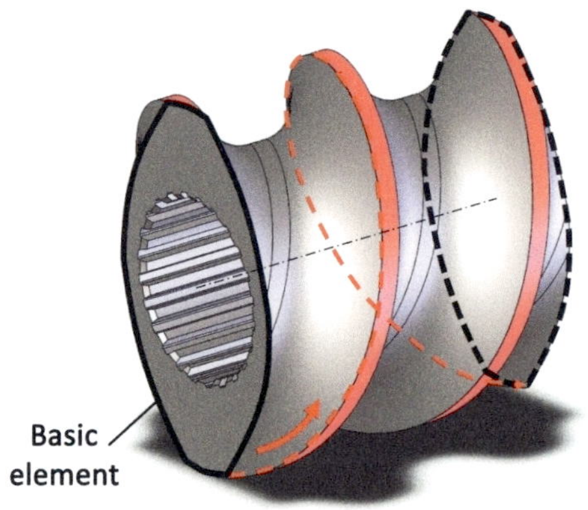

Figure 2.21
Twisted connection of a basic element

The pitch need not be constant along the element axis, but rather may follow any function. If the element is cut at a random point in a level vertical to the longitudinal axis, one will always have the basic profile of the element.

The facing elements of both drive shafts are synchronized such that during one rotation at least one element crest wipes the housing surface and, moreover, the elements are mutually tangent once at each point with little clearance (see Section 2.1). Since all surfaces that come into contact with product are wiped by a moving geometry, such a system is *self-cleaning*.

The overwhelming majority of today's standard screw elements are implemented symmetrically and congruently for reasons, among others, of ease of use and manufacturing (Figure 2.22):

- *Elements are symmetrical* when their geometry can be congruently reproduced by mirroring. Based on the symmetry, it does not matter which end face of the element is first placed onto the axis. This markedly reduces the risk of assembly errors.
- *Elements are congruent* when the forms of the right and left shaft elements are identical. These elements can be installed on each of the shafts, thus reducing the number of necessary spare parts and the risk of assembly errors.

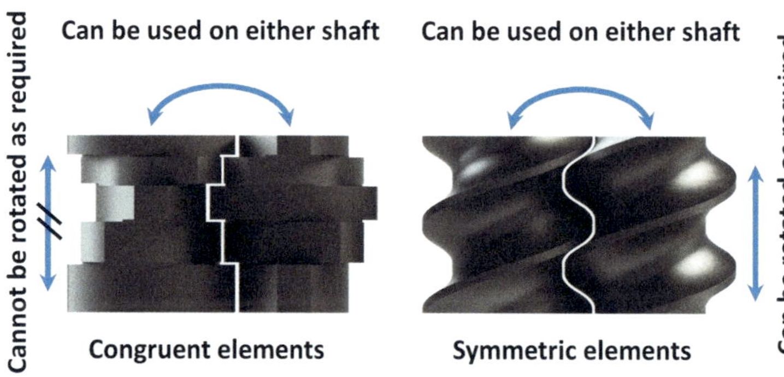

Figure 2.22 Arrangement of screw set elements

The Erdmenger patent also describes some non-symmetrical and non-congruent elements that are rarely used in practice. The transition elements by which a basic element transitions along the axial threaded connection into another cross section without abandoning the principle of self-cleaning are one particular exception. As an example, a transition element for the connection of a one- and two-flighted Erdmenger profile is displayed in Figure 2.23. With this cross-section transfer for the left and right shaft, there are different elements, because two-flighted elements – in contrast to one-flighted elements – must be arranged at a 90° offset angle to one another. In order to not hinder the rotation of the shaft, both elements must be assembled in the prescribed installation direction and position.

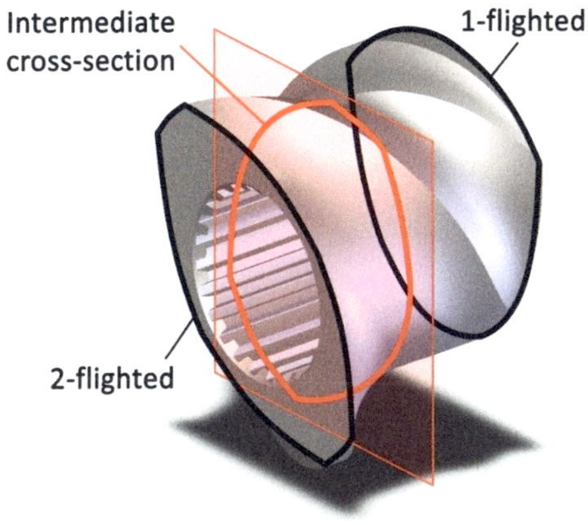

Figure 2.23
Transition element

The basic elements of the size of an extruder are explicitly defined by a few, characteristic parameters. On **conveying elements,** these parameters are:

- *Number of flights*
- *Pitch:* This is usually described by the axial length in millimeters that is necessary for a complete (i. e., a 360° rotation of the cross section) threaded connection (Figure 2.24).
- *Pitch direction:* The pitch direction is the direction of the rotation angle of the threaded connection. If the threaded connection is counter to the rotational direction of the screw shaft, then a screw element with positive conveying direction (a conveying element) results. If the threaded direction is in the direction of the screw shafts, then the element will have a backward-pumping effect (Figure 2.25).
- *Element length:* Element length is given in millimeters and is ordinarily equal to the pitch or to an integer value of multiples or dividers of the pitch, respectively.

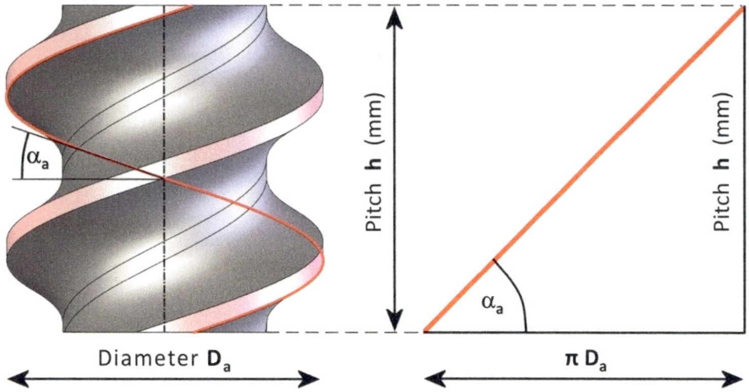

Figure 2.24 Pitch of a conveying element

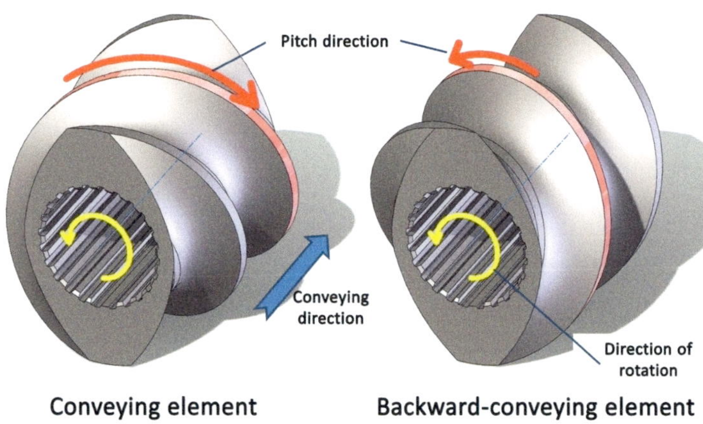

Figure 2.25 Pitch direction

Limiting element length to multiples or fractions of the element diameter is not physically necessary, but it does ease element kit assembly as well as up- or down-scaling machines.

An axially parallel body, a so-called kneading disc, results if the basic element along the element axis "extrudes" and does not skew. Such discs are usually not used individually but rather are included in combinations of up to 7 elements to form a **kneading element** (kneading block). This kneading element is defined using parameters similar to those of a conveying element:

- *Number of flights*
- *Number of kneading discs* that are included within an element.
- *Offset angle:* The offset angle describes the rotation of a kneading disc relative to the preceding disc. Information is usually provided in angle degrees (Figure 2.26).

- *Pitch direction:* The pitch direction describes the direction of the disc offset. Disc offset can occur in either clockwise or counterclockwise direction. Depending upon the rotational direction of the screw shaft, kneading elements thus result with either positive or negative conveying direction, i. e. conveying or sealing elements. At a disc offset of 90°, no conveying effect results; the mixing effect of the element predominates.
- *Element length* measured in millimeters.

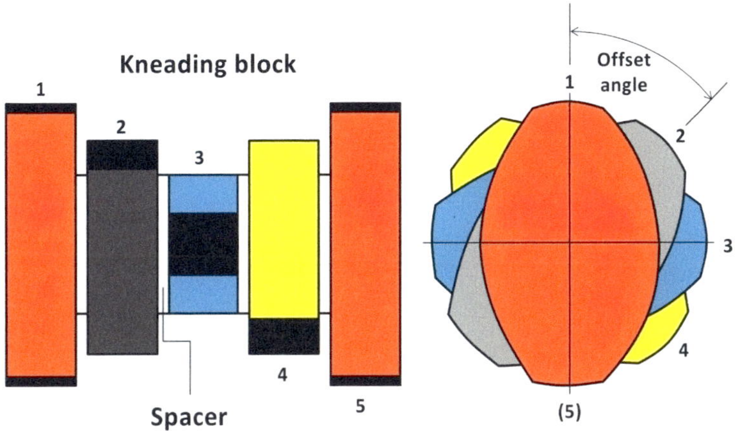

Figure 2.26 Pitch of a kneading element

In order to maintain the clearance necessary for operation between corresponding kneading blocks, a spacer is inserted between every disc of a kneading block. Thus, the actual disc width is somewhat less than the quotient of element length and the number of discs.

Supplemental to the standard elements, there is a large number of variations which do away with reciprocal cleaning or wiping of the housing wall in favor of other technical engineering options. The necessity of complete self-cleaning depends upon the procedure and must be tested in each individual case. With low-viscosity products that adhere strongly to metal and require an extreme degree of purity – i. e. in the pharmaceutical field – only self-cleaning elements are used. When processing non-adhesive or highly viscous products, such complete self-cleaning is frequently not necessary. The geometry that results from the self-cleaning principle can even be counterproductive if shear peaks or "hot spots" occur in narrow crevices which may lead to product damage.

New construction methods enable the development of intermeshing profile cross sections that are not composed exclusively of arcs in accordance with the Erdmenger principle. Thus, very complex cross sections can be created using skillful combinations of involute curves with classical arcs. The elements arising for both drive shafts in such cases are either entirely identical, or can be very different and not congruent. The contours of such new special elements are often described using nicknames such as “potato” or “camel” (Figure 2.27). Compared to classical elements, these new forms can be precisely modified to optimize individual characteristics of technical procedures, thus achieving a reduction in shear peaks or a marginal increase in the degree of pumping effect with certain profile forms. With tangential transitions or small transition angles between profile segments, converging crevices can be constructed which will support the generation of extensional flows for dispersed mixing.

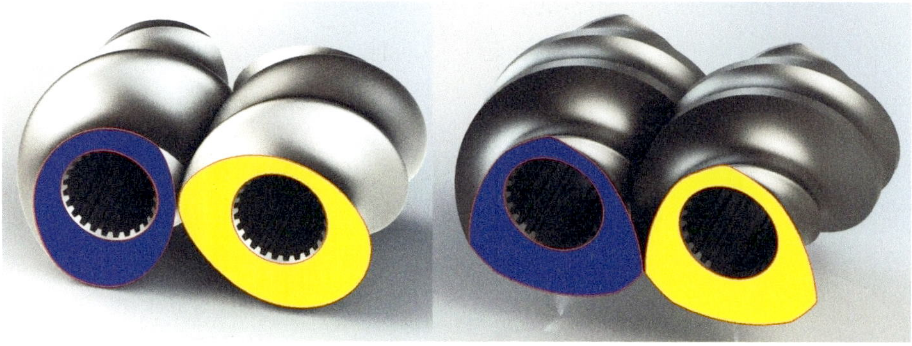

Figure 2.27 New intermeshing elements

2.2.2 Combining Screw Elements

Screw elements can be combined in any sequence reasonable for process-engineering, as long as a few requirements are met:

- The elements must rotate past each other without making contact, i.e. the elements must be correctly mounted in pairs along both shafts. To simplify installation, as a rule, symmetrical and congruent elements are used, i.e. the opposing elements on both screw drive shafts are identical.
- The installation position must be correct relative to the screw shaft. This is particularly important to observe when concurrently using elements of different flight counts. For this reason, one- and three-flighted elements are installed “profile-parallel” (without angular offset) and two-flighted elements are offset at 90° (Figure 2.28).
- Sufficiently large axial clearance between both elements of the shaft must be available. As a rule, this is the case with screw elements which are based upon the same basic element and are installed onto the previous element with no angular offset. Sufficiently large axial clearance has already been incorporated into

the construction of these elements (see Section 2.1). Thus, one can, for example, add two-flighted screw elements with different pitch flush with each other as long as their beginning and ending cross sections lie covering each other.
If, in contrast, elements of differing flight count, individual kneading discs, and special elements are combined with each other, or if standard elements are installed onto the previous element with angular offset, then the danger arises that the axial clearance needed for operation will not be available. In this case, special transition elements (for example, from two- to three-flighted) and/or spacer rings must be installed at the point of the profile transition. If this specification is not heeded, the danger will exist that elements of both drive shafts will come into contact, resulting in increased abrasion or even damage to the machine (Figure 2.29).

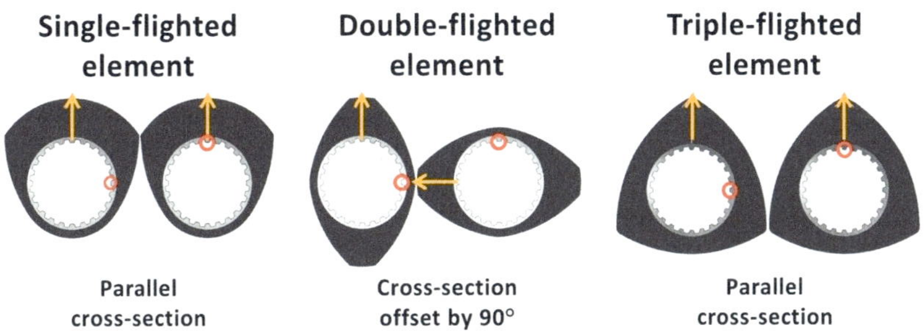

Figure 2.28 Assembly specification for screw elements of different flight counts

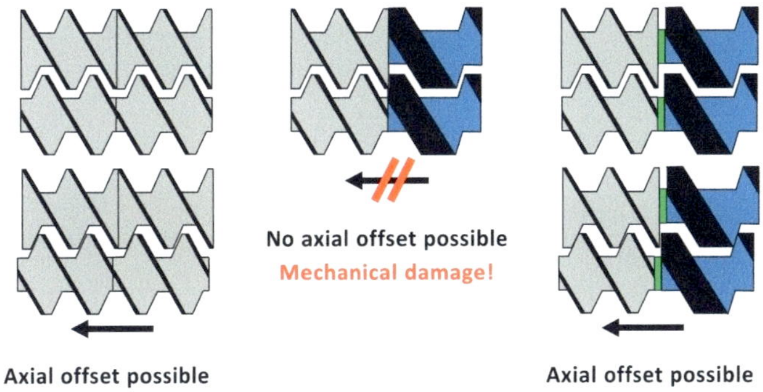

Figure 2.29 Combination of elements with different flight counts

For constructive reasons, it is not always possible to combine elements of different flight counts while maintaining the self-cleaning principle. The maximum diameter ratio for which three-flighted intermeshing profile elements can be manufactured is $D_a/D_i = 1.36$ (see Section 2.1). This limit, resulting from the geometrical dependency of the elements which belong together, lies at a diameter ratio of

D_a/D_i = 2.46 for a two-flighted profile, but no such limit exists for a one-flighted profile. Accordingly, three-, two-, and one-flighted elements can be manufactured only for extruders with a relatively small diameter ratio of $D_a/D_i \leq 1.36$.

It must be noted that the free volume, meaning the space available for the product in the screw channels, reduces with the reduction of the flight count at constant diameter ratio. In the case of machines with a D_a/D_i diameter ratio of 1.55, the relative free volume (i.e., the free volume relative to the total volume of the housing bore) decreases from approximately 36% for two-flighted elements to approximately 32% for one-flighted elements.

Conversely, the free volume increases with the diameter ratio. Processes which require large free volume are therefore carried out on machines exhibiting a large diameter ratio of 1.8 for example. Older machines from the early days of twin-screw extruders, with the then-customary D_a/D_i diameter ratio of 1.22 and three-flighted elements, achieved a free volume of approximately 20%.

In the case of high volume, for two-flighted extruders (D_a/D_i > 1.4), which are overwhelmingly in use today, only one- and two-flighted elements *with seal profiles* are possible. Elements with higher flight counts inevitably have larger radial clearances and are therefore not completely self-cleaning (Figure 2.30).

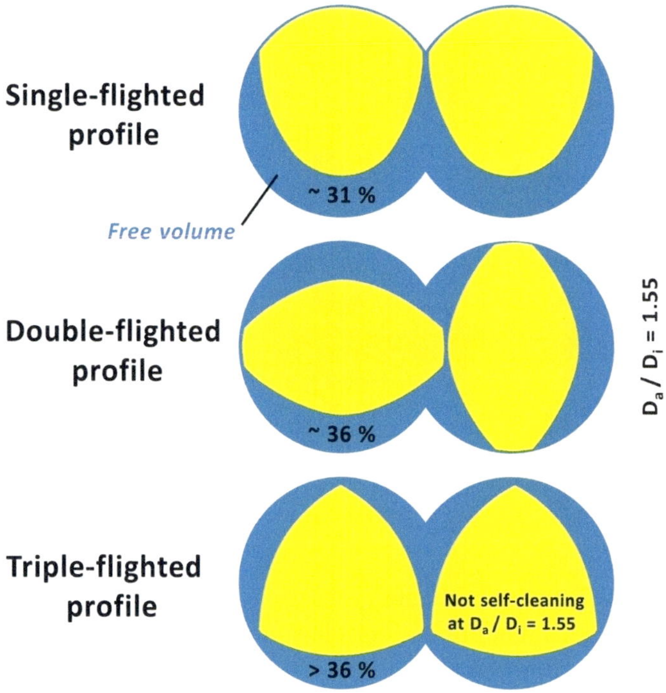

Figure 2.30 Free volume of one- and multiple-flighted screw elements

2.2.3 Screw Elements and Their Operating Principles

2.2.3.1 Conveying Elements

Conveying elements draw the product - such as pellets, powder, ground material, molten material, etc. - into the extruder, transport it downstream to the first processing zone, and compress it as needed. They then take over transport between the individual processing zones, create retention time in the devolatilization space, and finally generate the necessary pressure at the end of the extruder to press the product through the discharge sections.

Conveying elements transport the product downstream in a figure-eight-shaped double helix and transfer it to the intermeshing area from one screw shaft onto the other. The drag flow outside the apex obeys the laws of all screw feeders, i. e. the degree of conveying effect is dependent upon the friction characteristics between the product and the screw surface or, respectively, between the product and the housing surface. In the intermeshing area, i. e. in the transfer area between the two screws, the interplay of the screw elements reduces the free volume upon the transferring shaft (Figure 2.31, green) and increases the volume on the receiving shaft (Figure 2.31, red). As a result of this displacement effect, a type of axial force feed results which differentiates the twin-screw extruder from the single-shaft variety.

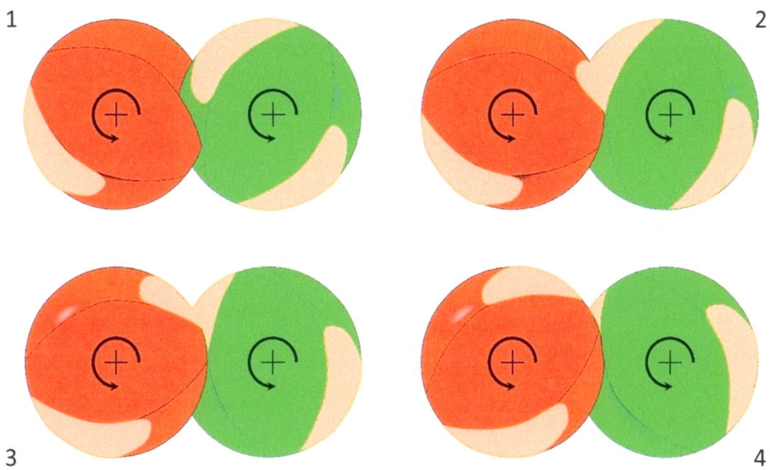

Figure 2.31 Displacement effect in the intermesh area

With one-flighted elements, the change in volume and, along with it, the displacement effect, is markedly larger than with two-flighted elements. Here, the wide screw crests hinder product transition onto the neighboring shaft and lead to a redirection of the material stream (Figure 2.32).

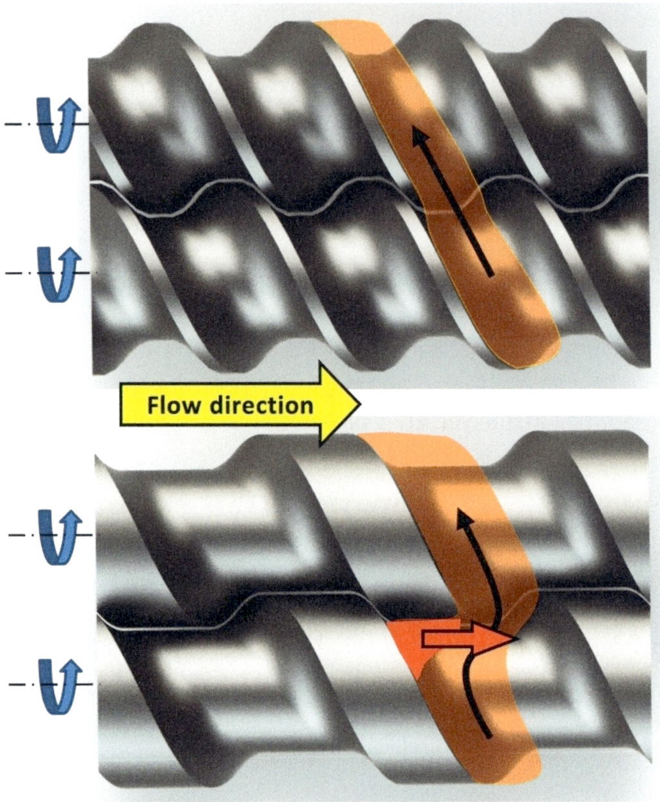

Figure 2.32 Conveying mechanism of one- and two-flighted screws

Conveying elements are available with various pitches and lengths. They are usually constructed such that the form and position of both transverse sections of the element are identical. Typically, two-flighted elements whose length encompasses an entire or a half pitch, i. e. a complete or a half profile turn, are used (Figure 2.33). This principle enables elements to be installed in a continuous profile with no angle offset or discontinuity to the neighboring element.

Pressure-building conveying elements scrape the screw housing almost exclusively with little clearance to minimize leakage over the crests and contact zone of the element. If these clearances are set too high, then too much material will flow back against the desired flow direction because of the counter pressure, and the conveying effect of the element will be lost.

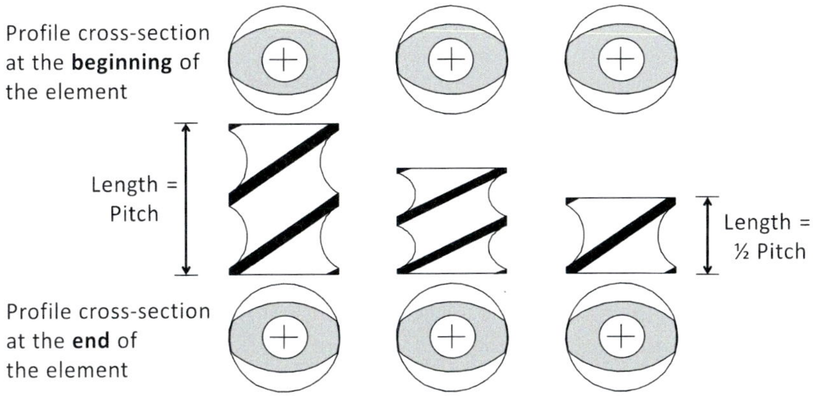

Figure 2.33 Correlation of pitch and length in conveying elements

If any profile corrections to axial clearance are neglected, the transverse section of a self-cleaning screw element will always show the same cross-section area, independent of the executed screw pitch or pitch direction. If an increase of the free cross section is desired for better product addition, then one must deviate from the contour of the seal profile. Thus, shear edge or box section elements (Figure 2.34) offer an increased free volume of 15 to 30% as opposed to the standard element. Still, they harbor the risk that product which is no longer scraped from the counter profile will remain on the flanks of the element.

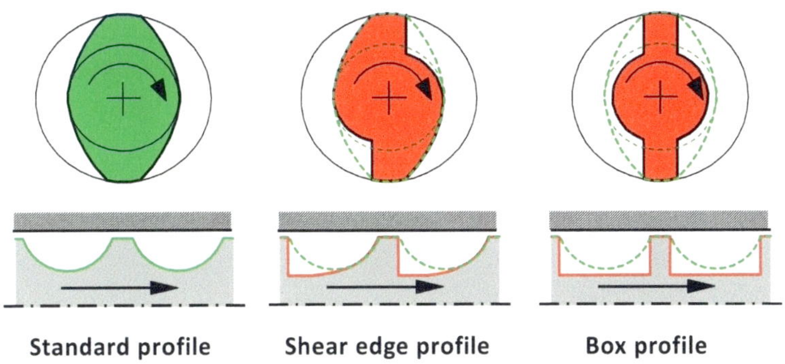

Figure 2.34 Conveying elements with increased free volumes

Conveying elements with one- or two-flighted seal profiles, as well as with shear edge or box profile, differ significantly in factors that are important for the conveying performance of the extruder – the *free volume* and the *conveying efficiency* (Figure 2.35). One-flighted elements have the smallest free volume and demonstrate the best conveying efficiency. With shear edge elements, a larger quantity of product can be conveyed (largest free volume); because of the open cross section, however, they still have a low conveying effect which corresponds to that of a single-shaft extruder with the same profile.

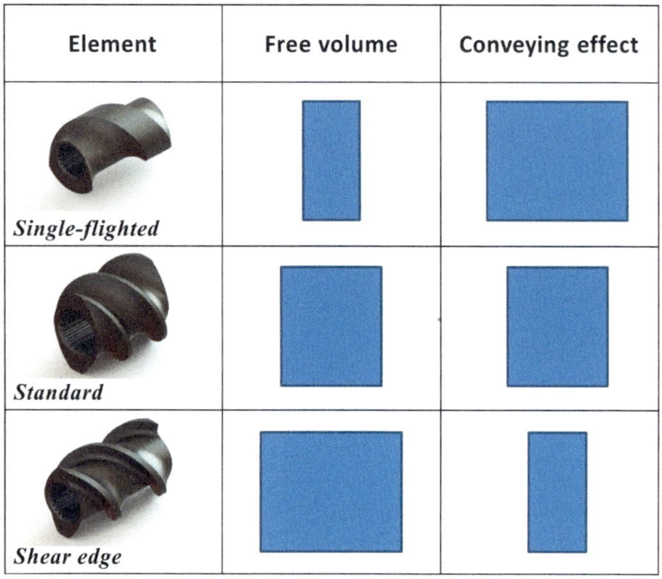

Figure 2.35 Operating principle of conveying elements

The cross-section area of the seal profile, and with it, the free volume of a screw channel, is independent of the pitch of the element. In contrast, with increasing pitch, the conveying output of the element passes through a maximum. As limit values of a minuscule or a theoretically infinitely large pitch, the element possesses absolutely no conveying effect. At the prescribed product throughput, the degree of filling in the screw channel thus depends upon the pitch of the element. The back-pressure length – i. e. the length of the processing area in which pressure is built up – is markedly larger at very small or very large pitches than at an optimal mid-range pitch (Figure 2.36, above).

Even the degree of pumping efficiency – the relationship of the pressure build output to the overall output – passes through a maximum with increasing pitch.

If a screw element is chosen for feeding that has a too small pitch, it will not work at its optimal operating point (Figure 2.36, below). Possibly, a clearly lesser degree of pumping efficiency will be achieved, as would be the case using an element with greater pitch. If the operating throughput lies far beyond of the optimal throughput of the element, beyond the area of active conveying, then the element may dam up the product flow and would necessarily be overrun by conveying elements located further upstream.

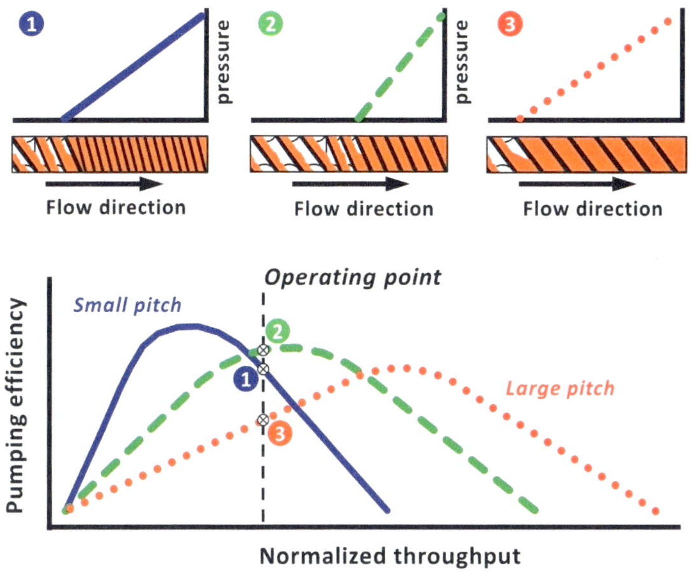

Figure 2.36 Conveying effect as a function of pitch

The free volume and free cross section are independent of the element pitch vertical to the axes, but this does not apply to the cross sections of a screw channel. With decreasing screw pitch *H*, the cross section *A* decreases, while the length of the helical mount (i.e., the length of the coil) increases (Figure 2.37). This has a direct influence upon materials transport in the helical mounts, which is hindered by a narrow cross section and longer transport routes. Thus, for example, devolatilization in which gas is drawn off through the helical mounts between the product, is better (i.e., at low pressure loss) realized using elements with greater pitch.

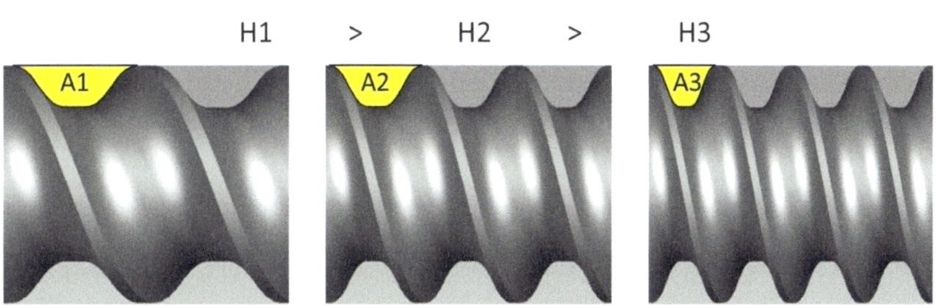

Figure 2.37 Channel transverse section depending upon element pitch

For the pressure build-up section, an optimal screw pitch may be calculated for the desired extruder operation method, insofar as all necessary material substance data, as well as the process engineering limiting conditions, are known. In this context, "optimal" means that the necessary head pressure is established using the best efficiency, i.e. with minimal energy consumption. In reality, however, the number of element pitches available for use is limited, so that one element which comes closest to being the optimum is chosen from the "element kit".

Elements with very high pitch, such as > 6 *D*, are no longer used as classical conveying elements (Figure 2.38). They can be implemented in conveying or reverse conveying, or as plasticizing elements in the melting segment or downstream as homogenizing elements.

Figure 2.38 Screw elements with high pitch

Calculation of optimum pitch is markedly more difficult for the product feed section than for the discharge section where the material is available, already fused and ideally homogeneously mixed. As a rule, a pellet- or powder-form material is conveyed in the feed section; such a material can be deformed or even compacted in the transport process and partially melted. Additionally, transport in the feed zone can be influenced by gas that is siphoned off via the feed hopper. At this time, selection of the appropriate pitch for product feeding is based essentially on experience. A screw pitch of 1.5 to 2 *D* has proven to be optimal.

2.2.3.2 Kneading Elements

Kneading elements, or kneading blocks, are used primarily for plasticization of polymers and dispersing of fillers and reinforcing materials, but also for mixing processes.

In kneading blocks, several individual discs - usually with a seal profile - are assembled into a single element. The discs are offset against each other at a constant angle from kneading disc to kneading disc. Standard kneading elements are, analogous to conveying elements, symmetrically (point symmetry) constructed, meaning that here the form and placement of both transverse sections are identical. The elements are defined by the number of discs, the offset angle of the discs, and their length (Figure 2.39).

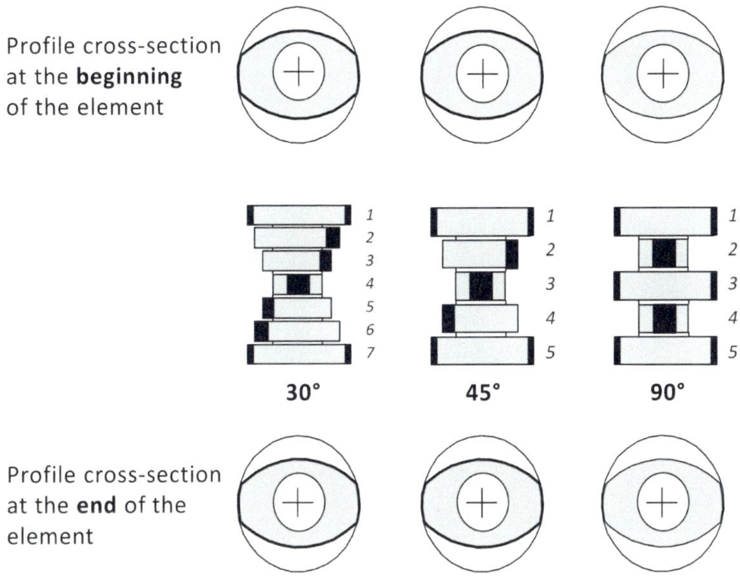

Figure 2.39 Offset angle and number of discs in kneading elements

Offset angle, number of discs, and width of discs (or element length) ultimately determine the operating principle of the kneading element. The greater the offset angle between the kneading discs, the more open the element is in the axial direction. Inevitably, the feeding performance of the kneading element is reduced, while the axial mixing effect increases. With a disc offset of 60° for triple-flighted or 90° for double-flighted, feed-neutral kneading elements result (Figure 2.40).

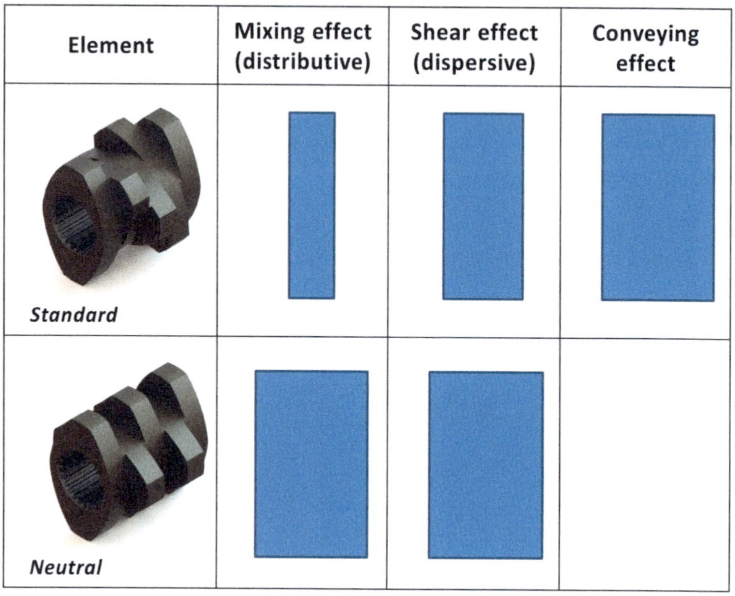

Figure 2.40 Effect of kneading elements – offset angle (double-flighted)

The width of a kneading disc is responsible for the dispersive effect of the kneading element. With the width of the element, the likelihood grows that a particle will be forced into the shearing gap between the crest of a disc and the housing wall. In other words, wide kneading discs have a pronounced dispersive effect versus the more distributive effect of a narrow kneading disc (Figure 2.41).

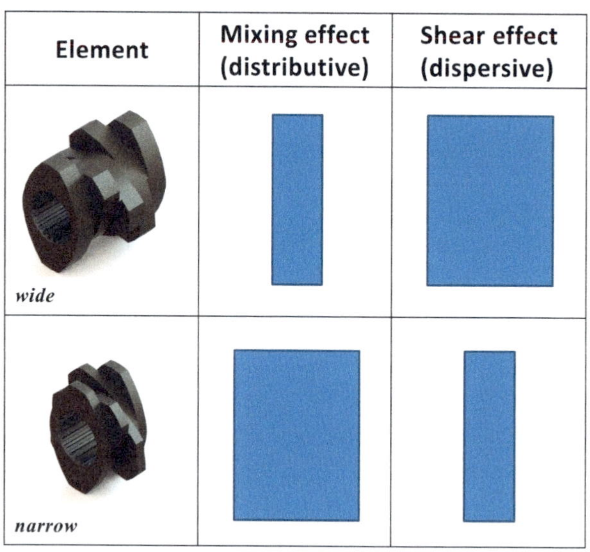

Figure 2.41 Effect of kneading elements – disc width

Common to all *standard kneading elements* is intensive product shear stress between the corresponding flanks of the kneading discs on both screw shafts. The discs move in opposing directions past each other with narrow clearance, thereby creating a narrow shearing gap. Because of the movement of the discs relative to one another and the larger side area of the disc compared to the crest area, the resulting overall shear stress for the product is, as a rule, markedly greater than, for example, that in the crest gap of a two-flighted element.

This condition led to the development of so-called *shoulder kneading discs* which can be implemented with reduced disc width. With increased axial clearance between the corresponding kneading discs, shearing in the axial gaps is decreased.

Increased axial clearance can be created for these elements in different ways. Either the outer diameter on one part of the disc is reduced to a prescribed mid-range diameter (Figure 2.42), or this disc section is reduced completely to the interior diameter, corresponding to a combination of a narrower kneading disc with a spacer. On the shoulder kneading discs, the front surfaces may be vertical - as with standard kneading elements - or at an angle to the shaft axes.

Figure 2.42 Kneading element with modified kneading discs

Three-flighted kneading elements that are implemented in current high-volume extruders have very narrow, pointed crests, only two of which - conditioned upon the construction principle - can wipe the housing. A symmetrical, self-cleaning, three-flighted profile is, as already described, not possible for large D_a/D_i diameter ratios.

With the narrower crest width and the enlarged radial clearance on one, if not an all three ridges, the shear stress for the product is clearly reduced.

In contrast to the two-flighted profile, whereby only three product channels result, five melt channels are formed with the three-flighted profile. This causes a more consistent product stress and a concurrently more intensive product mixing effect. Therefore, installation of such elements in the melting section is preferred.

The disadvantage of a lack of self-cleaning can be balanced out by eccentrically arranging the three-flighted kneading discs on their rotational axis (patents: Coperion and Extricom). Using this technical trick, at least one crest of one disc will clean the housing wall as well as the screw lying opposite to it (Figure 2.43).

Figure 2.43 Three-flighted kneading element

2.2.3.3 Sealing Elements

In the case of *counter-rotating* extruders and intermeshing screws, closed chambers are created using their special geometry along the screw. In other words, there is an axially closed system. *Co-rotating*, intermeshing screws, in contrast, create an axially open system. Most processing tasks, however, require sections that must operate under conditions that are independent and different from each other. An intensively kneading plasticizing section, a partially filled ventilation section under atmospheric pressure, the adding and mixing section for fillers, or a vacuum devolatilization section must work separately from one another while the extruder is in operation. Such a division is enabled, for example, by counter-conveying sealing elements. Although, geometrically speaking, these elements are also axially open, during operation they lead to a filling of elements located upstream and thus separate the different process segments with "melting plugs" (Figure 2.44).

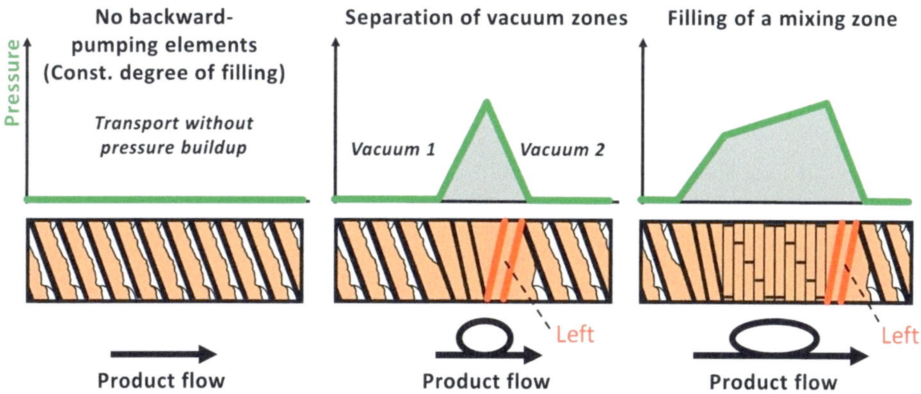

Figure 2.44 Operating principles of counter-conveying elements

Aside from elements with counter-conveying pitch (the so-called "left elements"), all feed-neutral elements, as well as those elements whose active conveying output is smaller than the product stream to be conveyed, belong to the sealing elements. In particular, these are conveying elements with very low pitch.

Pressure generation, length of the filled section, and the accompanying energy consumption in this section can be influenced by the selection of the sealing element (Figure 2.45). Counter-conveying elements always have a sealing effect, independent of operating conditions. The sealing effect of a counter-conveying kneading element is, conditioned upon the axial openings between the kneading discs, weaker than that of a counter-conveying screw element.

The sealing effect of feed-neutral elements or elements with less pitch always depends upon throughput and the characteristics of the product, as well as the screw rotation speed. As a rule, they are less suited to reliably separate the processing sections.

For flow restriction elements, discs that reduce the entire cross section of the housing boring down to one narrow gap are also an option. However, these are used only in specialized applications and rarely for separating process sections; this is described in more detail in the "Special Elements" Section 2.2.3.5.

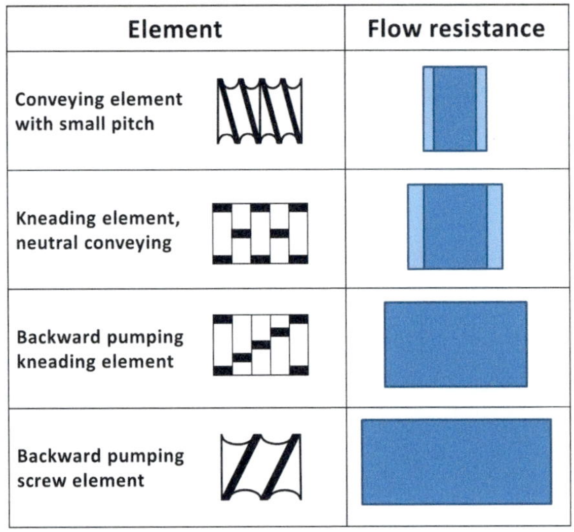

Figure 2.45 Comparison of the effect of counter-conveying elements

2.2.3.4 Mixing Elements

Kneading elements - even those with narrow, modified discs - mix the product and are predominantly self-cleaning. Alongside the desired mixing effect, with kneading elements, significant product shearing always results. Material is more or less strongly sheared between the kneading discs as well as in the crest section of the element, and is thus both mechanically and thermally stressed.

Mixing elements, in contrast, are constructed such that shear-intensive areas are avoided whenever possible, facilitating a distributive mixing of the product stream. Usually, one achieves this goal by increasing the gap between the corresponding elements and between the element and housing wall, as well as by forming new element openings. All of these measures run counter to the self-cleaning principle. Mixing elements which are based upon conveying, feed-neutral, or counter-conveying basic elements are thus only partially or non-self-cleaning.

2.2.3.4.1 Conveying Mixing Elements

- *Segment screw elements* (Figure 2.46) were first implemented in their simplest form in 1975 for melt homogenization at the screw tip. With these elements, a standard conveying element was divided into individual segments which then were arranged offset at certain angles to one another, both conveying and counter conveying, comprising the actual mixing element.
- *Screw mixing elements* (SME) (Figure 2.47) are based upon conveying seal profiles in which the flanks of the basic element are penetrated by notches in order to lead a partial product flow back into the upstream screw channel.

Conveying mixing elements work most effectively in filled extruder sections, meaning that they should work against pressure. Only when counter pressure is present parts of the product will flow back through the gaps in the screw crest and mix with product located in the screw channel lying upstream.

Figure 2.46 Segment mixing element

Figure 2.47 Screw mixing elements (SME)

2.2.3.4.2 Feed-Neutral Mixing Elements

- *"Igel" elements* (Figure 2.48) result when a conveying basic element overlaps a counter-conveying profile with the same pitch and flight depth. The remaining element body exhibits very sharp-edged spikes which give the element its name ("Igel" is German for "hedgehog"). The spikes periodically break the melt stream into individual partial streams and lead to a mixing of product from the creation of localized transverse flows.
- *TME elements* (turbine mixing elements) are created from cylindrical discs whose outer diameter D_a corresponds to that of the screw. Into the flank of the disc, uniformly distributed notches are milled so that a type of impeller results as remaining geometry. Ultimately, form and orientation of the blades are defined by form and milling direction of the notches relative to the screw channel. Depending upon the placement of the blades, conveying, feed-neutral, or counter-conveying elements will result. However, the conveying effect fails due to the low element length and open geometry, such that these mixing elements are better counted as feed-neutral elements.

  A disc with a core diameter D_a should be set as a partner on the countershaft for the TME element. Often, the tooth block is implemented as a combination with this disc (Figure 2.49).

Feed-neutral mixing elements can be installed at any point in the screw configuration. They are always completely filled and must be overrun by upstream elements.

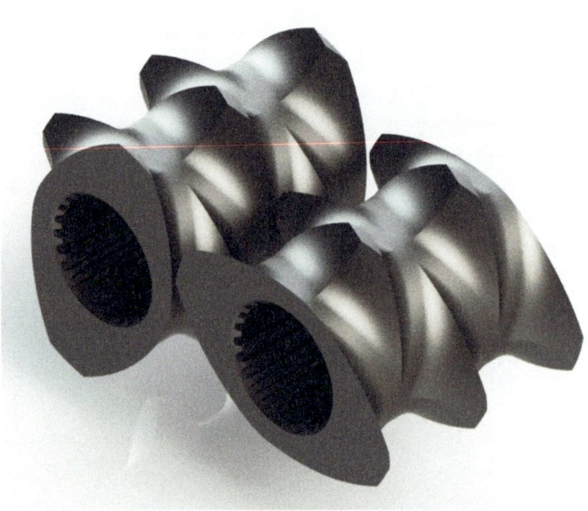

Figure 2.48 "Igel" elements

Figure 2.49 TME elements

2.2.3.4.3 Counter-Conveying Mixing Elements

ZME elements (Figure 2.50) are based upon a counter-conveying, one-flighted seal profile element with little pitch and which likewise has notches milled into its screw flank. These counter-conveying mixing elements can likewise be installed as desired along the screw set; they are always completely filled and must be overrun by upstream elements.

Figure 2.50 ZME elements

2.2.3.5 Special Elements

Feeding, kneading, and reverse-conveying elements comprise the basis of the screw kit for a co-rotating twin-screw extruder. With ongoing technical developments, this processing system has become useful even for procedures whose requirements deviated significantly from those that were common prior thereto and that required development of specialized elements. Over time and based on development engineers' knowhow, a variety of special elements came into existence.

Today, these special elements are preferred for applying extensional flows or creating defined shearing fields. In these flow fields, it is then possible to low-shear plasticize material in volume or even to hold back unmelted particles that remain following plasticization.

If stress points must be avoided during the plasticization procedure, then large-volume, low-shear elements should be installed. These can be single-flighted kneading discs, for example, with a narrow crest angle that are then combined with different offset angles in conveying or counter-conveying form into one element (Figure 2.51).

Figure 2.51 One-flighted kneading elements

Modified mixing elements with high pitch are also used whose outer diameter has been reduced so much that the elements no longer reach completely into one another but are instead only tangent to one another at some points. Such elements can also be combined with each other as pairs, conveying or counter conveying as desired, without risk of elements colliding with each other (Figure 2.52).

Figure 2.52 Mixing elements with high clearance

Bimodal polymers that are distinguished by high-molecular particles embedded within a low-molecular matrix can be broken up in shear fields only to a certain size. If highly viscous droplets are to be further fragmented, then a flow field is necessary where these particles will be sufficiently extended. Such an extensional flow is achieved, for example, when product is fed through a tapering cross section, thus raising the flow velocity. A change to the cross section can be executed both in an axial as well as peripheral direction. An axial reduction in the flow cross section was realized using *barrier screws* (Figure 2.53), representing a further development of segment screws. Between the individual segments, a defined bar is mounted as a barrier which leaves a ring gap open on the entire periphery with the preferred half-flight depth. Every particle must pass this barrier gap in the axial direction.

Figure 2.53 Barrier elements

In the case of *eccentric discs* (Figure 2.54) or *one-flighted kneading discs with an integrated extensional channel*, an extensional flow in the peripheral direction occurs. Eccentric discs are cylindrical discs that are arranged eccentrically to the screw shaft. The product is drawn into the tapering eccentric gap by the rotational movement of the discs and is thus extended. However, the flow is not led in the axial direction so that parts of the product can deviate up- and downstream and particles are not subject to any defined extension.

Figure 2.54 Eccentric discs

Figure 2.55 Single-flighted kneading discs with extension channel

In one-flighted kneading discs, the eccentric disc is contained on both sides by a one-flighted profile disc. The polymer that is drawn into the extension channel can no longer escape to the sides and thus is subject to the full extension flow as defined by the geometry.

In similar fashion to the kneading discs, both disc implementations can be combined into larger element units.

In the *screw shear elements* (Figure 2.56), shear gaps are worked into the screw crest section by section. In these gaps, a portion of the melt is exposed to a defined shear field in order to disperse higher molecular polymer content, for example. One should bear in mind with these elements that the stress of the material does not occur quantitatively across the entire product stream.

Figure 2.56 Screw shear elements

In their basic form, the majority of the elements previously described are based on the Erdmenger patent in their basic form, and are more or less self-cleaning. In the following, those elements which clearly deviate from this principle will be introduced.

HME elements consist of a cylindrical disc with an outer scroll diameter, upon which notches protrude in the axial direction from the outer edge (Figure 2.57). The element is combined with a geometrically equal counter element which is installed in mirror fashion in the axial direction. The notches rotate inside of the contact area of both elements behind each counter notch and pass the intersection area through a gap between two consecutive notches. Using this form of movement, intensive mixing is realized that is necessary, for example, in the compounding of glass fibers.

Figure 2.57 HME elements

Sealing and blister discs impede the melt flow over the entire cross section of the housing bore. *Blister discs* (Figure 2.58) are cylinders that are centrically arranged to the screw shaft; they are furnished with numerous bores and have a definite gap with the housing wall. As with the TME elements, a ring with a core diameter constitutes the counterpart on the second shaft. This combination is repeated in exchanged configuration downstream, whereby the complete housing bore cross section is closed off. On both shafts, the same element is used in a mirrored arrangement.

The product must flow either through the holes in the discs, the ring gap in the housing wall, or in the apex between both blister discs. With correct dimensioning of the holes and gaps, the size of unmelted particles is determined which this element can still pass. Larger particles are held back – as with a sieve – until they are sufficiently fused by the addition of energy in the shear field.

Figure 2.58 Blister discs

Sealing discs are constructed similarly to blister discs but with no through holes in the discs. Hence, they close off the cross section nearly completely and must be adjusted product specifically exactly at the operating point of the extruder. Because the pressure loss with these elements can drastically increase, even with small throughput changes, these discs destabilize the process and are therefore seldom used.

Theoretical investigations have shown that an infinite number of different intermeshing screw profiles can be constructed. For that reason, an all-encompassing overview of possible forms and their areas of use is simply not possible. Therefore, only a small selection of the most basic elements have been listed and described here. Most certainly, in the future a number of new elements, which more or less clearly deviate from the basic forms presented here, will be developed for specialized processes or products.

References for Section 2.2

[1] VDI-Gesellschaft Kunststofftechnik, *Der Doppelschneckenextruder.* VDI-Verlag, Düsseldorf (1998)

[2] *Der gleichläufige Doppelschneckenextruder*, Carl Hanser Verlag, Munich (2007)

[3] Sämann, H.-J., *Polyolefine Extrusion Process Training*, internal training, Stuttgart (2011)

2.3 Overview of Patented Screw Elements

Michael Bierdel

In the history of twin-screw extruders, which has lasted for over 70 years, a large number of screw elements have been developed and patented. This section gives an overview. To allow easy access to the patents, they are divided into seven categories, whereby the classification within the categories takes place according to the increasing number of flights:

Construction of screw profiles (Sections 2.3.1–2.3.3): Although a large number of special screw profiles were already known from the literature and patent applications, rules for the general construction of screw profiles have only been published in recent years.

Erdmenger profiles (Sections 2.3.4–2.3.9): Even today, most of the screw elements used are based on the classic, self-cleaning Erdmenger profiles. A detailed derivation of these profiles is shown in Section 2.1.

Eccentric screw profiles (Sections 2.3.10–2.3.13): Here, the screw profiles are arranged eccentrically on the shafts. Such screw elements have a lower energy input, since all other tips except one have an increased clearance to the barrel. The self-cleaning of the barrel is ensured by the one tip with a narrow barrel clearance.

General screw profiles (Sections 2.3.14–2.3.25): In addition to the classic screw profiles, further completely self-cleaning profiles have been developed in which the tip angles are different or in which asymmetrical profiles are present.

Transition elements (Sections 2.3.26–2.3.30): Usually, spacers are used between elements with different flight numbers to compensate for the axial play of the shafts. However, it is also possible to use transition elements whose screw profiles change continuously along the axis from one flight number to the other while maintaining complete self-cleaning.

Non-self-cleaning profiles (Sections 2.3.31–2.3.43): This includes, in particular, all screw elements that are used as mixing elements. In addition, for kneading discs, the energy input and product stress can be reduced.

Optimization of clearances (Sections 2.3.44–2.3.49): Every twin-screw extruder requires a specific clearance between the screws and between the screw and barrel. By skillful selection of the clearances, the processing conditions in the extruder can be improved.

The individual patents are only explained in a few sentences, supplemented by the most important figures. The figures of the screw geometries are much more illustrative than detailed descriptions of the geometry. Anyway, each reader must carry out a comprehensive analysis and assessment of the claims for himself on the basis of the original patent documents, since much depends on each individual task or question. The patents listed do not claim to be complete either. Even with carefully planned patent searches, it must (unfortunately) always be expected that some patents will be overlooked. All patents mentioned can be obtained free of charge in full text from the internet at *www.epo.org/searching-for-patents.html* or *www.depatisnet.de*.

2.3.1 WO 2009152910, EP 2291277, US 20110110183

Filing date: 2008-06-20

Company: Bayer Technology Services, now Covestro

This patent shows universally how one- to four-flight, self-cleaning screw profiles can be constructed from circular arcs using symmetries. Furthermore, the patent discusses how complete, self-cleaning screw profiles can be created from circular arcs using a transition element as an example.

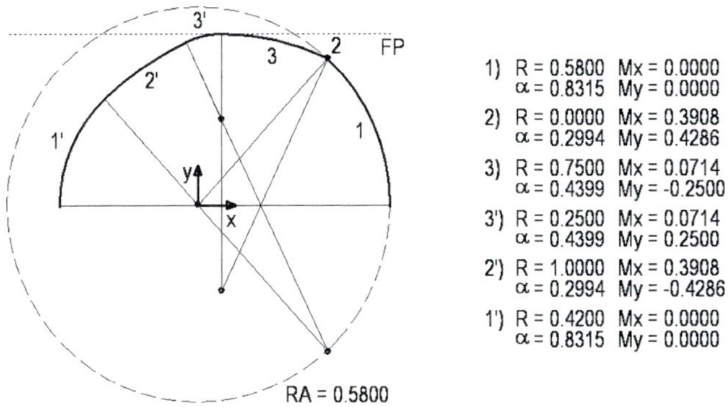

Figure 2.59

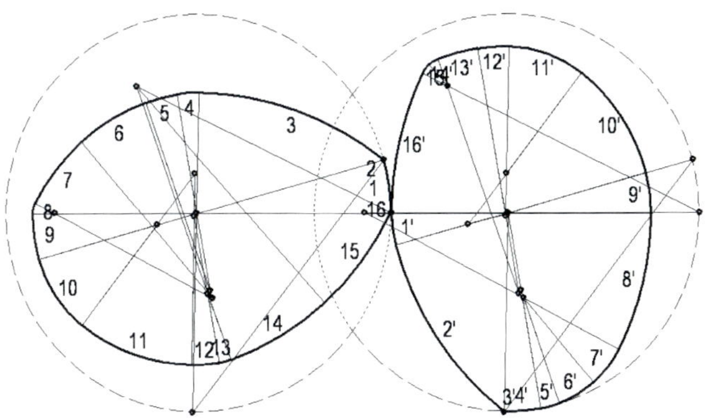

Figure 2.60

2.3.2 WO 2011039016, EP 2483051, US 20120320702

Filing date: 2009-09-29

Company: Coperion, formerly Werner & Pfleiderer

In this patent, the section of a screw profile is obtained over an evolute E, which consists of a set of points P(1) to P(*n*). The involute of a point-shaped evolute is a circle (arc), so that in the end, the design rule shown is also based on circular arcs.

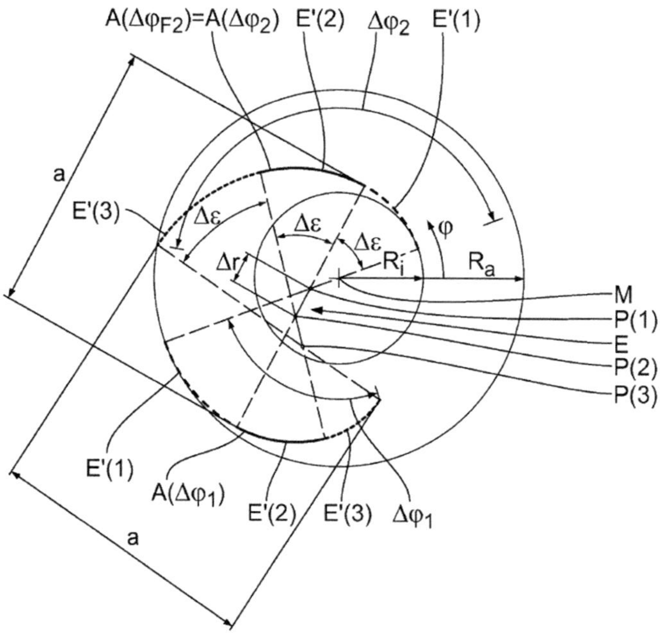

Figure 2.61

2.3.3 WO 2011069896, EP 2509765, US 20120281001

Filing date: 2009-12-08

Company: Bayer Technology Services, now Covestro

In the two previous patents, the screw profiles are based on a sequence of more or less circular arcs. This patent generally shows how the corresponding screw profile q is obtained based on a screw profile p that is given by a continuously differentiable curve. In the example shown, the screw profile p is an ellipse. This results in a corresponding screw profile q, which resembles a football.

$$\vec{p}(s)=\begin{pmatrix}29\cos(s)\\19\sin(s)\end{pmatrix}, s\in[0...2\pi] \tag{2.21}$$

$$\vec{q}(s)=\begin{pmatrix}\dfrac{\cos s\left(-912+29\sqrt{841-480\cos^2 s}\right)}{\sqrt{841-480\cos^2 s}}+48\\-\dfrac{\sin s\left(-1392+19\sqrt{841-480\cos^2 s}\right)}{\sqrt{841-480\cos^2 s}}\end{pmatrix} \tag{2.22}$$

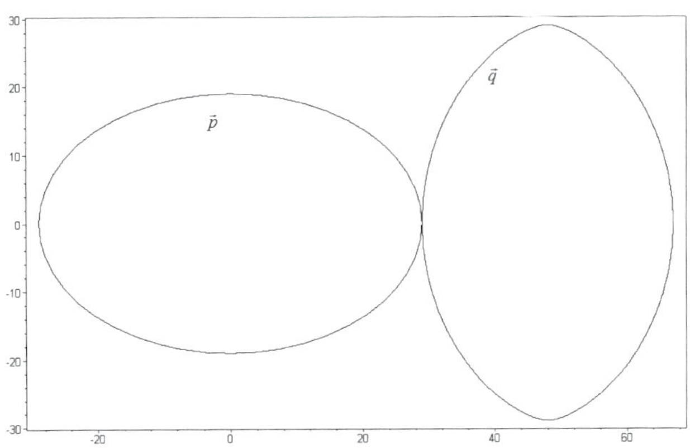

Figure 2.62

2.3.4 DE 00813154, US 2670188

Filing date DE: 1949-09-29

Company: Farbenfabriken Bayer

This patent by Mr. Erdmenger with regard to one- and two-lobe, self-cleaning kneading discs is regarded as one of the basic patents for twin-screw extruders. In addition to the classic profiles, the eccentric disc is also described.

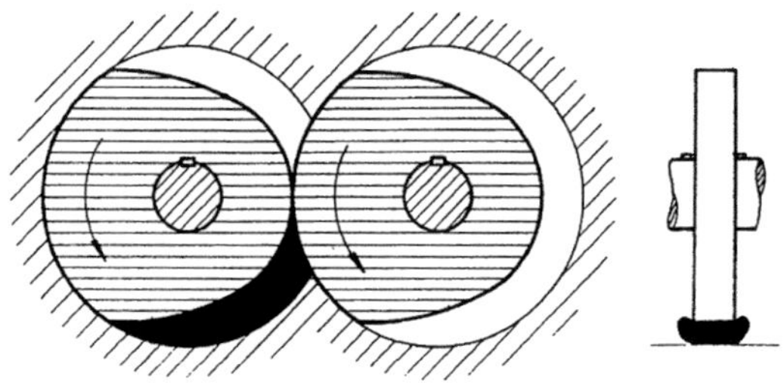

Figure 2.63

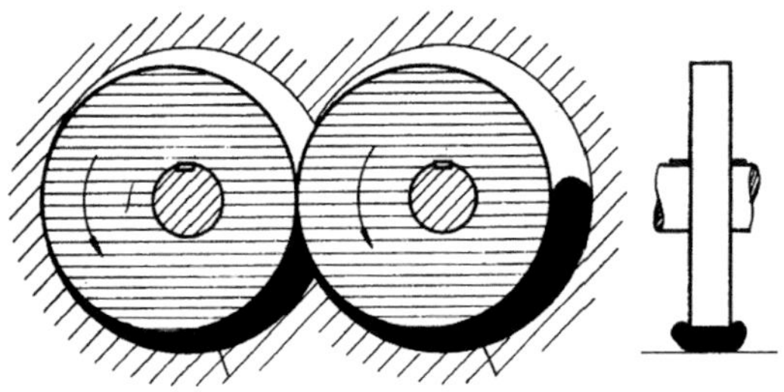

Figure 2.64

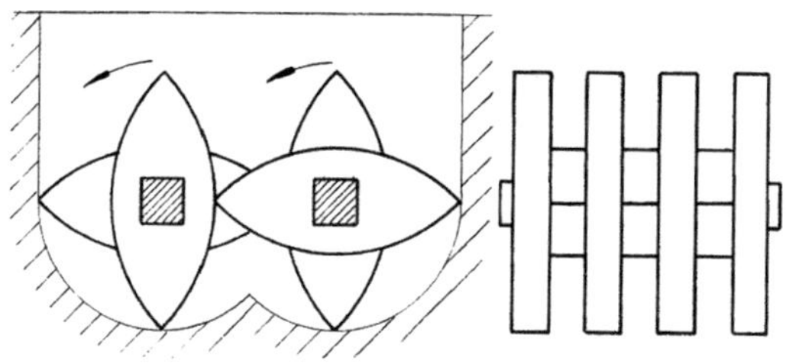

Figure 2.65

2.3.5 DE 19947967, EP 1121238, WO 2000020188

Filing date: 1999-10-05

Company: Krupp Werner & Pfleiderer

Typically, kneading discs are always offset in one direction. In this case, the single-flight screw profile is offset once clockwise and once counterclockwise, whereby the polymer is enclosed between two kneading discs in the intermeshing zone. When turning the two screws, the polymer is pressed through the tight clearances.

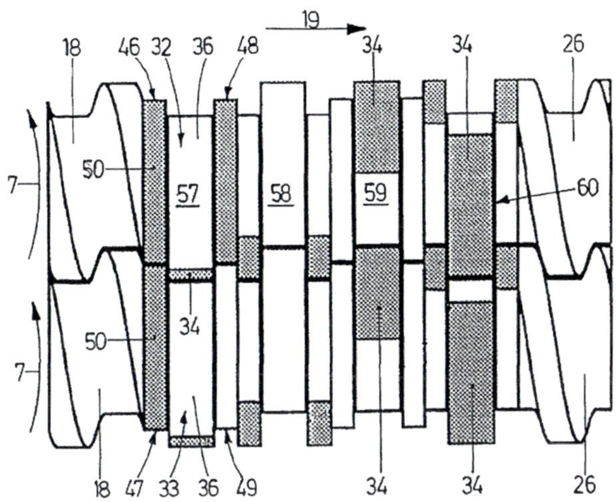

Figure 2.66

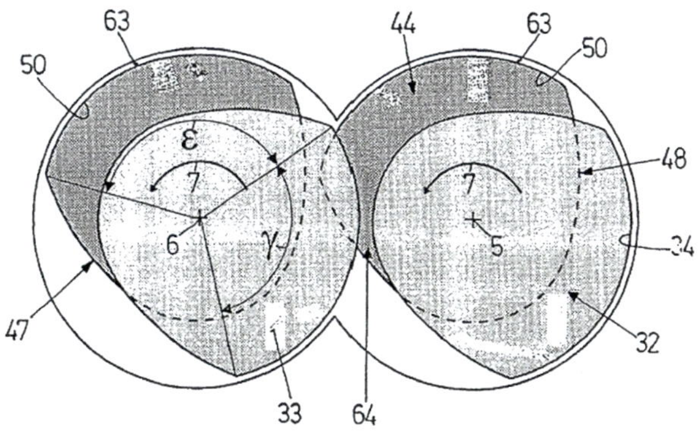

Figure 2.67

2.3.6 US 1868671

Filing date: 1931-06-04

Company: Universal Gypsum and Lime Co.

Interestingly, double-flighted kneading discs were described long before Mr. Erdmenger's time.

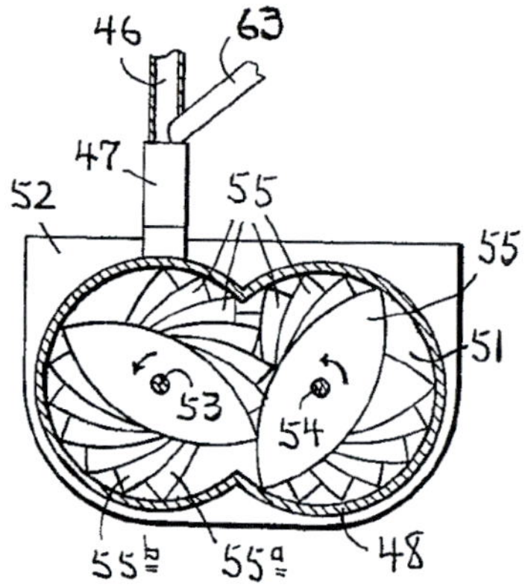

Figure 2.68

2.3.7 DE 10207145, EP 1476290, US 20050152214

Filing date: 2002-02-20

Company: Blach and Extricom, respectively

This patent describes short one- and two-lobe conveying elements, which are arranged with an offset to each other. In principle, this is a hybrid between conveying and kneading elements.

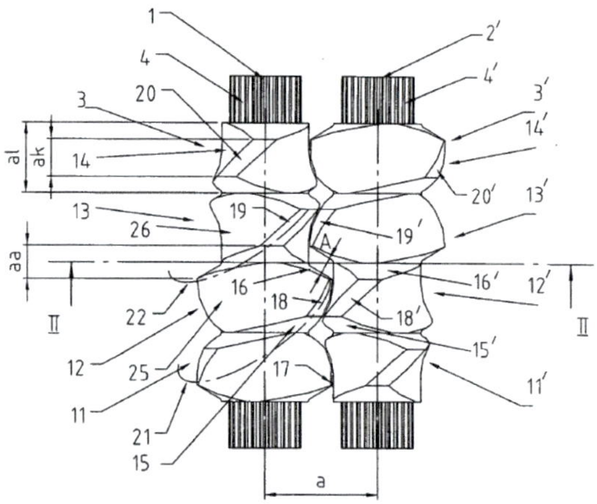

Figure 2.69

2.3.8 DE 00940109, US 2814472

Filing date: 1953-07-27

Company: Farbenfabriken Bayer AG

This is another patent of Mr. Erdmenger's: self-cleaning, three-lobe kneading discs.

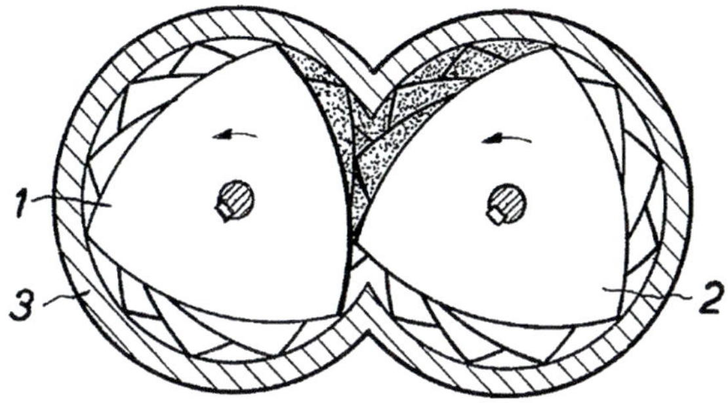

Figure 2.70

2.3.9 US 5713209

Filing date: 1996-10-24

Company: General Mills

One- or two-flighted screw elements have the disadvantage that the product conveyed in the twin-screw extruder can only be cooled very poorly due to the deep channels. For (partial) freezing of a food product, a flat-cut five-flighted conveying element is therefore proposed.

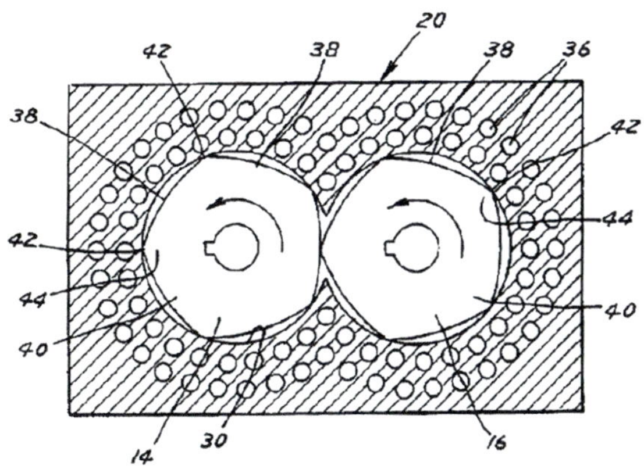

Figure 2.71

2.3.10 US 3717330, DE 2128468

Filing date: 1970-06-08

Company: Du Pont, Canada

This patent describes a screw mixer with eccentric three-lobe spirals.

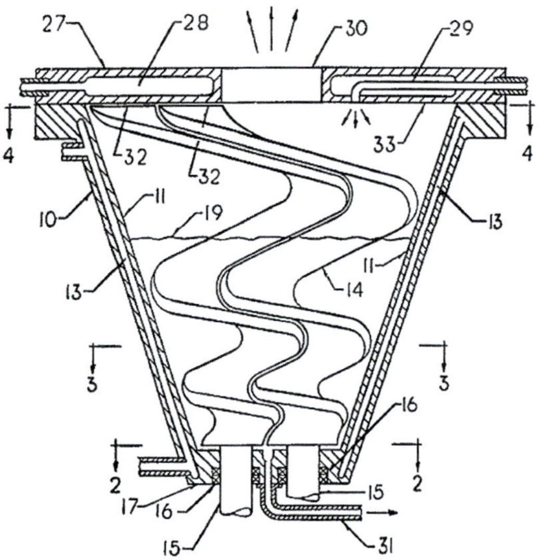

Figure 2.72

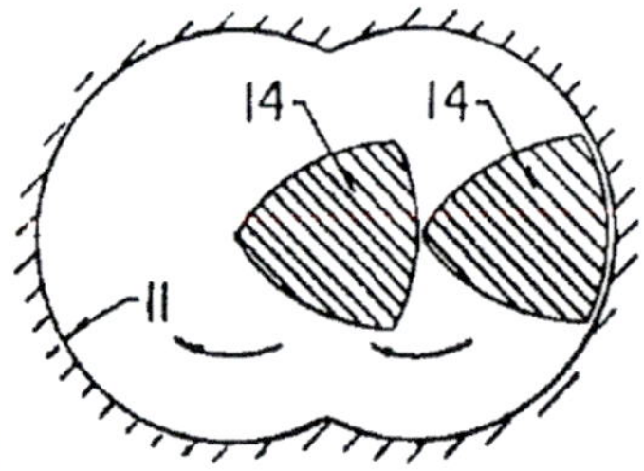

Figure 2.73

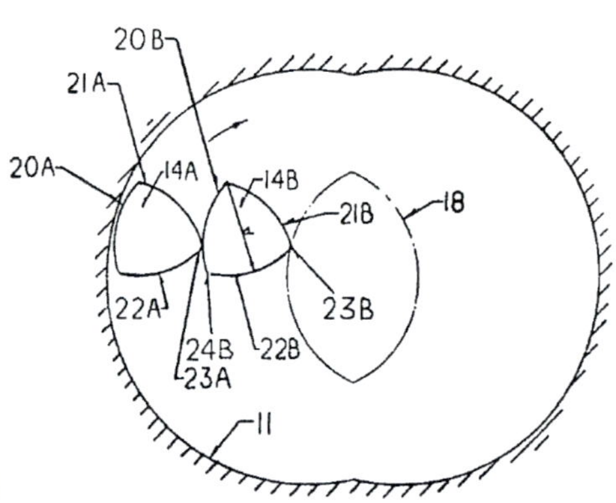

Figure 2.74

2.3.11 DE 4118530, EP 516936, US 5338112

Filing date: 1991-06-06

Company: Werner & Pfleiderer

Here, too, a screw mixer with eccentric three-lobe spirals is described. The similarity to the previous patent is striking.

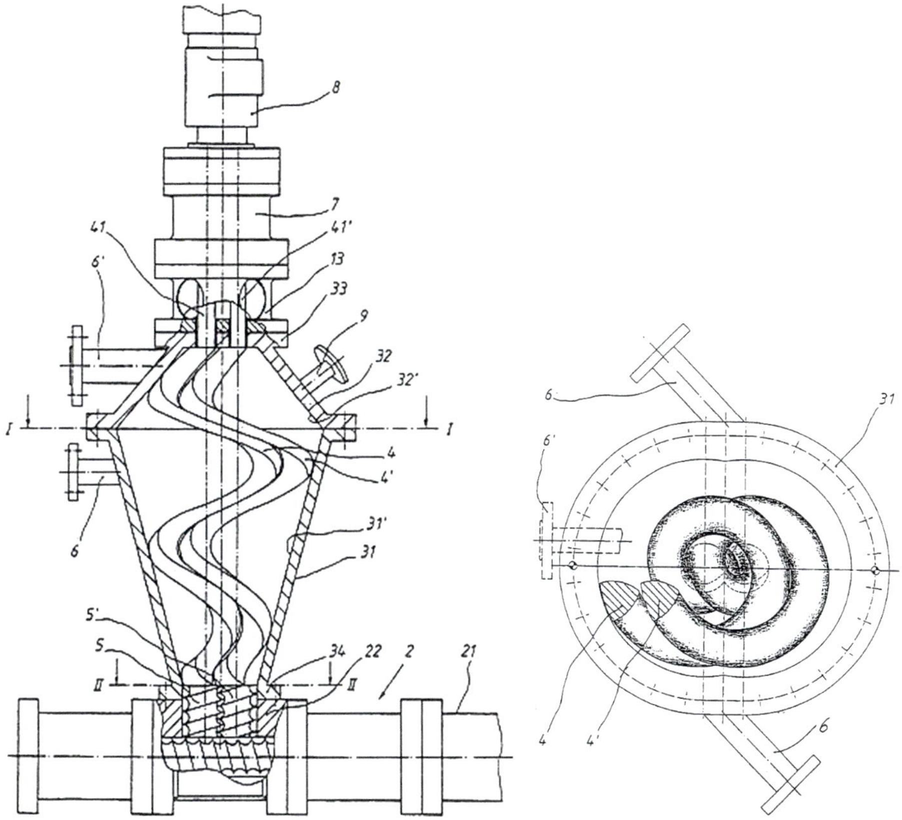

Figure 2.75

2.3.12 US 4131371

Filing date: 1977-08-03

Company: E.I. Du Pont de Nemours and Company

This patent claims eccentric screw profiles in a twin-screw extruder. The eccentricity is selected so that only one tip cleans the barrel.

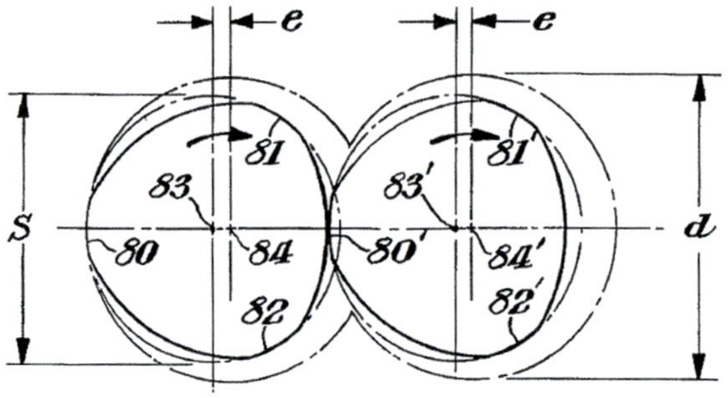

Figure 2.76

2.3.13 DE 03412258, US 4824256

Filing date: 1984-04-02

Company: Werner & Pfleiderer

In contrast to the previous patent, eccentricity is achieved by shifting the screw profiles within a significantly larger barrel clearance.

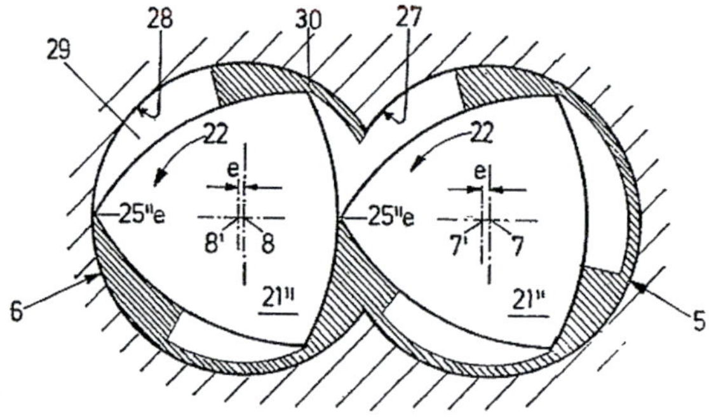

Figure 2.77

2.3.14 DE 1180718, US 3254367

Filing date: 1962-04-11 (DE)

Filing date: 1963-03-26 (US)

Company: Farbenfabriken Bayer AG

This is a single-flighted, asymmetrical screw profile, which essentially consists of a circular disc with a small tip and small groove area.

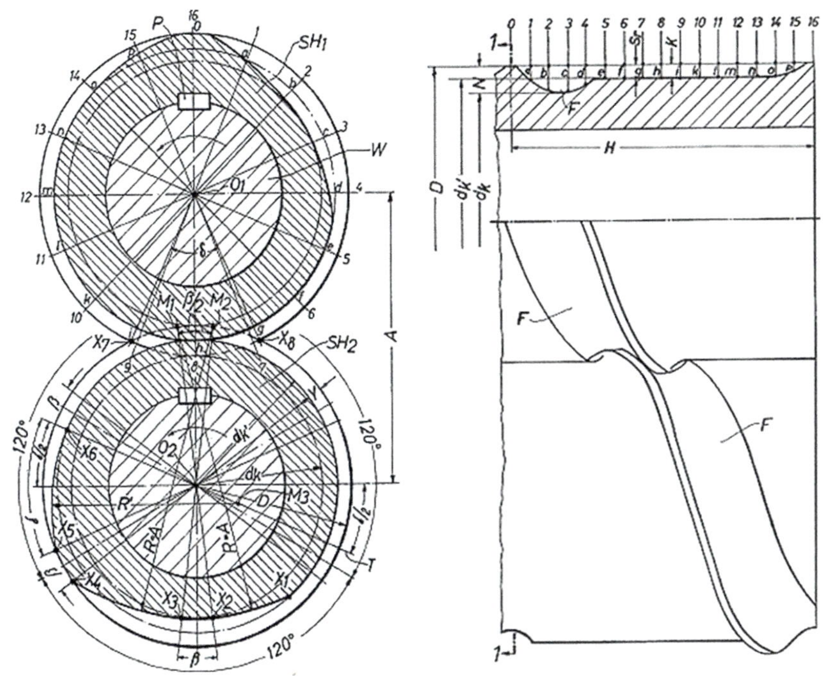

Figure 2.78

2.3.15 US 3900187

Filing date: 1973-10-29

Company: Baker Perkins

As is well known, a lot of energy is dissipated in the tip area of a one-lobe screw profile. This patent succeeds in significantly reducing the tip angle and thus the energy input.

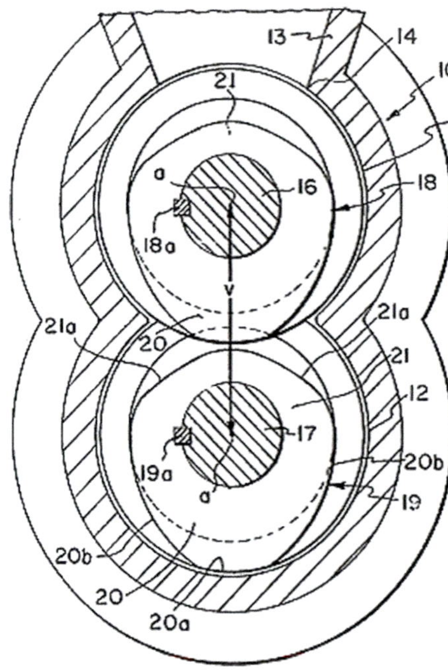

Figure 2.79

2.3.16 WO 2009153003, EP 2303544, US 20110112255

Filing date: 2009-06-12

Company: Bayer Technology Services, now Covestro

In the design specification of the screw profile of the previous patent, the centers of certain circular arcs must lie on the perpendicular of the symmetry axis, which leads through the center of rotation of the screw profile. This patent avoids this restriction and thus enables an even more flexible design of one-lobe screw profiles with a small tip angle.

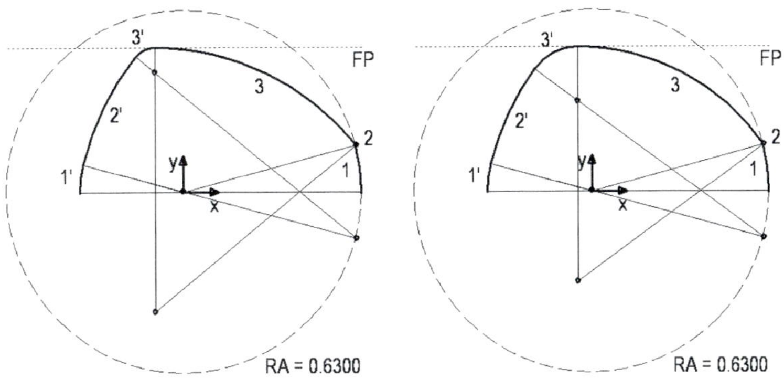

Figure 2.80

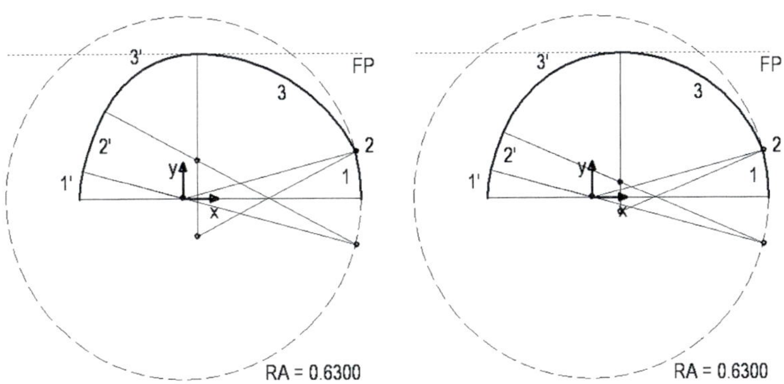

Figure 2.81

2.3.17 WO 2009152974, EP 2291279, US 20110180949

Filing date: 2008-06-20

Company: Bayer Technology Services, now Covestro

The screw profiles shown here represent a variation of the familiar single-flighted Erdmenger profile. In principle, "material" is cut out of the middle of the screw tip and some "material" is added again to ensure self-cleaning in the screw channel. The screw profile allows a high pressure build-up with small tip angles.

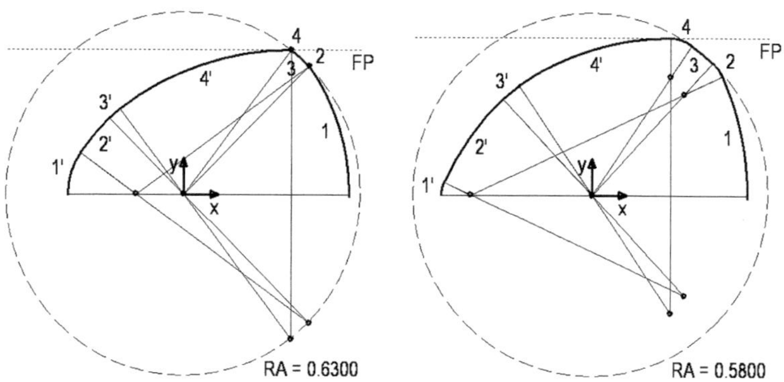

Figure 2.82

2.3.18 US 3216706

Filing date: 1963-08-21

Company: Baker Perkins

This patent on one-lobe screw profiles makes it clear that the tip angles of a pair of screws do not have to be the same.

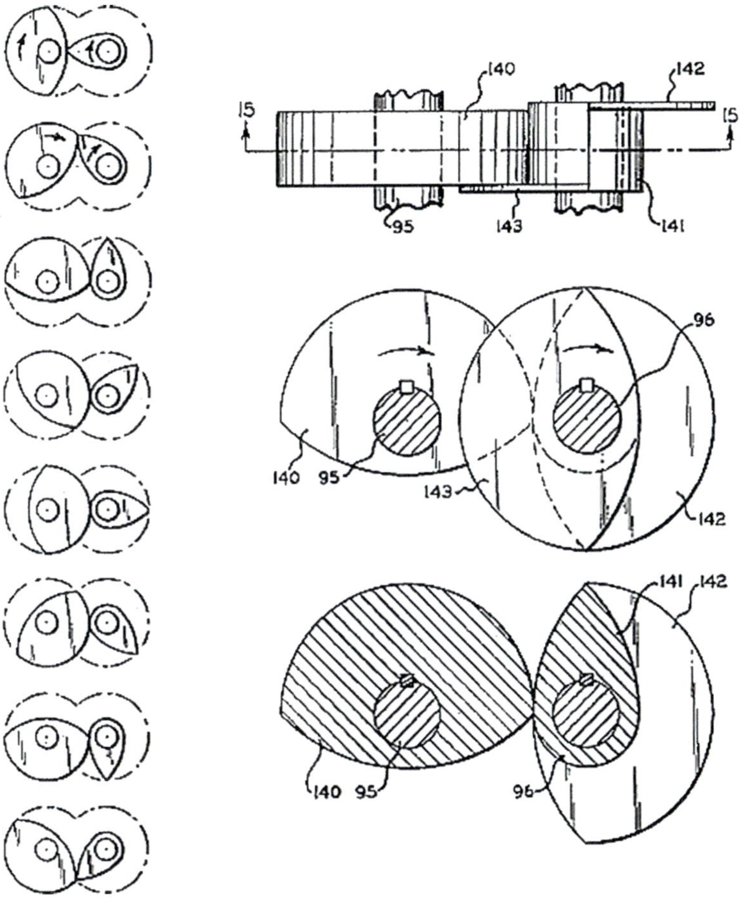

Figure 2.83

2.3.19 WO 2009152968, EP 2303531, US 20110158039

Filing date: 2008-06-20

Company: Bayer Technology Services, now Covestro

The previously known two- or multi-lobe screw profiles always had several kinks in the profile, mostly on the screw tips. In this patent, self-cleaning screw profiles that are completely continuously differentiable are shown, that is, they do not show any kinks at any point.

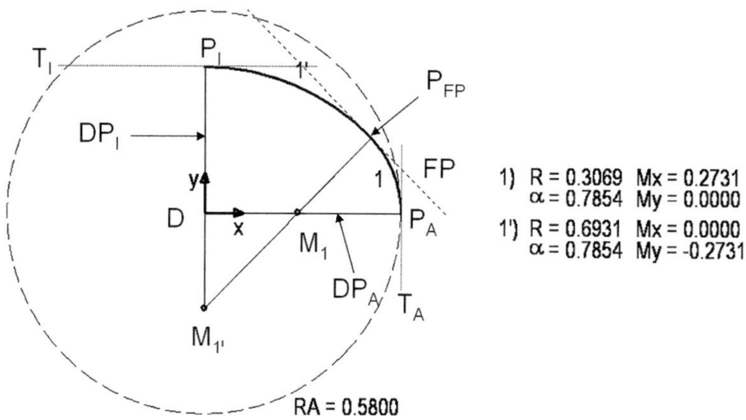

Figure 2.84

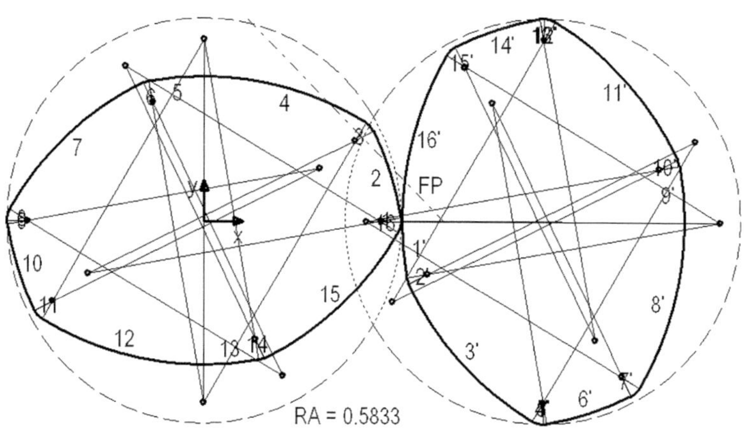

Figure 2.85

2.3.20 WO 2013045623, EP 2760658

Filing date: 2011-09-28

Company: Lanxess

Lanxess uses screw elements with a large pitch and large barrel clearances for degassing thermoplastics and rubbers whose screw profile has a maximum of 30° kinks, preferably a maximum of 10° kinks before the screw tips.

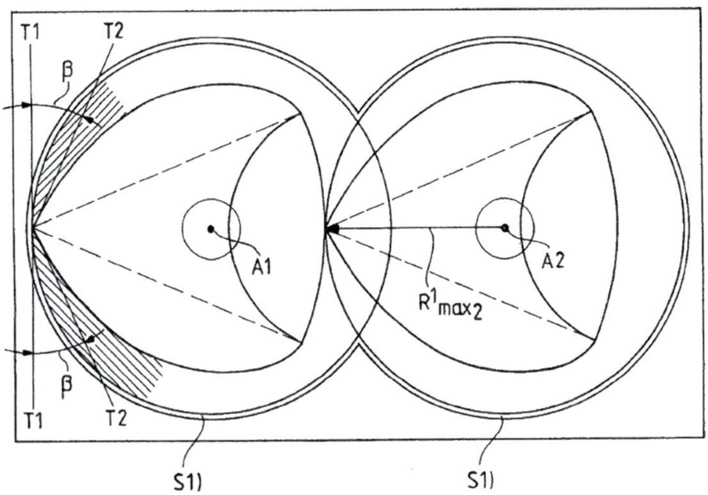

Figure 2.86

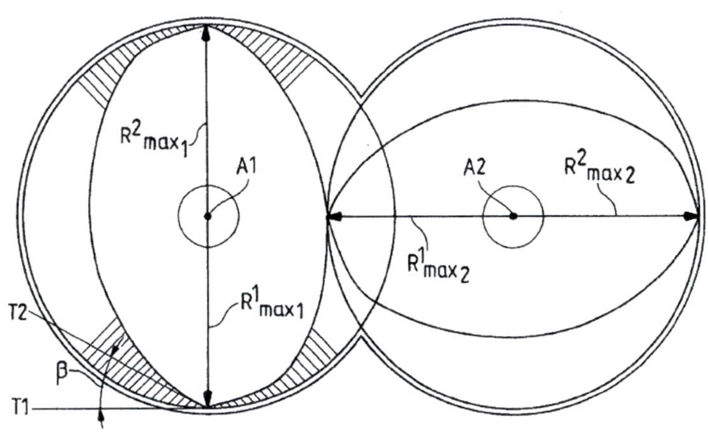

Figure 2.87

2.3.21 WO 2009152973, EP 2291270, US 20110141843

Filing date: 2008-06-20

Company: Bayer Technology Services, now Covestro

In contrast to the classic two-lobe Erdmenger profile, it is shown here how the tip angle can be reduced. The size of the tip angle is flexibly adjustable. The advantage is that there is a smaller energy input between the screw tip and the barrel, thus reducing local temperature peaks.

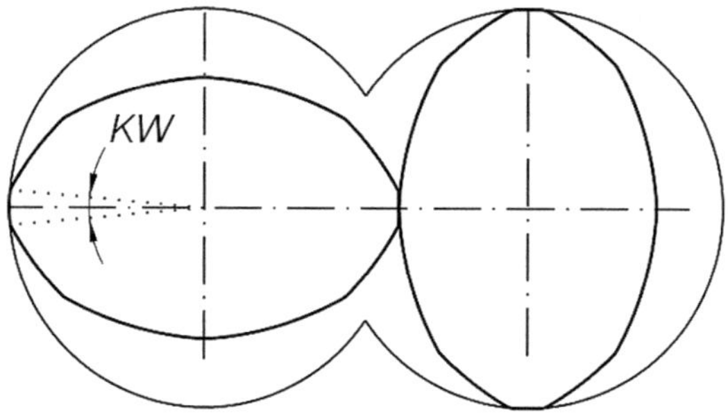

Figure 2.88

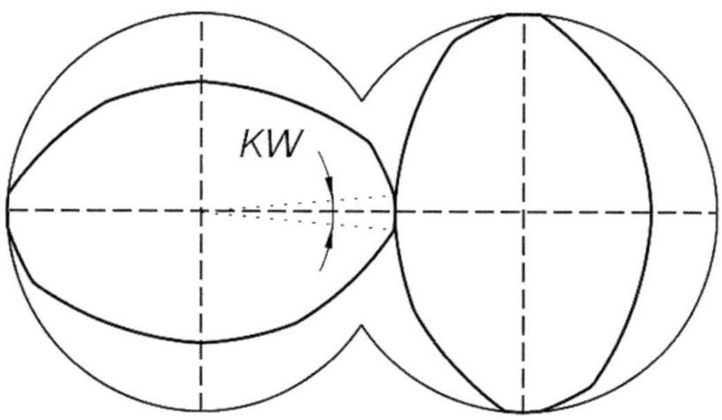

Figure 2.89

2.3.22 WO 2009153002, EP 2307182, US 20110096617

Filing date: 2009-06-12

Company: Bayer Technology Services, now Covestro

In flat-cut two-flighted screw elements there are already very large tip angles, which can lead to temperature damage to the polymer. Analogous to the patent in Section 2.3.17, "material" is cut out from the middle of the screw tips and some "material" is added again to ensure self-cleaning in the screw channels. The resulting larger barrel "clearances" lead to lower local temperature peaks.

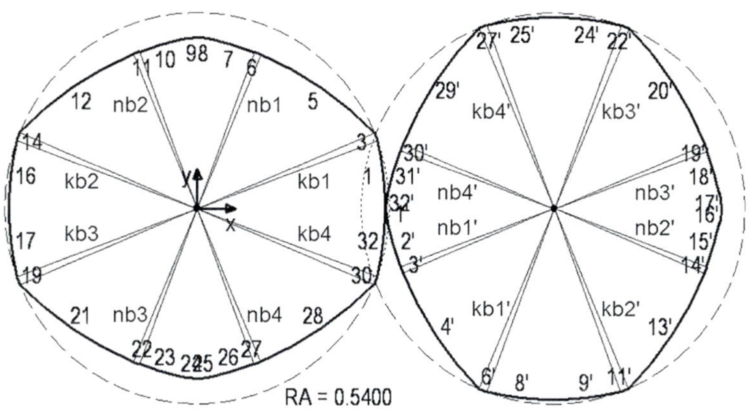

Figure 2.90

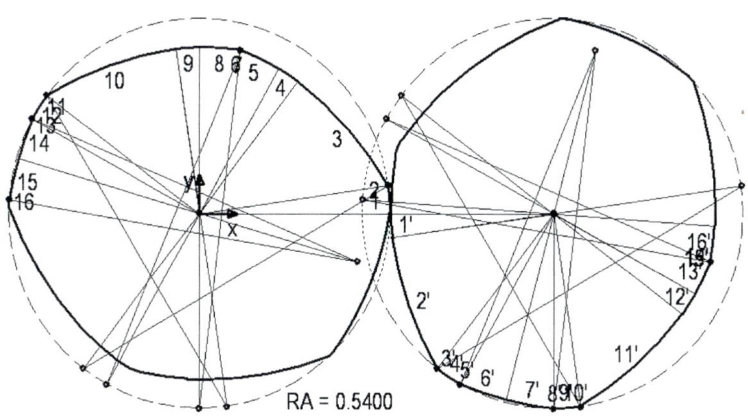

Figure 2.91

2.3.23 EP 0002131, JP 54072265, US 4300839

Filing date: 1977-11-19

Company: Sekisui Kagaku Kogyo Kabushiki Kaisha

In this patent, n-lobe screw profiles with different tip and core diameters are claimed. At least one tip cleans the barrel. A similar transition from a screw tip to a larger core diameter is already shown for a one-lobe screw profile in Section 2.3.14.

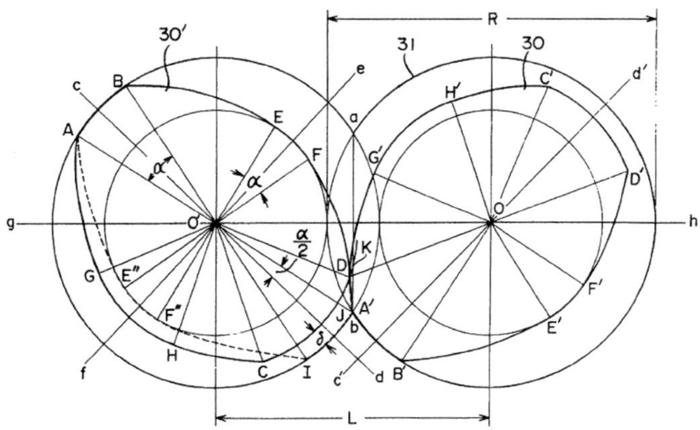

Figure 2.92

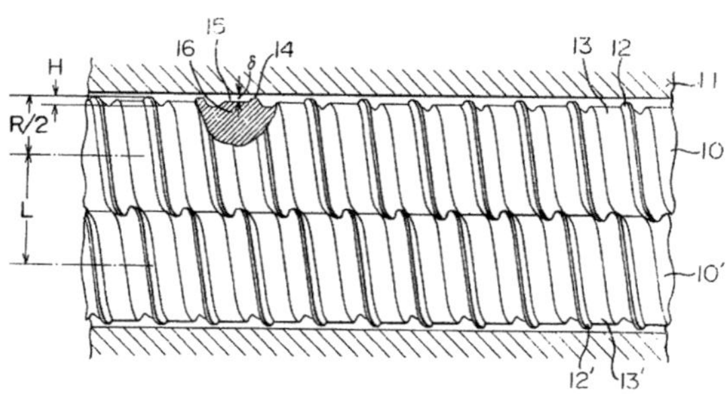

Figure 2.93

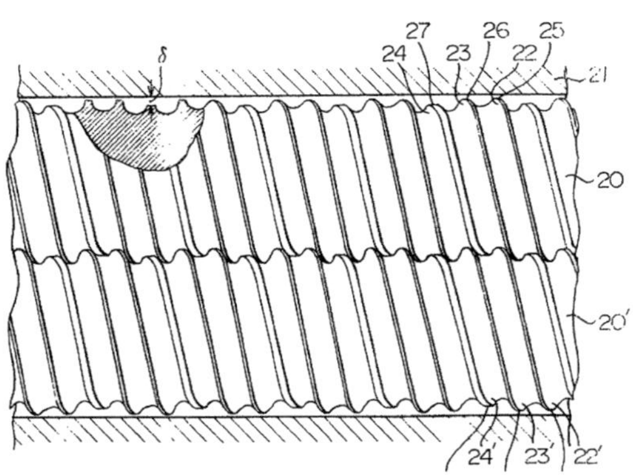

Figure 2.94

2.3.24 DE 19718292, EP 0875356, US 6048088

Filing date: 1997-04-30

Company: Krupp Werner & Pfleiderer

Similar to the previous patent, this is a two-lobe screw profile with different tip and core diameters. From the different tip radii, it follows that only one tip cleans the barrel.

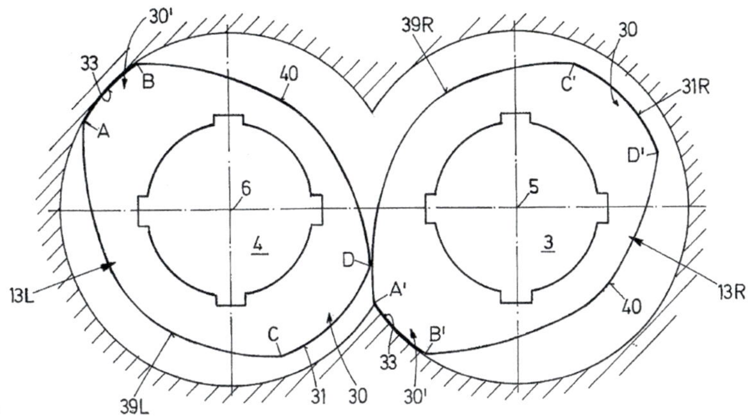

Figure 2.95

2.3.25 DE 04239220

Filing date: 1992-11-21

Company: Blach and Extricom, respectively

A self-cleaning three-lobe screw profile with two or three different tip widths is claimed. Only the broadest tip cleans the barrel.

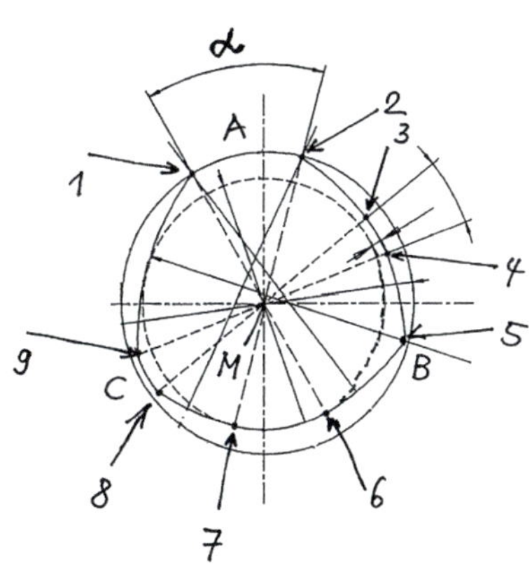

Figure 2.96

2.3.26 DE 01529919, US 3288077

Filing date: 1964-09-11

Company: Farbenfabriken Bayer AG

This patent deals with one-lobe conveying elements with alternating periodic profiles.

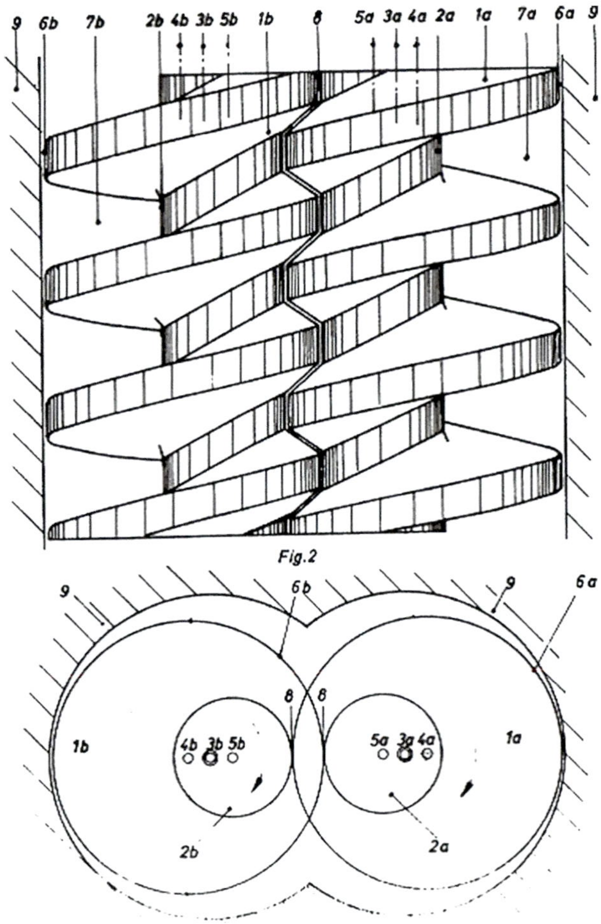

Figure 2.97

2.3.27 EP 0330308, US 5048971

Filing date: 1988-02-24

Company: APV PLC

This patent claims the transition from an x-flighted to a y-flighted screw profile. In the transition area, both screws have a different screw profile.

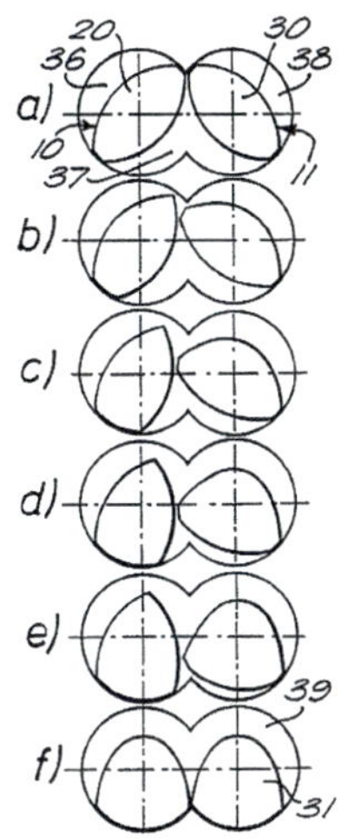

Figure 2.98

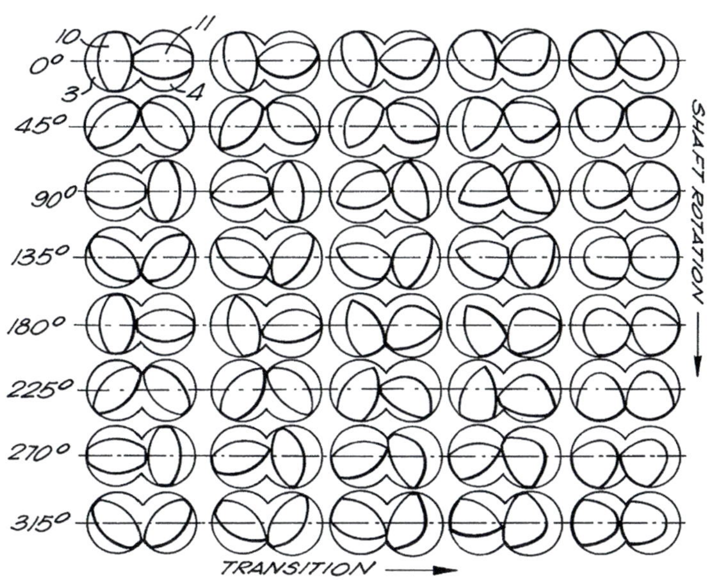

Figure 2.99

2.3.28 DE 10114727, US 6974243, WO 2002076707

Filing date: 2001-03-22

Company: Berstorff, now KraussMaffei Berstorff

Here, a transition from a one-lobe to a three-lobe and back to a one-lobe screw profile is described. The barrel is not cleaned in the three-lobe region.

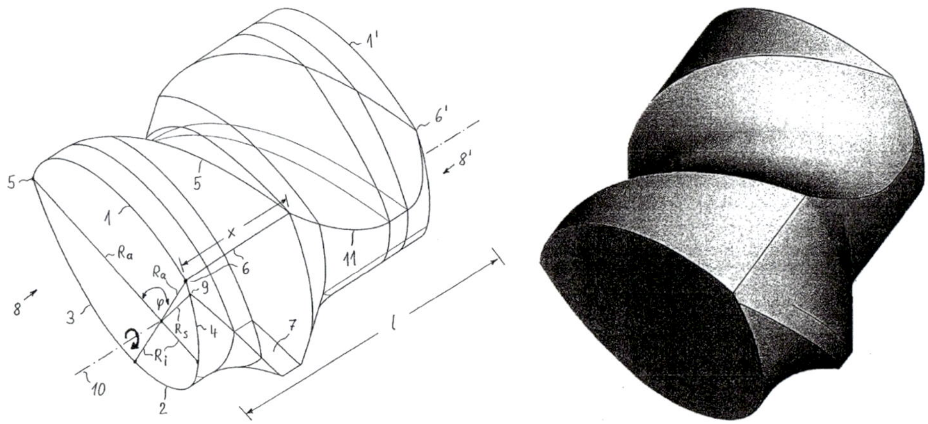

Figure 2.100

2.3.29 US 6783270, WO 2002009919

Filing date: 2000-07-31

Company: Steer Engineering Ltd

This patent claims the transition from an x-flighted to a y-flighted screw profile. In contrast to the previous patent, the screw profiles in the transition area are identical for both screws after mirroring. It is also interesting to note that profile 1.2.75 is very similar to the screw profile in Section 2.3.23.

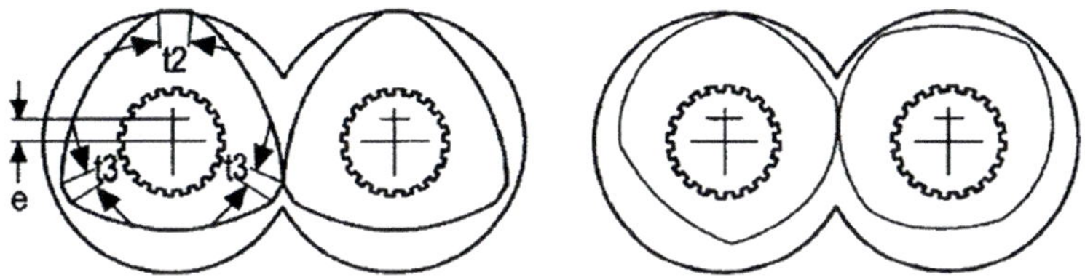

Figure 2.101

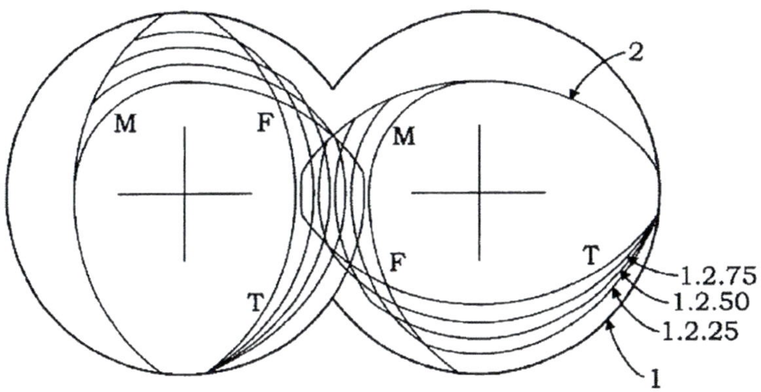

Figure 2.102

2.3.30 WO 2013128463, EP 2747980, US 20140036614

Filing date: 2012-04-16

Company: Steer Engineering Ltd

The previous patent describes various transition screw profiles. From the profiles 1.2.100 to 1.2.50 described there, a conveying element is constructed here, which consists of the profiles 1.2.100 to 1.2.50 and back again twice. The aim is to achieve a better mixture of the product.

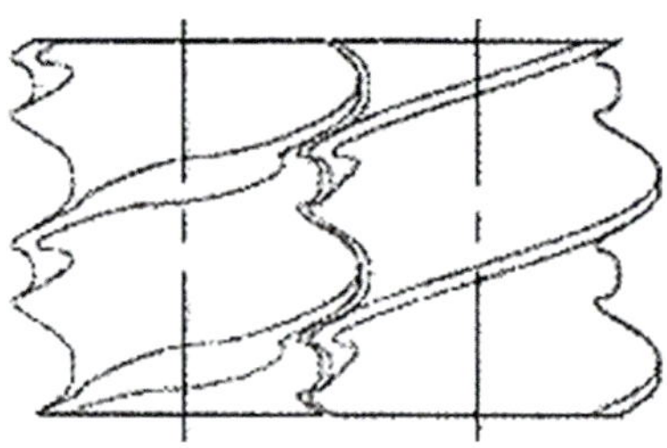

Figure 2.103

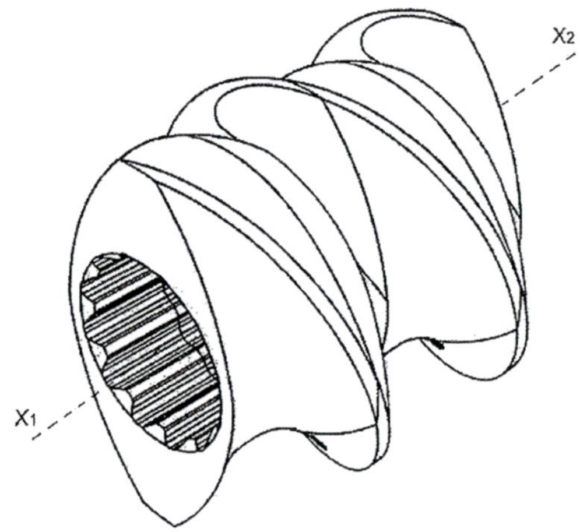

Figure 2.104

2.3.31 JP 2008183721, DE 102007055764, US 2008181051

Filing date: 2007-01-26

Company: Kobe Steel Ltd

The one-lobe screw tip of kneading discs has been modified by neglecting self-cleaning in such a way that a hydrodynamic lubrication bearing results. This prevents contact of the screw tip with the barrel and thus metallic abrasion.

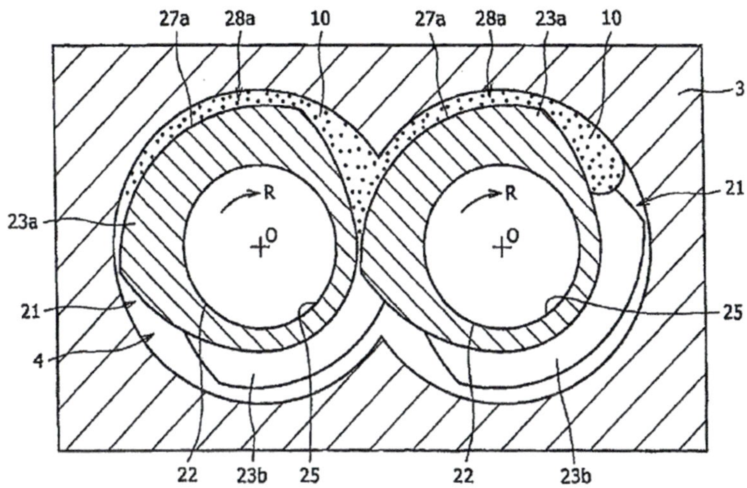

Figure 2.105

2.3.32 DE 4329612, EP 641640, US 5573332

Filing date: 1993-09-02

Company: Werner & Pfleiderer

This mixing element is based on a classical single-flight conveying element in which the tip angle on both screws is significantly reduced. The reduction of the tip angle is such that a conveying element is created on one shaft, but a backward conveying element is created on the second shaft. This increases the mixing effect.

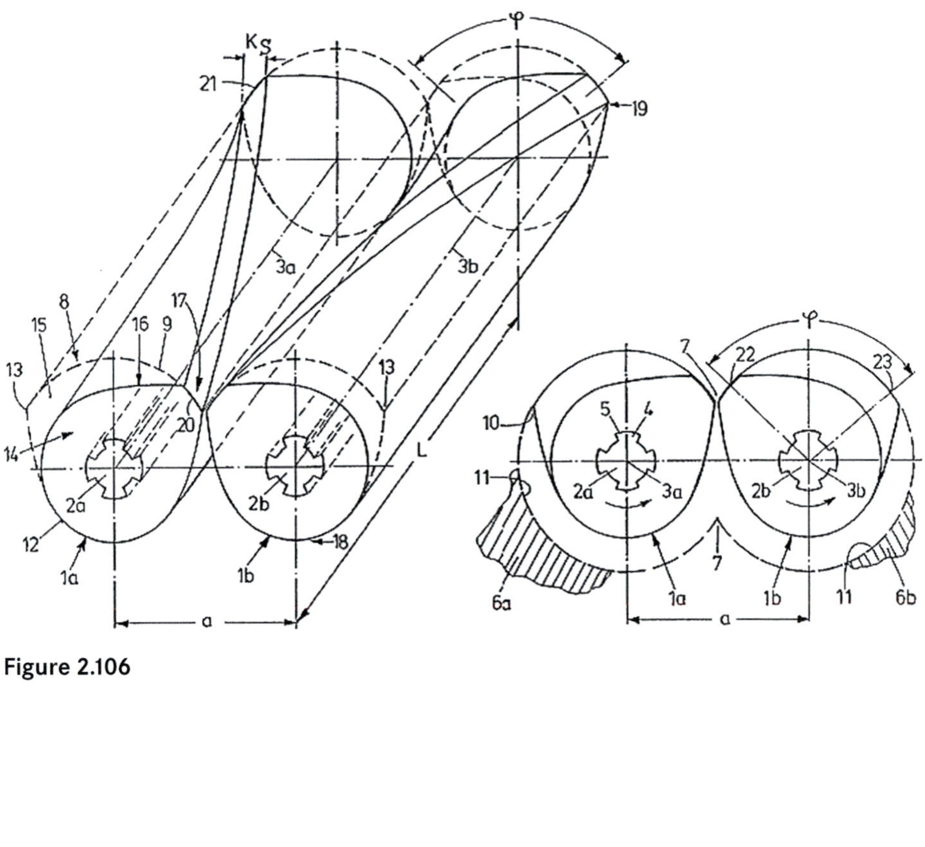

Figure 2.106

2.3.33 DE 19860256, EP 1013402, US 6179460

Filing date: 1998-12-24

Company: Krupp Werner & Pfleiderer

In comparison to the previous patent, one-lobe kneading discs with a reduced tip angle are described here.

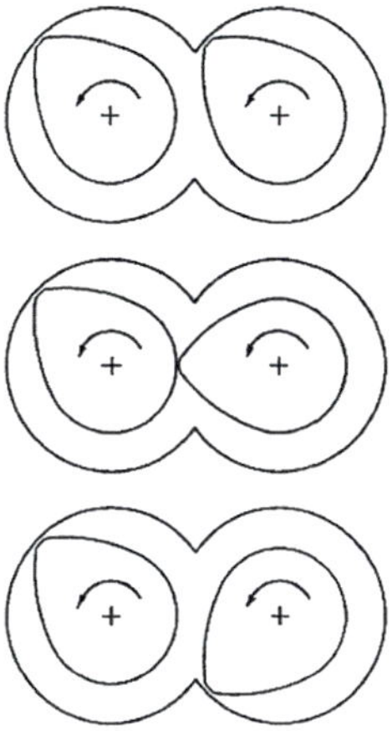

Figure 2.107

2.3.34 DE 04134026, EP 0537450, US 5318358

Filing date: 1991-10-15

Company: Werner & Pfleiderer

This patent describes the already classical tooth mixing element. The one-lobe basic geometry is conveying backwards. By selecting an appropriate pitch of the grooves, the mixing element can still exhibit slightly positive conveying properties.

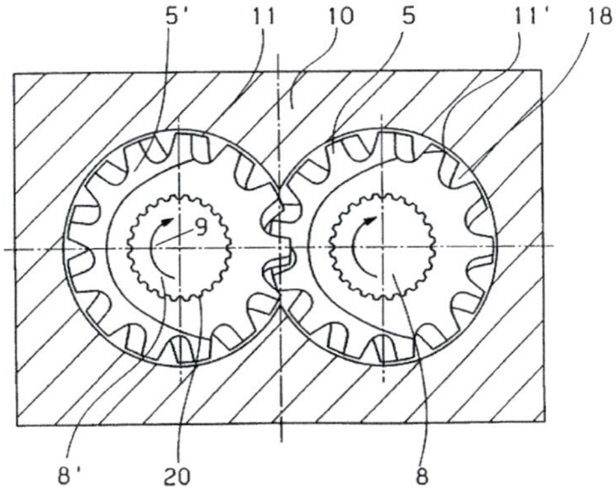

Figure 2.108

2.3.35 DE 19706134

Filing date: 1997-02-08

Company: Friedrich Theysohn

In contrast to the previous patent, one-lobe kneading discs with teeth are described here.

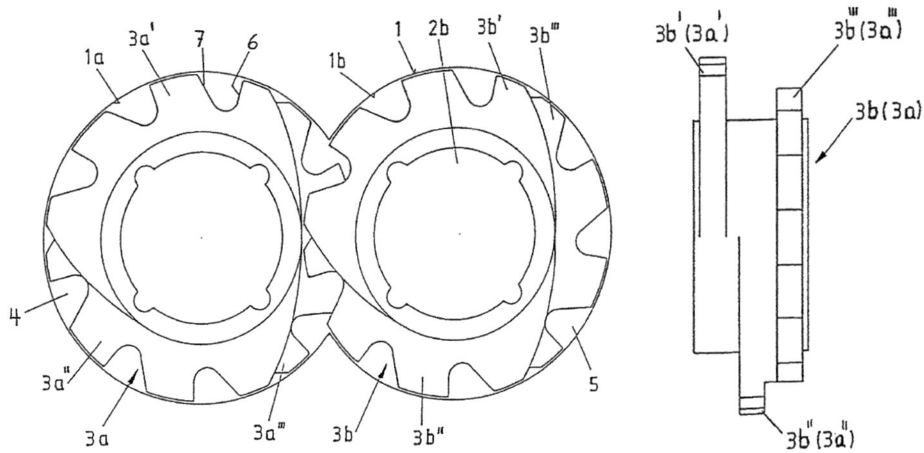

Figure 2.109

2.3.36 JP 2013028055

Filing date: 2011-07-28

Company: Japan Steel Works Ltd

By skillful removal of material, undercuts are possible in the screw elements of this patent, whereby a better mixing behavior is to be achieved. This is reminiscent of the high viscosity reactors of List, Switzerland, and Buss-SMS-Canzler, Germany.

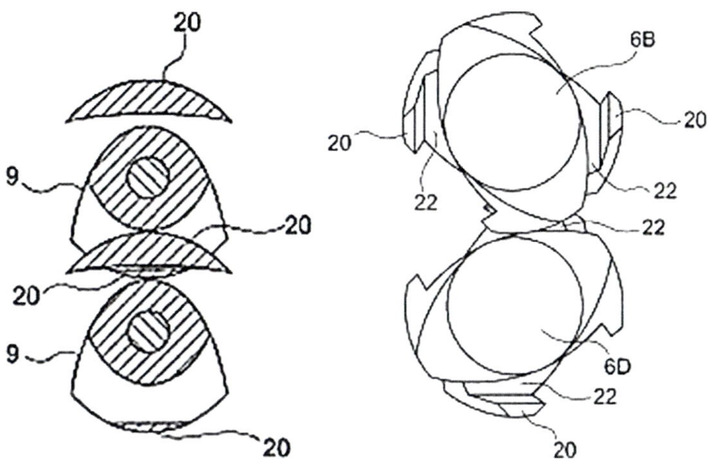

Figure 2.110

2.3.37 WO 1998013189, US 6022133, EP 934151

Filing date: 1996-09-24

Company: Dow Chemical Comp.

This patent describes a mixing element in which adjacent screw elements do not have the same number of tips.

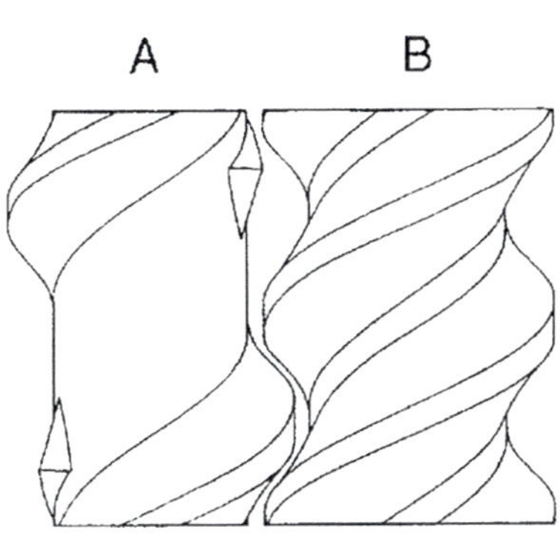

Figure 2.111

2.3.38 WO 1999025537, EP 1032492

Filing date: 1997-11-19

Company: Dow Chemical Comp.

In the previous patent, individual screw tips are completely removed. In this case, the tip height of additional screw elements is also reduced.

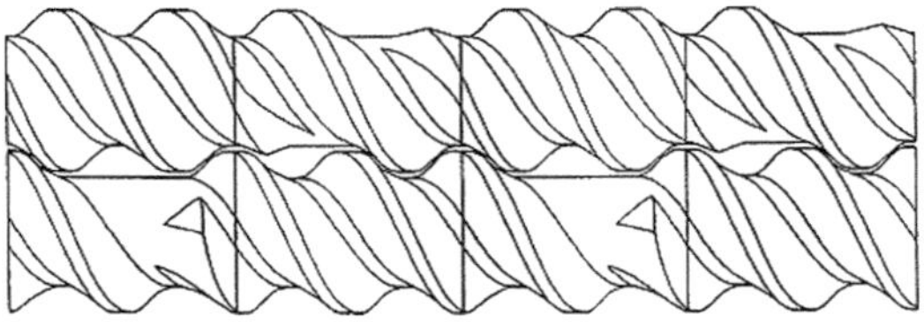

Figure 2.112

2.3.39 US 6116770, EP 1035960, WO 2000020189

Filing date: 1998-10-02

Company: Krupp Werner & Pfleiderer

In the case of this two-lobe kneading element, the sealing surface is reduced by material removal. This results in low product stress and a lower energy input.

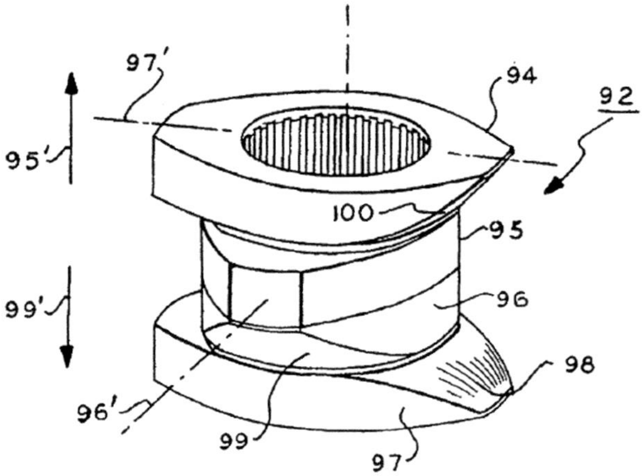

Figure 2.113

2.3.40 DE 29901899 U1

Filing date: 1999-02-04

Company: Friedrich Theysohn

In the case of this two-lobe kneading element, the sealing surface is reduced by material removal. This results in low product stress and a lower energy input.

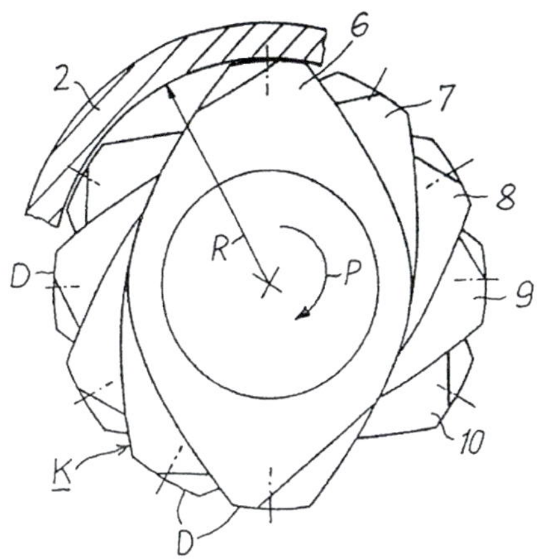

Figure 2.114

2.3.41 US 6170975, WO 2000047393

Filing date: 1999-04-21

Company: Krupp Werner & Pfleiderer

In this patent, two-lobe and three-lobe shoulder elements are claimed in a twin-screw extruder. This results in low product stress and a lower energy input.

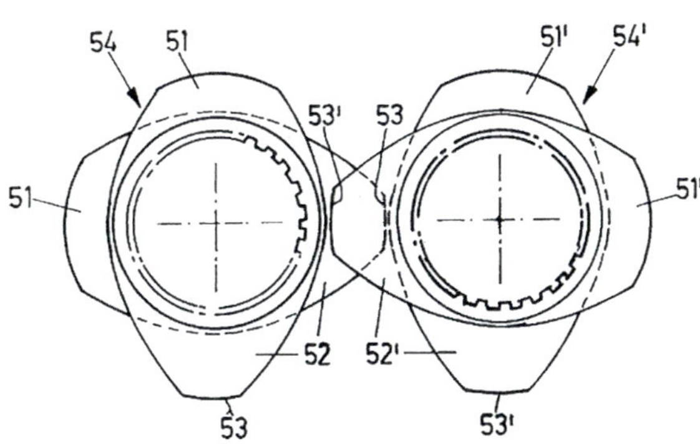

Figure 2.115

2.3.42 DE 10150006, EP 1434679, US 7080935

Filing date: 2001-10-11

Company: Bühler AG

In this patent, two-lobe and three-lobe shoulder elements are claimed in a multi-shaft extruder, especially in a ring extruder. This results in low product stress and a lower energy input.

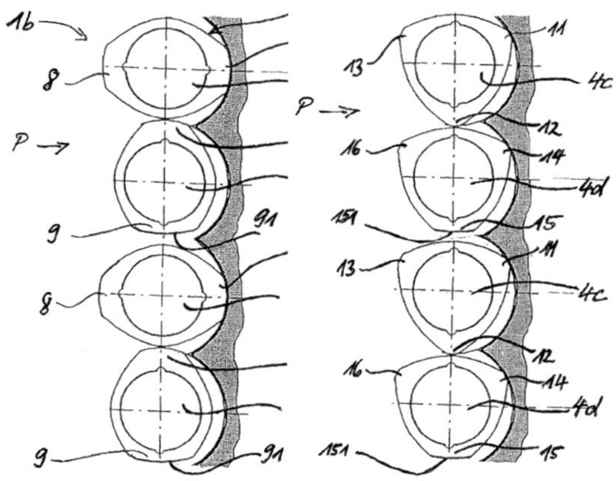

Figure 2.116

2.3.43 DE 4202821, US 5267788, WO 1993014921

Filing date: 1992-01-31

Company: Rockstedt

This patent shows polygonal kneading elements, especially hexagonal polygonal elements. This profile shape ensures gentle kneading of the polymer in the intermeshing region.

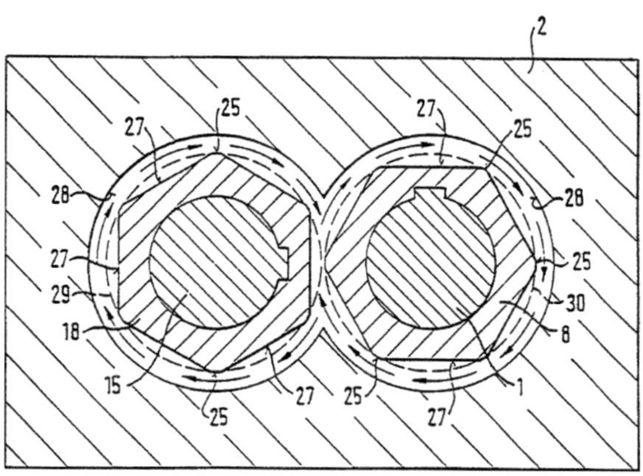

Figure 2.117

2.3.44 DE 03014643, EP 0037984, US 4352568

Filing date: 1980-04-16

Company: Bayer AG

This patent describes an eccentric disc whose screw clearance is smaller than the barrel clearance. This reduces the energy input. Furthermore, the pellets are not jammed between the eccentric disc and the barrel during melting.

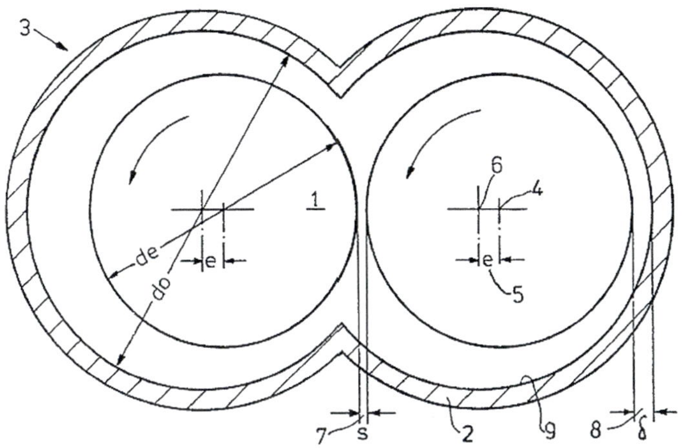

Figure 2.118

2.3.45 DE 02611908, US 4162854

Filing date: 1976-03-20

Company: Bayer AG

Here, clearance strategies for two-lobe and three-lobe conveying elements are described, which lead to a maximum pump efficiency during pressure build-up. The figure shows the optimum pitch for three-lobe conveying elements depending on the flow rate and the barrel clearance.

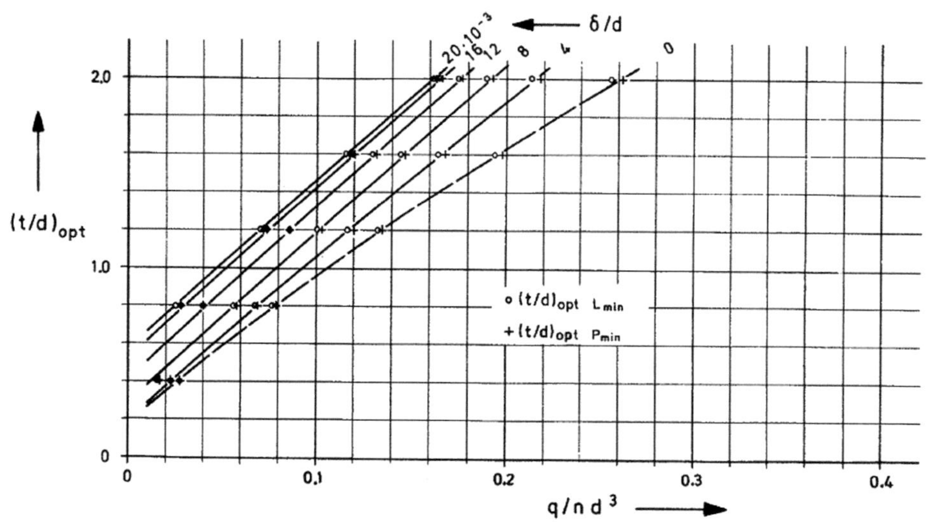

Figure 2.119

2.3.46 WO 1995033608, US 5487602, EP 764074

Filing date: 1994-06-03

Company: Farrel Corp.

The screw profiles are asymmetrical and exhibit large, uneven clearances.

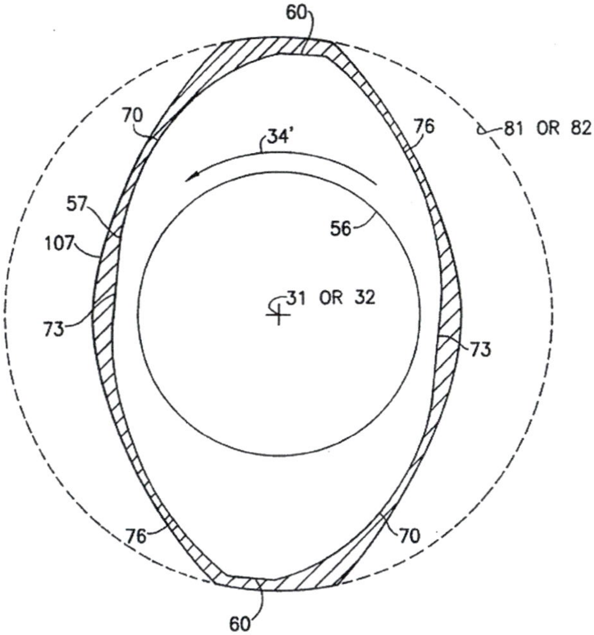

Figure 2.120

2.3.47 DE 102004010553

Filing date: 2004-03-04

Company: Coperion Werner & Pfleiderer

This patent describes screw elements with a narrow clearance between the screws and a large clearance between the screw and barrel. This measure reduces product stress and energy input.

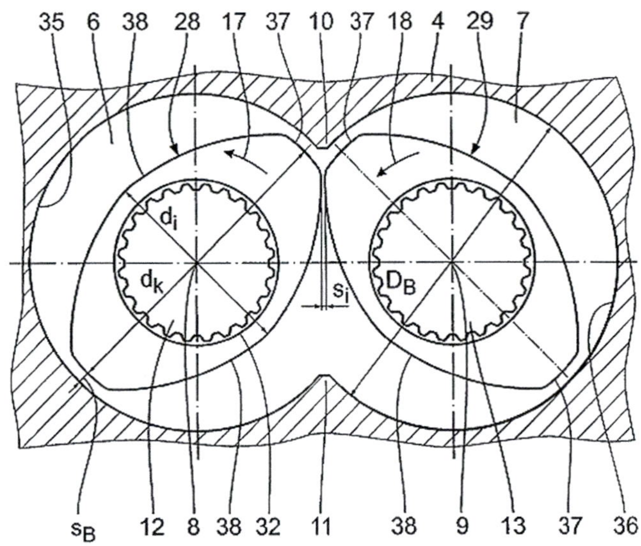

Figure 2.121

2.3.48 DE 04115591, EP 0513431

Filing date: 1991-05-14

Company: Blach and Extricom, respectively

Another way to increase the clearance between the screw and the barrel is to increase the bores in the barrel.

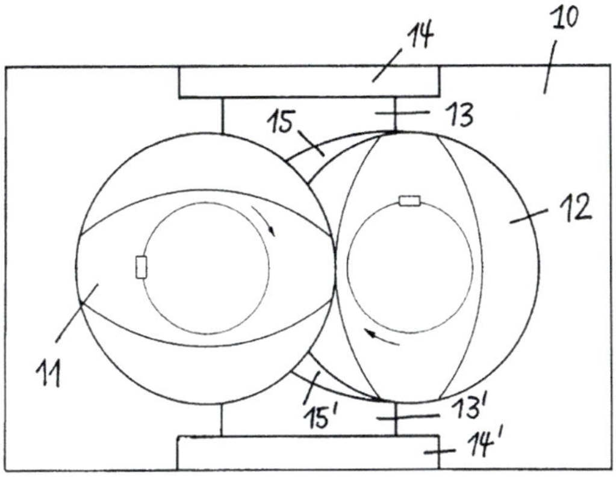

Figure 2.122

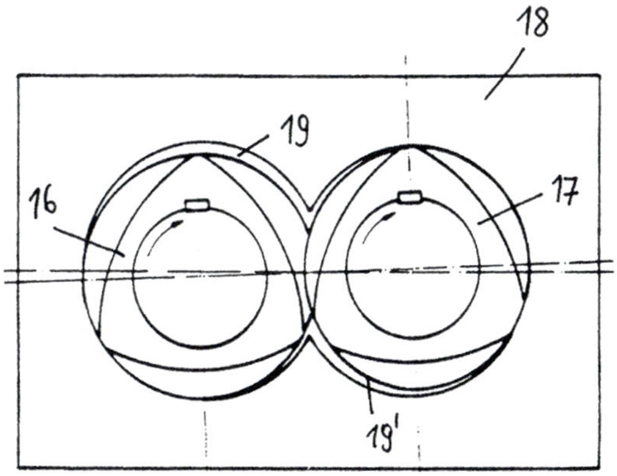

Figure 2.123

2.3.49 WO 2011073181, EP 2512776, US 20120245909

Filing date: 2009-12-18

Company: Bayer Technology Services, now Covestro

This patent claims a method of constructing a data-based model from a multitude of single simulation calculations with different barrel and screw clearances, but also with different pitches T and axial distances A, which enables a prediction of the operating behavior.

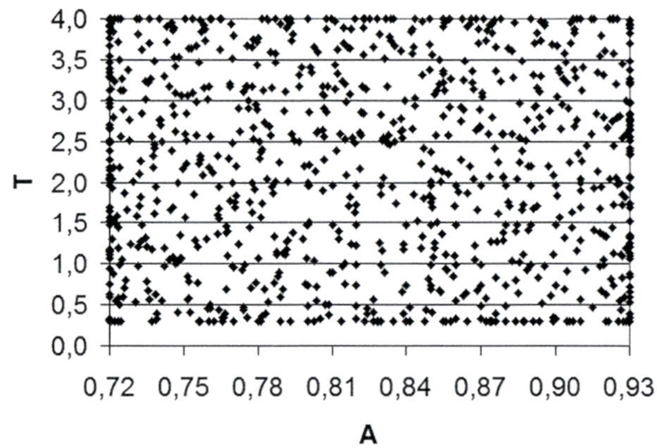

Figure 2.124

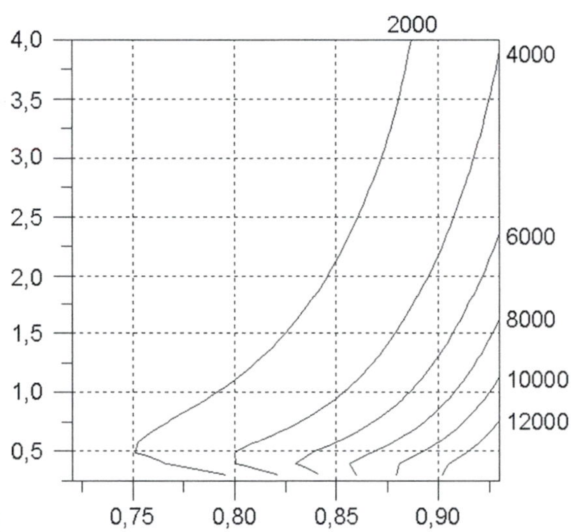

Figure 2.125

3 Material Properties of Polymers

3.1 Rheological Properties of Polymer Melts

3.1.1 Introduction and Motivation

Jens Hepperle

As pellets are frequently used in plastics processing as starting materials, polymers pass initially through the following aggregate conditions in the extruder: solid → solid/liquid → liquid. While for liquid polymers one-dimensional modeling has been available since the 1970s in the form of a dimensionless description of the pressure build-up for filled and partially filled zones in the twin-screw extruder (see Section 4.5) and we have since succeeded in producing a three-dimensional description of the speed, pressure, and temperature fields of filled screw elements (see Section 4.7), it was only in recent years that models of the actual plastification process and the melting behavior (in other words the transition from solid to solid/liquid and to liquid) were developed. This is due to the fact that plastification in the extruder is a very complex process, dominated by viscous energy dissipation. In this process, friction and plastic deformation of the individual granules on the walls (of the housing and the screws) and between individual granules produces heat which causes melting. On the other hand, the quantity of heat introduced through the housing wall by thermal conduction is usually quite small in practice [1, 2].

Since homogeneous melts are covered in a later account of pressure build-up and power input in the extruder (Section 4.5), this section confines itself to the flow behavior of homogeneous unfilled polymer melts and on the introduction of the most important rheological parameters such as viscosity, shear thinning, elasticity, and extensional viscosity. The influence of these rheological properties on simple pressure- and drag flows is demonstrated, while the influence of rheological parameters on pressure build-up and power input in the extruder is described in more detail in Section 4.5.

3.1.2 Classification of Rheological Behavior of Solids and Fluids

Even at a phenomenological level, the flow behavior of polymers differs significantly from that of low-viscous liquids such as water. The flow behavior of polymer melts is markedly dependent on the applied shear stress σ, the shear rate $\dot{\gamma}$, and time. Elastic effects and normal stresses also occur in practice, and their effects can either help or hinder the processes underway.

A fundamental rheological value is the viscosity function, i.e., the shear viscosity as a function of the shear rate $\eta = f(\dot{\gamma})$. The relationship between η and $\dot{\gamma}$ is determined by experiments and is described by means of heuristic functions. The viscosity measurement of fluids is shown here by shearing a fluid between two parallel plates (Figure 3.1). The lower plate is fixed $\left[u(x_2 = 0) = 0\right]$ and the upper plate is drawn across it at a speed of u_0 $\left[u(x_2 = H) = u_0\right]$, for which a force F is required. The shear rate is defined as a time derivation of the deformation δL relative to the height H in a time period dt, or as a quotient:

$$\dot{\gamma} = \frac{d\gamma}{dt} = \frac{d}{dt}\left(\frac{\delta L}{H}\right) = \frac{u_0}{H} \tag{3.1}$$

The viscosity is derived from the shear stress $\sigma = F/A$ and the shear rate $\dot{\gamma}$:

$$\eta(\dot{\gamma}) = \frac{\sigma}{\dot{\gamma}} \tag{3.2}$$

At infinitesimally small shear rates, the shear viscosity is described as zero shear viscosity:

$$\lim_{\dot{\gamma} \to 0} \eta(\dot{\gamma}) = \eta_0 \tag{3.3}$$

When interpreting measurement data it is important to note that the shear viscosity is dependent on temperature, pressure, and time as well as on the shear rate.

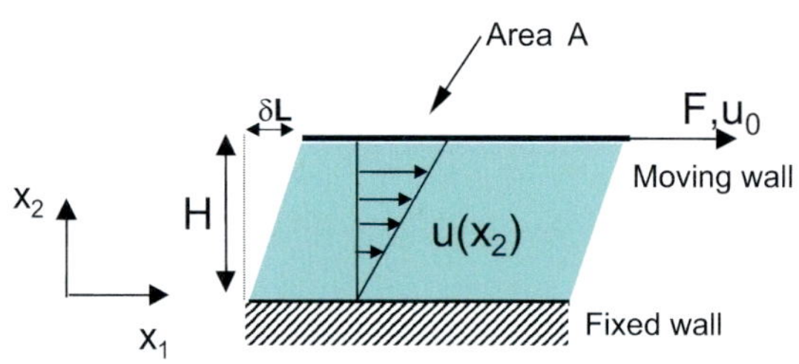

Figure 3.1
Simple shear of a fluid between two parallel plates

Figure 3.2 illustrates a classification of the rheological behavior of solids and fluids. Examples of different flow behaviors are shown in the lowest boxes. Figure 3.2 also illustrates the resulting shear stress as a function of the (shear) deformation γ or, for fluids, the shear rate $\dot{\gamma}$. The two most important material properties for our discussion in this chapter are the viscoelastic and the Newtonian fluid circled in the figure.

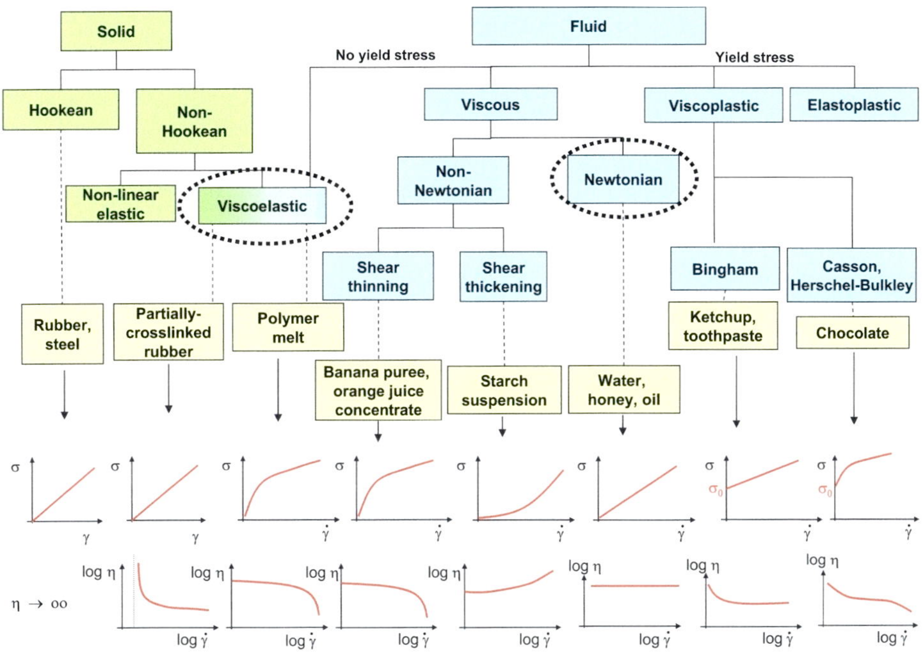

Figure 3.2 Different types of rheological behavior of solids and fluids with examples of materials [7]. Bottom: shear stress as a function of the deformation γ or shear rate $\dot{\gamma}$ and shear viscosity η as a function of the shear rate $\dot{\gamma}$ (double-logarithmic axes)

In the case of solids it is evident that deformation is either linear elastic – like a Hookean solid (most solids including steel and rubber) – or non-linear elastic or viscoelastic. In the case of liquids, fluids differ between those without yield stress and those with yield stress (so-called plastic materials). Fluids without yield stress will flow if subjected to even slight shear stresses, while fluids with yield stress start to flow only above a material-specific shear stress which is indicated by σ_0.

In the case of fluids *without* yield stress, viscous and viscoelastic fluids can be distinguished. The properties of viscoelastic fluids lie between those of elastic solids and those of Newtonian fluids. There are some viscous fluids whose viscosity does not change in relation to the stress (Newtonian fluids) and some whose shear viscosity η depends on the shear rate $\dot{\gamma}$ (non-Newtonian fluids). If the viscosity increases when a deformation is imposed, we define the material as a shear-thickening (dilatant) fluid. If viscosity decreases, we define it as a shear-thinning fluid.

In the case of fluids *with* yield stress, viscoplastic fluids differ from elastoplastic fluids. With the application of a shear stress σ above the yield strength σ_0, Bingham fluids show a linear dependence of shear stress on shear rate, whereas Casson and Herschel-Bulkley fluids show a nonlinear dependence on these parameters.

Newtonian fluids display the simplest rheological behavior. They show a constant viscosity η and there is a direct proportionality between the shear rate $\dot{\gamma}$ and shear stress σ:

$$\sigma = \eta \cdot \dot{\gamma} \tag{3.4}$$

Non-Newtonian fluids whose viscosity depends on shear rate can be described using a power law:

$$\sigma = K \cdot \left(\dot{\gamma}\right)^n \tag{3.5}$$

Equation (3.5) contains a constant factor K and a varying factor n, which specifies the slope of the of the viscosity function. For Newtonian fluids, K corresponds to the shear viscosity η and $n = 1$. For $0 < n < 1$, the fluid is a shear thinning fluid. For shear-thickening fluids, i. e. liquids, whose viscosity increases with shearing, $1 < n < \infty$.

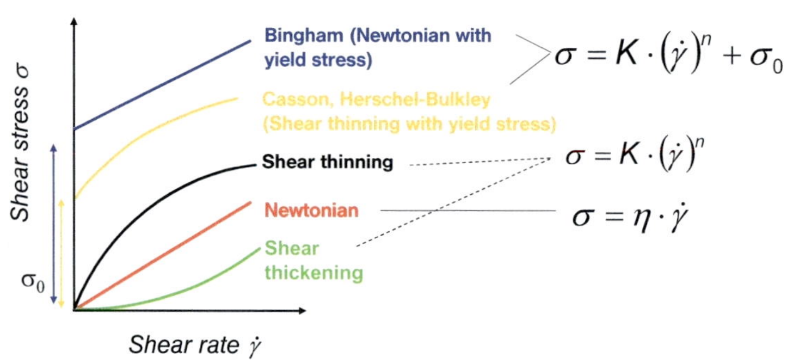

Figure 3.3 Graphical illustration of Equation (3.6)

Figure 3.2 and Figure 3.3 show the sub-division of fluids into those with and without yield stress. Fluids with yield stress require a shear stress σ_0 in order to flow. We are all familiar with this characteristic from tomato ketchup, which requires a certain "minimum force" before it starts to flow out of the bottle. Below σ_0 it is still a solid – in rheological terms. This behavior is described in the model by adding a shear stress σ_0 to Equation (3.5). We then obtain a Herschel-Bulkley correlation:

$$\sigma = K \cdot (\dot{\gamma})^n + \sigma_0 \quad (3.6)$$

Figure 3.3 illustrates the special cases of Equation (3.6) used to describe Herschel-Bulkley fluids and, depending on the flow exponent and yield stress values, Newtonian fluids, shear thinning, shear thickening, and Bingham fluids. The values for Equation (3.6) are given in Table 3.1.

Table 3.1 Special Cases of the Herschel-Bulkley Model, Equation (3.6)

Fluid	Factor *K*	Flow exponent *n*	Yield stress σ_0
Newtonian	> 0	1	0
Shear thinning	> 0	$0 < n < 1$	0
Shear thickening (dilatant)	> 0	$0 < n < \infty$	0
Herschel-Bulkley fluid	> 0	$0 < n < \infty$	> 0
Bingham fluid	> 0	1	> 0

The viscosity function is often approximated using the "Carreau equation", which uses three variable parameters:

$$\eta(\dot{\gamma}) = \frac{\eta_0}{\left(1 + \frac{\dot{\gamma}}{\dot{\gamma}_c}\right)^m} = \frac{\eta_0}{\left(1 + \Theta \cdot |\dot{\gamma}|\right)^m} \quad (3.7)$$

It contains the free parameters zero shear viscosity η_0, a critical shear rate $\dot{\gamma}_c$, at which shear thinning starts, and the gradient m at high shear rates. The reciprocal value of the critical shear rate $\dot{\gamma}_c$ gives a time constant Θ. At high shear rates, Equation (3.7) can be approximated by Equation (3.5) with a constant gradient of $m = 1 - n$.

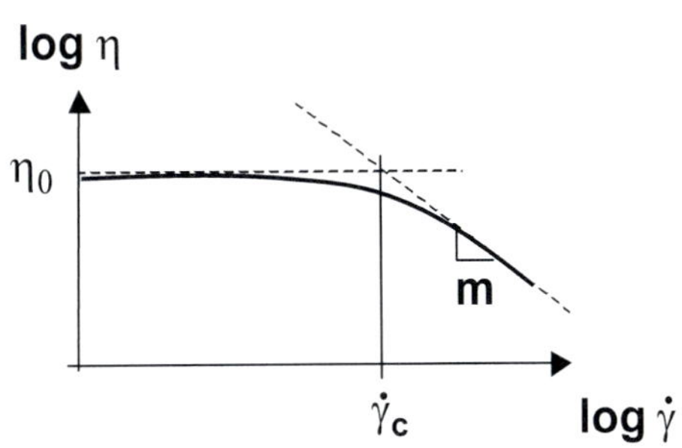

Figure 3.4
Carreau equation for approximating the viscosity function

The form of the viscosity function is of practical relevance for processing plastic melts and other shear thinning materials. Shear rates can differ greatly depending on the process (Figure 3.5). Sedimentation and trickle processes under shear force give shear rates of far less than 1 s^{-1}. In the case of extrusion and in pipe flows, shear rates can reach several hundreds of reciprocal seconds, particularly if a portion of the volume is flowing over the tip of the screw. In this case, shear rates can reach several thousand reciprocal seconds.

In processes with very small liquid dimensions and high speeds, such as spraying or coating processes, shear rates can reach up to 10^5 s^{-1}. If shear thinning or shear thickening materials are used, the viscosity can therefore vary greatly depending on shear rate.

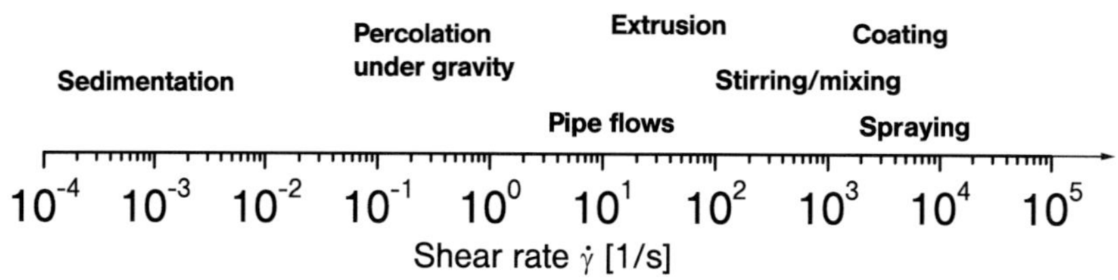

Figure 3.5 Shear rates for different applications

In practice, the complete viscosity functions are often not specified for polymer melts. Instead, the melt flow rate (MFR) and/or the melt volumetric rate (MVR) is used. These parameters specify how much polymer mass and/or volume will flow out of a normalized flow channel within a specific period (e. g., [cm^3/10 min]) when subjected to stress (see DIN EN ISO 113, ASTM D1238). Low-viscosity, easily flowing polymers have high MFR or MVI values, while high viscosity/slow flowing polymers have low MFR or MVI values. These values are listed e. g. in plastics databases such as CAMPUS® [3].

Here are a few examples of viscosites at low shear rates: the viscosity of air is 10^{-5} Pas and the viscosity of glycerin is around 1 Pas. At processing temperature, polymer melts have (zero shear) viscosities of somewhere between 100 Pas and 10^6 Pas. Even amorphous glass has a viscosity, which is around 10^{21} Pas [6]. Processing polymer melts involves different viscosity ranges, which can generally be achieved by adjusting the molecular mass. Highly viscous types with viscosities between 10^3 Pas and 10^4 Pas (at processing temperature) are used for extrusion and blow molding, because a certain degree of stability of the melt is required at the die outlet. Low-viscosity types with viscosities ranging from 10^1 Pas to 5×10^2 Pas are used for injection molding because of the need to overcome small flow cross sections and long flow paths in the tool at a moderate pressure. Medium-viscosity types with viscosities of between 10^2 and 10^3 Pas are used, for example, in the laminating process or for manufacturing plastic films.

3.1.3 Comparison of Viscous Fluid and Viscoelastic Fluid

In order to illustrate the specific material properties of polymers, we compare a viscous fluid (silicone oil) with a viscoelastic shear thinning fluid (aqueous polyethylene oxide solution). These fluids are used as model fluids in order to show the flow behavior limits for polymer melts, which corresponds to the behavior of a viscous fluid at very low shear rates and to the behavior of a shear thinning fluid at very high shear rates.

Both cases are referred to in Chapter 4 with different numerical values for the material constants. For instance, calculating the pressure build-up in the extruder:

- for Newtonian fluids, see Sections 4.5.1 to 4.5.5.
- for shear thinning fluids, see Section 4.5.6.

3.1.3.1 Viscous Fluids

Silicone oils whose molecular weight is not too high show a Newtonian flow behavior in the lower shear speed range. Figure 3.6 illustrates silicone oils of different viscosities ranging between 1 and 2000 Pas in glass test tubes after rotation through 90° for different time periods t. While the low-viscosity oils have already achieved a horizontal state, the high-viscosity oils have only just begun to flow.

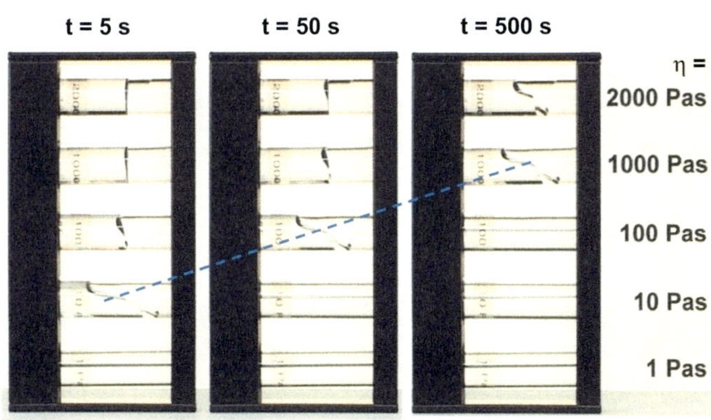

Figure 3.6 Flow behavior of silicone oils of different viscosities as a function of time period as indicated. The blue line connects flow statuses corresponding to $\eta \propto t$

If we assume laminar flow (Re << 1), the time necessary for the liquid surface to reach a specified position can be derived by dimensional analysis (ρ: fluid density, D: characteristic length, g: gravity):

$$t \propto \frac{\eta}{\rho g D} \tag{3.8}$$

In Figure 3.7, the time t^* taken by the oil to flow to a specific position inside the test tube is shown as a function of the shear viscosity η for silicone oils of different viscosities. The experiment therefore confirms the proportionality $t \propto \eta$.

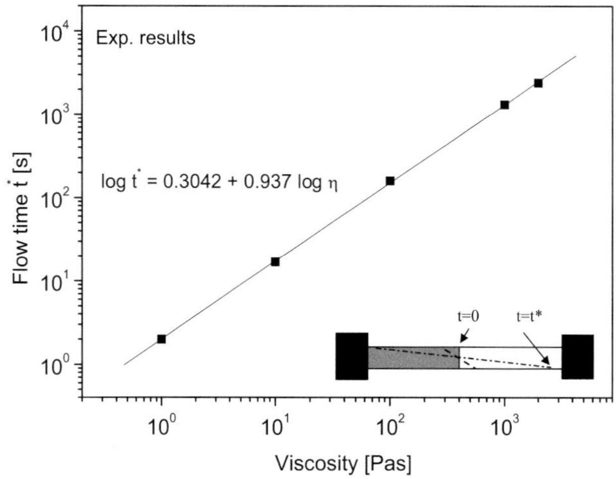

Figure 3.7 Experimentally determined flow times in test tubes (D = 10 mm) as a function of shear viscosity η for different silicone oils

Figure 3.6 shows the same effect if the same liquid levels for different viscosities at the corresponding times are compared (dotted blue line). Absolute values of the liquid level in the test tube cannot be determined with a dimensional analysis, because other parameters (the test tube wall) would have to be taken into account.

3.1.3.2 Viscoelastic Fluids

Figure 3.8 shows the shear viscosity functions for viscous and viscoelastic fluids. In the range of shear rates investigated, the silicone oil displays a constant viscosity, i.e., Newtonian flow behavior. The viscosity of the PEO solution depends significantly on the shear rate in the investigated range and shows strong shear thinning behavior. The viscosity function of the PEO solution can be approximated using a power law (Equation (3.5)), while the viscosity function of the silicone oil can be determined using Equation (3.4).

In the case of elastic fluids and for simple shear flow, the first normal stress difference is $N_1 = \sigma_{11} - \sigma_{22}$. When shearing a fluid between two plates (x_1 direction), the first normal stress difference N_1 forces the plates apart (x_2 direction). The first normal stress difference N_1 is shown together with the measured shear stress τ as a function of the shear rate in Figure 3.9. In the range of shear rates investigated, the shear stress in the case of silicone oil is substantially greater than the normal stress difference and we see substantially greater normal stress differences for viscoelastic PEO solution than for viscous silicone oil.

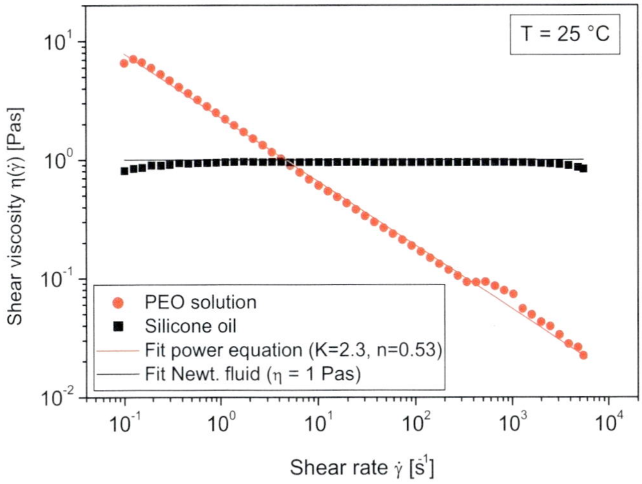

Figure 3.8 Shear viscosity of Baysilone M 1000 silicone oil and aqueous PEO solution (c = 1 wt.-%, Polyox WSR 301)

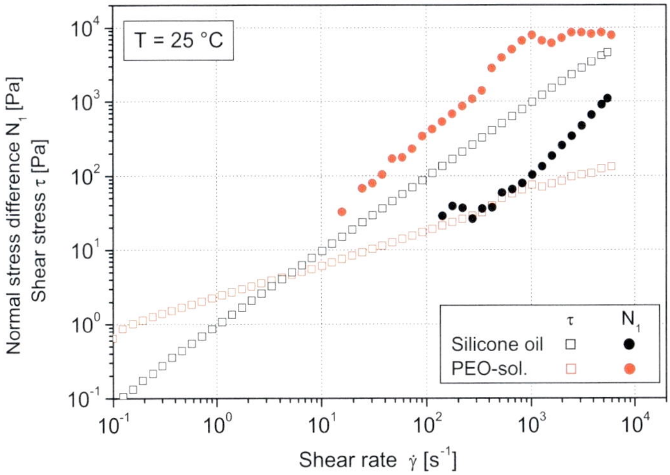

Figure 3.9 First normal stress difference N_1 and shear stress σ as a function of shear rate for Baysilone M 1000 silicone oil and aqueous PEO solution (c = 1 wt.-%, Polyox WSR 301)

The high normal stress differences in comparison to the shear stress cause flow phenomena which may influence many technical processes. One example is the *Weissenberg* effect (see Figure 3.10), which arises when a shaft rotates within a viscoelastic fluid. The first normal stress difference leads to a pressure distribution which causes the fluid to climb up the stirrer shaft. This effect occurs when processing polymer color dispersions or mixing cake dough.

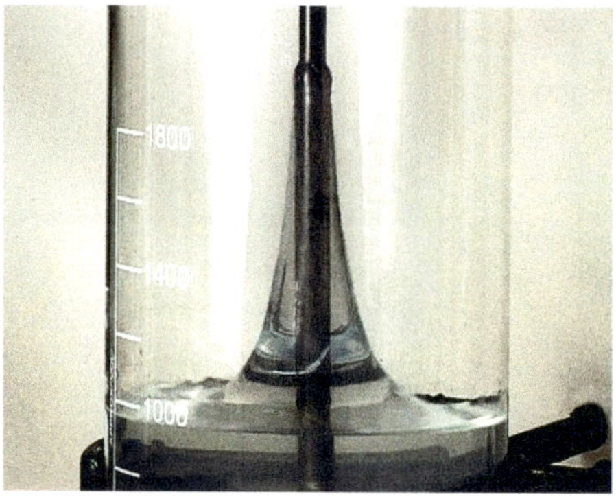

Figure 3.10 Example of the "Weissenberg Effect" for a viscoelastic solution. Stirrer is rotated (n = 200 s^{-1}) in a glass beaker containing aqueous PEO solution (c = 1 wt.-%, Polyox WSR 301). The solution climbs up the stirrer shaft due to normal stress effects

Normal stress differences also have practical relevance for extrusion: They assist in centering the shaft in the extruder when processing thermoplastics as the rotation of the screw generates radial forces (from the shaft to the wall).

Figure 3.11 shows the difference between a viscous and a viscoelastic fluid flowing out of a die. The silicone oil (case A) is only slightly different in diameter compared to the one of the die, while the emerging strand of PEO solution (case B) is considerably wider.

According to *Dealy,* the extrudate swell can be described in terms of the ratio of the die diameter D_0 to the strand diameter D_e after relaxation (due to tempering) [4]:

$$B_{ex} = \frac{D_e}{D_0} \tag{3.9}$$

For small Reynolds numbers (Re ≤ 2), Newtonian fluids also show a slight extrudate swell and the measured values of B_{ex} must be corrected accordingly [4].

$$B = B_{ex} - 0.13 \tag{3.10}$$

Based on an integral model, it can be shown that the corrected strand widening B can be quantitatively correlated with the first normal stress difference N_1, if the values for B and N_1 are determined with the same wall shear stress σ_w (see, e.g., Table 3.3) [4, 8]:

$$N_1^2 = 8 \cdot \sigma_w^2 \left[B^6 - 1\right] \tag{3.11}$$

Although Equation (3.11) does not fit with all experimental data in literature, the dependence of the extrudate swell on the first normal stress difference is nevertheless shown by $B \propto \sqrt[3]{N_1}$. Figure 3.11 was used to determine the extrudate swell of the PEO solution, where $B \cong B_{ex} = 1.6$. For polymer melts, B can even reach values of 2 or more. The extrudate swell has practical relevance for extrusion, e.g., in pipe or sheet extrusion.

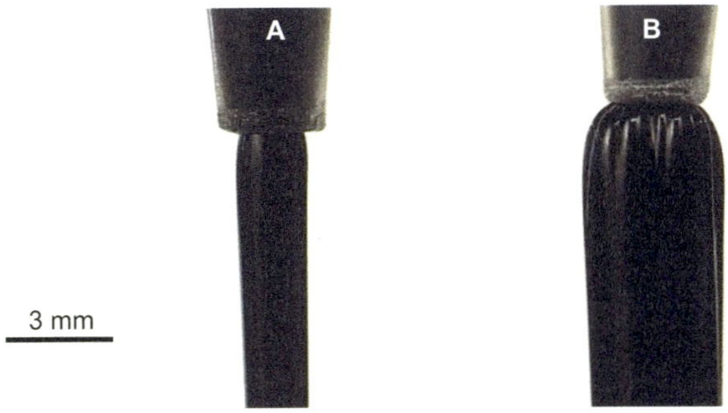

Figure 3.11 Extrudate swell for silicone oil (Baysilone M 1000) (A), and for viscoelastic PEO solution (aqueous 1% PEO solution Polyox WSR 301) (B) when extruded from a die

During extrusion of polymer melts with high throughputs, the elastic melt properties can also lead to elastic instabilities which can result in surface distortions of the extrudate. One example is the wavy distortions also described as sharkskin. Depending on the polymer, this can also lead to helical extrudate structures (stick-slip effect) or to very irregular extrudate structures (melt fracture) at even higher throughput rates [10].

3.1.4 Temperature Dependence of Shear Viscosity

The viscosity functions of homopolymers and compatible polymer blends measured at different temperatures can be shifted together by displacement along a 45° axis to form a single curve *(mastercurve)* by time-temperature-superposition (see Figure 3.12). This corresponds to a displacement along trajectories with a constant shear stress σ. The *shape* of the viscosity function does not change with temperature. One advantage of time-temperature-superposition is the availability of data outside a limited experimental window for the shear rates or frequencies.

From time-temperature superposition, the shift factors a_T can be obtained by Equation (3.12) and Equation (3.13).

$$\dot{\gamma}(T) = \frac{1}{a_T} \cdot \dot{\gamma}(T_0) \tag{3.12}$$

$$\eta(T, \dot{\gamma}(T)) = a_T \, \eta\left(T_0, \frac{\dot{\gamma}(T_0)}{a_T}\right) \tag{3.13}$$

Time-temperature superposition works for homopolymers and miscible blends but not for immiscible blends, filled systems (e.g., glass fiber reinforced plastics) or reactive or unstable polymers.

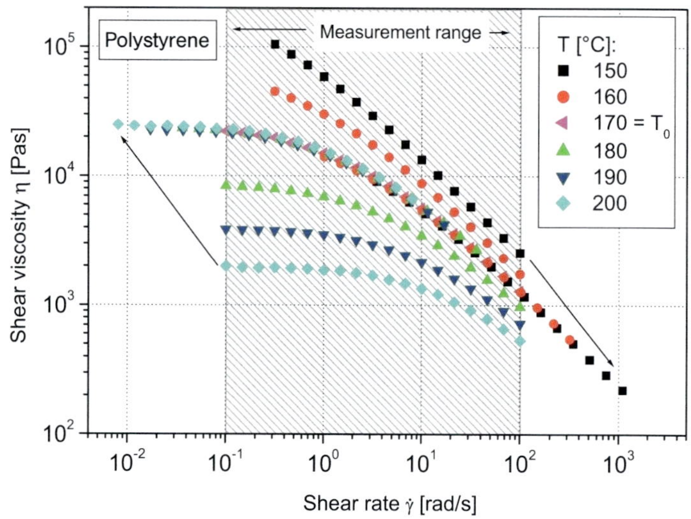

Figure 3.12 Time-temperature superposition along 45° trajectories (T = const.) to a reference temperature T_0 = 170 °C for the viscosity function of the polystyrene melt whose viscosity functions were measured between 150 and 200 °C

3.1.4.1 Temperature Dependence for Semi-Crystalline Polymers

For semi-crystalline polymers with melting points of more than 100 °C above the glass transition temperature and for amorphous polymers far above the glass transition temperature T_g (at around $T = T_g + 190$ °C), the shift factors obtained from time-temperature superposition can be plotted in the form of an *Arrhenius* plot for thermally activated processes:

$$\log a(T,T_0) = \frac{E_a}{2.3R}\left[\frac{1}{T} - \frac{1}{T_0}\right] \quad (3.14)$$

R is the gas constant and E_a the flow activation energy. The latter is a material-specific factor that is not dependent on the molar mass of the polymer and for thermorheological simple polymers also is not dependent on the shear stress. The activation energy E_a for polymer melts varies between 25 and 80 kJ/mol. It can be determined from the slope of the line in the Arrhenius plot.

Figure 3.13 shows the shift factors a_T determined from time-temperature superposition as a function of temperature for melts of two semi-crystalline thermoplastics as well as the Arrhenius plot. For the two polyethylenes (HDPE, LDPE), the progression of log a_T can be described with the Arrhenius equation. The activation energies can be determined from the slope as E_a(LDPE) ≈ 60 kJ/mol E_a(HDPE) ≈ 28 kJ/mol Along with polyethylenes (HDPE, LDPE, LLDPE), other significant semi-crystalline polymers are polypropylene (PP), polytetrafluoroethylene (PTFE), and polyamide (PA).

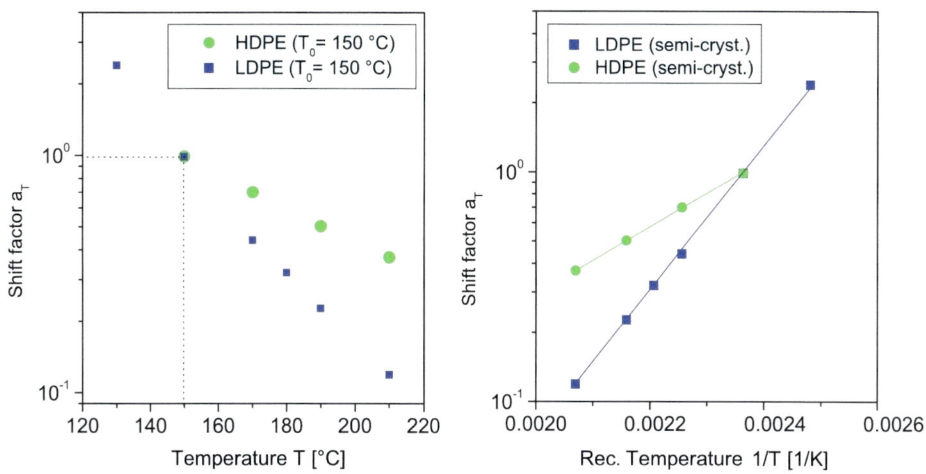

Figure 3.13 Left: Shift factors a_T from time-temperature superposition of two semi-crystalline thermoplastics [13]. Right: Arrhenius plot $a(T)$ = f(1/T). Lines: Arrhenius Equation (3.14) with $E_{a,HDPE}$ = 28 kJ/mol and $E_{a,LDPE}$ = 60 kJ/mol

3.1.4.2 Temperature Dependence for Amorphous Polymers

For amorphous polymers which melt above their glass transition temperature T_g, the WLF equation (according to *Williams, Landel, Ferry*, Equation (3.15)) with two material-specific parameters c_1 and c_2 gives a better description for the shift factors a_T than the *Arrhenius* function according to Equation (3.14).

$$\log a(T, T_0) = -\frac{c_1 (T - T_0)}{c_2 + T - T_0} \tag{3.15}$$

The WLF equation is based on the assumption that the segment mobility of polymer chains close to glass transition temperature is predominantly determined by the free volume which increases linearly with the distance from glass transition temperature. At high temperatures far above the glass transition temperature T_g ($T > T_g$ + 190 °C), the WLF relation approaches the Arrhenius equation (see line in Figure 3.14, right). In this case, the segment mobility of the molecules is no longer limited by the presence of free volume.

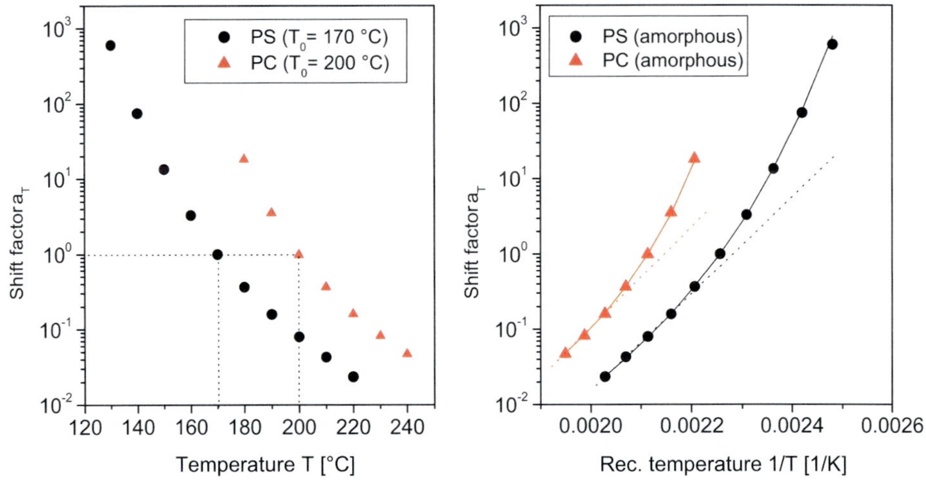

Figure 3.14 Left: Shift factors a_T from time-temperature superposition of two amorphous thermoplastics [9]. Right: Arrhenius plot $a(T)$ = f(1/T). Lines: WLF Equation (3.15) with $c_{1\text{-PC}}$ = 4.2 and $c_{2\text{-PC}}$ = 88 K for polycarbonate (PC) and with $c_{1\text{-PS}}$ = 5.3 and $c_{2\text{-PS}}$ = 113.4 K for polystyrene (PS)

The approximated lines at the right side of Figure 3.14 correspond to an Arrhenius activation energy of approximately 120 kJ/mol, which is significantly higher than the flow activation energy of the semi-crystalline polyolefin melts shown in Figure 3.13. The temperature dependence of the melt viscosity for amorphous thermoplastics is substantially higher than that of semi-crystalline polymers and increases dramatically as the temperature approaches the glass transition temperature.

3.1.5 Influence of Molecular Parameters on Rheological Properties of Polymer Melts

Important molecular parameters that influence the melt rheology of polymers are molar mass, molar mass distribution (polydispersity), and the degree of branching (molecular topology). A general empirical relation has been found for the dependence of (zero shear) viscosity on molar mass in the case of polymer melts, which applies for almost all polymers and for concentrated polymer solutions [5]. For higher molar masses $M > M_c$ a double logarithmic plot of η_0 = f(M_w) gives a straight line with a gradient of 3.4. This corresponds to a power law:

$$\eta_0 = K \cdot M_w^{3.4} \qquad M > M_c \tag{3.16}$$

K is a polymer-specific constant which also depends on the reference temperature T_0. The critical molar mass M_c is the equivalent of approx. 2 to 3 times the entanglement molar mass M_e. It is also a polymer-specific value and is insensitive to temperature. The molecular concept is that longer polymer molecules entangle themselves and the resulting increased interaction is responsible for the increasing zero shear viscosity with increasing molar mass. The molar masses of thermoplastics generally have molecular weights of above M_c.

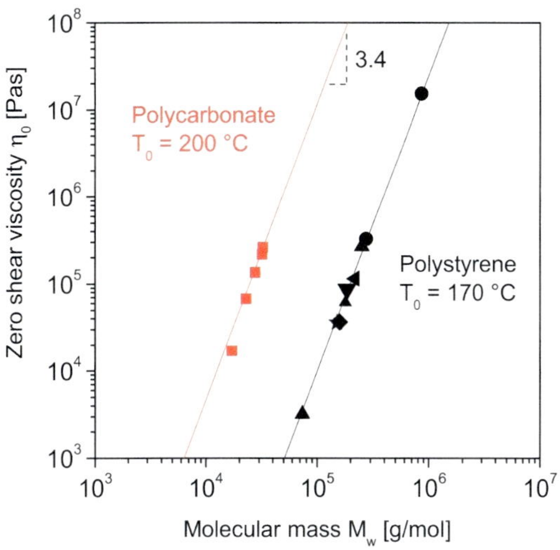

Figure 3.15
Influence of molecular mass on zero shear viscosity for polystyrene and polycarbonate melts at different reference temperatures T_0 [9, 11, 12]

The significant influence of molecular weight on viscosity becomes evident during processing: if, for example, polymers degrade and the molar mass decreases, the viscosity and therefore the pressure build-up in the extruder can decrease sharply.

The influence of molar mass distribution on the viscosity function is shown in Figure 3.16 on the basis of dynamic viscosities of different polystyrenes (PS), which were normalized with respect to their zero shear viscosity. A wider molar mass distribution results in a higher shear thinning in the normalized viscosity function, i. e., the drop in viscosity starts at lower normalized angle frequencies and/or shear rates.

This can be clearly seen in the comparison of viscosity functions of polystyrene 1 and polystyrene 2. If we add a very high-molecular weight component to polystyrene 2, we obtain polystyrene 3. It is therefore evident that the bimodal molar mass distribution causes the shear thinning to increase further, although not as far as the polydispersity change of the molar mass distribution between polystyrene 1 and polystyrene 2. In the case of higher shear rates, all flow curves proceed to similar viscosity functions.

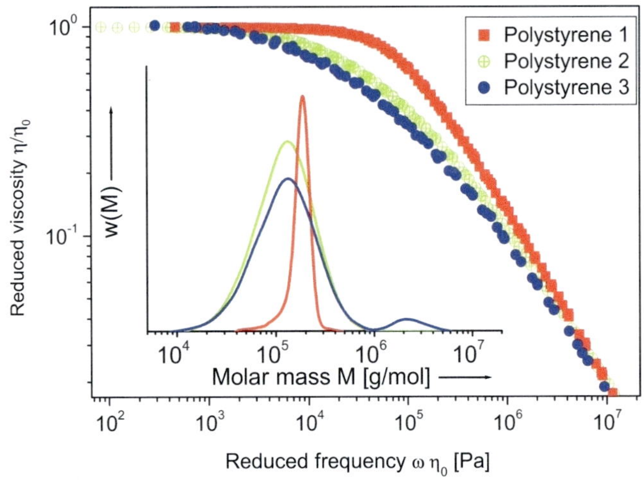

Figure 3.16 Influence of molar mass distribution on the normalized viscosity function (data from [9]). Inside figure: Molar mass distribution $w(M)$ of the three polystyrenes investigated

3.1.6 Shear Flows

In extrusion and polymer processing, shear and extensional flows occur. This section deals with basic flow shapes of pressure-driven pipe flows and simple drag flows, while the next section covers extensional flows.

In general, *shear flows* can be divided into *drag flows* and *pressure driven flows* (Figure 3.17). In addition, there are *extensional flows*, which are generated for instance by the deformation of polymer melts by compression or tensile deformation.

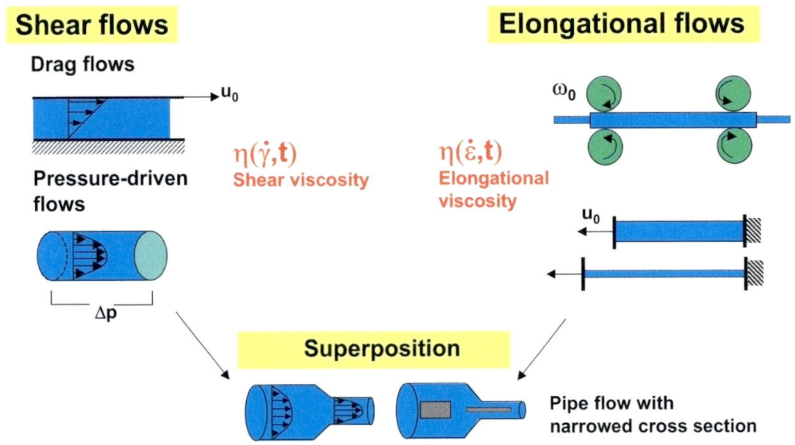

Figure 3.17 Simple shear and extensional flows and an example for overlapping shear- and extensional flow

3.1.6.1 Flow Profiles of Pressure-Driven Pipe Flow

For a *pressure-driven pipe flow* (laminar, stationary), force and mass balance are used to derive the relations summarized in Table 3.2. Different pipe flow profiles (Figure 3.18) are obtained, depending on the different rheological fluid properties. The profiles for shear rate and speed at constant volumetric flow and throughput depend substantially on the flow exponents: for Newtonian fluids (n = 1) we see a parabolic speed profile and a shear rate with linear dependence on the radial parameter r. The maximum shear rate is reached at the wall, because this is where the greatest speed gradients occur. At the center of the pipe, the shear rate is zero because there is no shearing at this point. For shear thinning fluids ($0 < n < 1$), the speed profile is flatter than for Newtonian fluids, while for shear-thickening (dilatant) fluids ($n > 1$) it increases much more sharply.

Table 3.2 Summary of the relations for Newtonian and shear thinning/shear thickening fluids in the case of a simple fully developed pipe flow (isothermal) [6, 14]

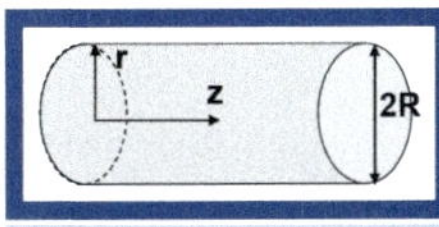	Newtonian fluid	Shear thinning/shear thickening fluid
Rheological constitutive equation	$\sigma = \eta \cdot \dot\gamma$	$\sigma = K \cdot \dot\gamma^n$
Shear stress at wall	$\sigma_w = \frac{\Delta p}{L} \cdot \frac{R}{2}$	
Shear stress	$\sigma(r) = \sigma_w \cdot \frac{r}{R}$	
Shear rate at wall	$\dot\gamma_w = \frac{4 \cdot \dot V}{\pi \cdot R^3}$	$\dot\gamma_w = \frac{3n+1}{n} \frac{\dot V}{\pi \cdot R^3}$
Shear rate	$\dot\gamma(r) = \dot\gamma_w \left(\frac{r}{R}\right)$	$\dot\gamma(r) = \dot\gamma_w \left(\frac{r}{R}\right)^{\frac{1}{n}}$
Max. velocity	$u_{max} = \frac{\Delta p \cdot R^2}{4 \cdot \eta \cdot L}$	$u_{max} = \frac{n}{n+1}\left(\frac{\Delta p}{2K \cdot L}\right)^{\frac{1}{n}} R^{\frac{n+1}{n}}$
Velocity	$u(r) = u_{max}\left[1 - \left(\frac{r}{R}\right)^2\right]$	$u(r) = u_{max}\left[1 - \left(\frac{r}{R}\right)^{\frac{n+1}{n}}\right]$
Pressure loss	$\Delta p = \frac{8\eta \cdot L \cdot \dot V}{\pi \cdot R^4}$	$\Delta p = \frac{2^{2+3n}\left(\frac{1+3n}{n}\right)^n K^n \cdot \dot V^n \cdot L}{\pi^n (2R)^{1+3n}}$

The residence time distribution also changes in the pipe in accordance with the flow profiles, depending on the rheological characteristics of the fluid: with increasing flow exponent n, the width of the residence time distribution increases, i. e., for $n \rightarrow 0$ the pipe flow practically is a plug flow with a very narrow residence time spectrum, while $n \rightarrow \infty$ results in a broad residence time distribution.

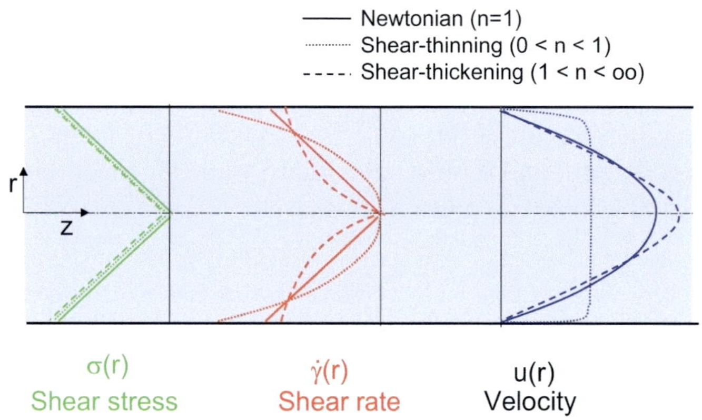

Figure 3.18 Radial profile of shear stress, shear rate, and speed through the cross-section for an isothermal pipe flow with Newtonian, shear thinning, and shear thickening fluids

When designing apparatuses for shear thinning fluids (such as polymer melts) it is important to note that pressure loss dependence changes if the throughput is changed (see formulas in Table 3.2): If the throughput is increased fourfold for a shear thinning fluid in which $n = 0.5$, pressure loss does not increase by a factor of 4, as in the case of a Newtonian fluid but by a factor of just 2. This is important when designing pressure consuming units such as those used downstream from extruders, e. g., pipes or die-plates.

3.1.6.2 Flow Profiles of Simple Drag Flow

Table 3.3 provides a summary of the relations for *simple drag flow* (laminar, stationary). Here, even if the fluids show different rheological behaviors, the speed profiles do not change (see Figure 3.19). The shear rate is also not dependent on rheological properties of the fluid.

Table 3.3 Summary of the Relations for Newtonian and Shear Thinning/Shear Thickening Fluids in the Case of a Simple, Fully Developed Shear Flow (Couette Flow, Isothermal) [6, 14]

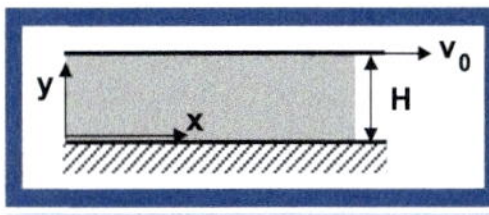	**Newtonian fluid**	**Shear thinning/shear thickening fluid**
Rheological constitutive equation	$\sigma = \eta \cdot \dot{\gamma}$	$\sigma = K \cdot \dot{\gamma}^n$
Shear stress	$\sigma_{yx}(y) = \frac{K \cdot v_0}{H} = const.$	$\sigma_{yx}(y) = K^{\frac{1}{n}} \left(\frac{v_0}{H}\right)^n = const.$
Shear rate at wall	$\dot{\gamma}_w = \frac{v_0}{H} = const.$	
Velocity	$v_x(y) = v_0 \cdot \frac{y}{H}$	

With respect to the shear stress distribution, shear thinning fluids ($0 < n < 1$) generally lead to lower shear stresses than shear-thickening fluids ($n > 1$), because the viscosities of shear thinning fluids in a stationary state are lower than in the case of Newtonian or shear thickening fluids at the same shear rate.

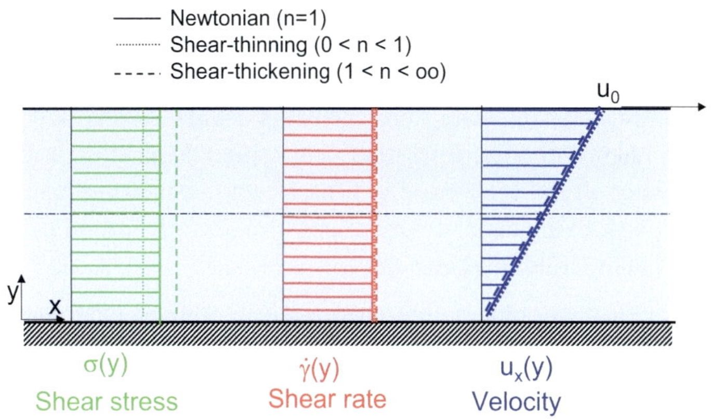

Figure 3.19 Profiles of shear stress, shear rate, and speed through the cross-section in the case of an isothermal, simple shear flow (Couette flow) with Newtonian, shear thinning, and shear-thickening fluids

To explain the influence of shear thinning on pressure build-up and on power input during extrusion, corresponding relations based on an overlap of the drag and the pressure-driven flows are developed. For more details, see Section 4.5.6.

3.1.7 Extensional Flows

In the extruder, not only shear flow is present, but also extensional flow occurs as well. This is illustrated in Figure 3.20 for the deformation of a fluid element. Wherever cross-sections narrow, such as at the tips or between kneading blocks and the wall, the fluid elements are compressed and extended. This effect is particularly relevant for non-homogeneous polymer melts, e.g., immiscible blends, in which the disperse phase can be split by extensional deformation.

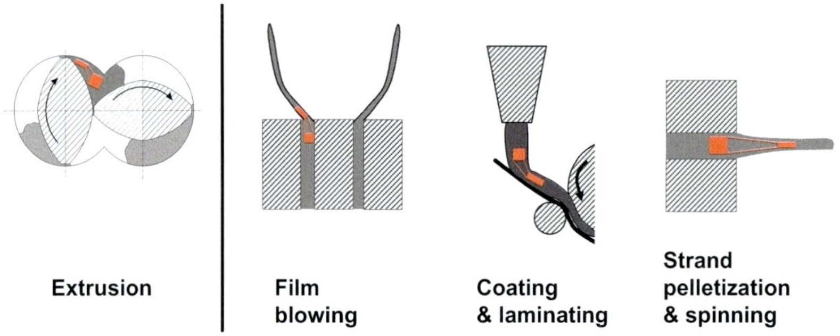

Figure 3.20 Extensional flows in the extruder and during processing of polymer melts

During polymer processing, extensional flows often occur; however, here generally those with free surfaces (see Figure 3.20). In the case of film blowing, the drawing of the film generates an extension. Here, the extensional viscosity plays a role with respect to a homogeneous film thickness distribution. In coating and laminating processes, drawing off the melt and the subsequent calendering results in extensional deformations. Extensional deformations also occur in strand pelletizing and spinning as the melt strand is drawn off.

Extensional viscosity can, depending on process conditions, therefore affect the homogeneity of the film thickness or the evenness of the layer thickness in coating processes. Extremely high extensional speeds can occur in spinning as the fibers are drawn from the die plates. The rheological properties play a role here for the maximum draw speed at which the fibers can be spun. In the case of extrusion and compounding, the extensional viscosity is relevant for pelletization because the polymer strands are drawn from the die-plate and then pelletized. The resulting strand diameter therefore influences the pellet form.

Figure 3.21 shows the influence of extensional viscosity on the flow behavior of an aqueous PEO solution. The high extensional viscosity in comparison with the relatively low shear viscosity can cause the upper container to empty when pouring, even when the liquid surface is below the glass rim. This is visible in Figure 3.21 by the liquid level markings. In the case of the viscous silicone oil, however, the liquid thread breaks as the pouring container is tipped back (not shown here).

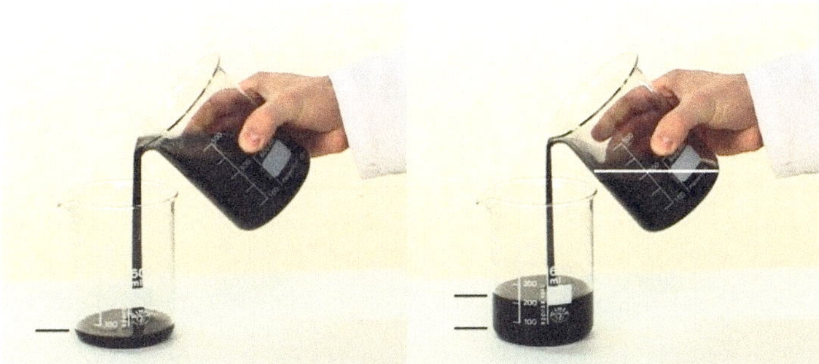

Figure 3.21 Pouring a viscoelastic solution with high extensional viscosity (open siphon effect). Aqueous PEO solution (c = 1% wt-%, Polyox WSR 301). Even tipping the solution out slightly (left) causes the container to empty: see liquid level change with respect to markings (right)

The extensional viscosity of the two fluids was determined using a punch that splits the liquid thread with a total length of 1 mm to 6.5 mm (method: see [15]). The resulting time-dependent thread diameter and the calculated extensional viscosities are shown in Figure 3.22.

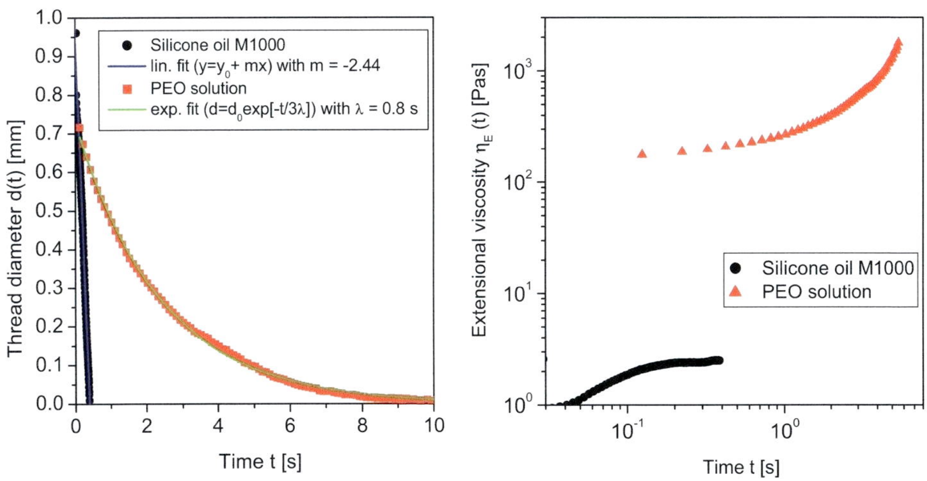

Figure 3.22 Measurement of the extensional viscosities of the PEO solution and of silicone oil Baysilone M 1000.
Left: Time-dependent thread diameter between two plates.
Right: Calculated extensional viscosities as a function of time

The exponential decrease in the thread diameter and the highly time-dependent increase in extensional viscosity is clearly visible in the case of the PEO solution. However, the silicone oil displays a linear drop in the diameter of the thread (Newtonian fluid) and no increase of the extensional viscosity over time; the extensional viscosity corresponds to approximately three times the shear viscosity.

References for Section 3.1

[1] Tadmor, Z., Klein, I., *Engineering Principles of Plasticating Extrusion*, Robert E. Krieger Publishing Company, Malabar (1985)

[2] Gogos, C. G., Tadmor, Z., Kim, M.-H., *Adv. Int. Polym. Tech.* (1998), 17(4), p. 1

[3] *http://www.campusplastics.com/*

[4] Macosko, C. W., *Rheology: Principles, Measurements and Applications*, VCH (1994)

[5] Ferry, J. D., *Viscoelastic Properties of Polymers*, 3rd ed., John Wiley & Sons (1980)

[6] Pahl, M., Gleißle, W., Laun, H.-M., *Praktische Rheologie der Kunststoffe und Elastomere*, VDI-Verlag, Düsseldorf (1995)

[7] Steffe, J. F., *Rheological Methods in Food Processing*, 2nd ed., Freeman Press (1996)

[8] Tanner, R. I., A theory of die swell, *J. Polym. Sci.* (1970), A-28, p. 2067

[9] Hepperle, J., *Einfluss der molekularen Struktur auf rheologische Eigenschaften von Polystyrol- und Polycarbonatschmelzen*, Shaker Verlag, Aachen, Germany (2003)

[10] Menges, G., Haberstroh, E., Michaeli, W., Schmachtenberg, E., *Werkstoffkunde der Kunststoffe*, Hanser Publishers, Munich (2002)

[11] Münstedt, H., Dependence of the elongational behavior of polystyrene melts on molecular weight and molecular weight distribution, *J. Rheol.* (1980) 24(6), pp. 847–867

[12] Graessley, W. W., Roovers, J., Melt rheology of four-arm and six-arm star polystyrenes, *Macromol.* (1979), 12(5), pp. 959–965

[13] Laun, H. M., Orientation of macromolecules and elastic deformations in polymer melts. Influence of molecular structure on the reptation of molecules, *Prog. Colloid Polym. Sci.* (1987), 75, pp. 111–139

[14] Fritz, H. G., Kunststofftechnologie - Einführung und Grundlagen. Teil A: Principles, lecture manuscript, Universität Stuttgart

[15] Stelter M., Das Zerstäubungsverhalten nicht-Newtonischer Flüssigkeiten, Dissertation, Universität Erlangen-Nürnberg (2001)

3.2 Material Behavior of Blends – Consideration of Polymer–Filler and Polymer–Polymer Systems

Jürgen Flecke

For the design of an extrusion or compounding line, especially of the process section of polymer processing, material data with rheological parameters, thermodynamic parameters, and densities are required. Today, the design of extruders is often based on mathematical methods and simulation calculations. In the case of processing a pure polymer without additives, the required material data can usually either be measured or extracted from databases.

In compounding technology, but generally during the processing of polymers with extruders, one often does not have to deal with “pure” polymers in practice. Multi-component systems are the reality in the plastics industry. Many processing steps include the filling or modification of polymers as well as the mixing and blending of different polymers or the blending of solids with the aim of improving the material performance of high-quality technical products [1–5].

The simplest case to be mentioned is mixtures with components of small parts by weight, such as color adjustments or the incorporation of fillers as UV stabilizers. In this case, the basic material properties of the polymer are changed only insignificantly, so that no special adjustments of the extruder screws or process parameters have to be considered for design studies. In addition to these simple modified products, however, polymer–polymer blends, highly filled solid compounds or even a combination of both are often used. In this section, the basics of handling modified polymers will be discussed.

Although many industrially produced compounds consist of more than two components, the following approaches refer to mixtures of two substances (polymer–filler or polymer–polymer systems):

(a) Mixing with fillers (polymer–filler systems)

During the compounding process of fillers, the filler, either in pure form or as a highly filled masterbatch, is added to the polymer during compounding. In the latter case, there is a kind of mixture of two polymers, since the properties of the highly filled masterbatch differ greatly from those of the origin polymer. In these systems, the material properties result as a function of the mixing partners and the composition of the mixture. Strictly speaking, the history of origin, i.e., the homogeneity of the mixture, must also be considered.

(b) Polymer blends (polymer–polymer systems)

During compounding of polymer–polymer blends, the mixture is formed in the extruder and during extrusion the material properties change depending on homogenization.

When describing the viscosity of polymer mixtures, the compatibility of the polymers used plays a decisive role. To simplify matters, one can differentiate between homogeneous mixtures of compatible polymers, partially and partially compatible, and heterogeneous mixtures of incompatible polymer blends [6–11].

Considering the process in an extruder, which is divided into different process sections and stages (feeding, solids conveying, compacting, melting, dispersing, degassing, conveying, and discharging), the respective process steps also depend on the mixture in each stage or in each process section. For a simulation calculation, therefore, for the mathematical basis, the material behavior description by equations in the zones or according to the individual sections must be determined. All mathematical methods for the design of extruders depend on how adequately and precisely the material properties are described (cf. [12]).

3.2.1 Material Properties of Two-Substance Systems

3.2.1.1 Introduction to Mixed Systems

In this chapter, some simple determination equations are presented and discussed, which can be used to determine the material parameters of a mixture of two substances from the parameters of the starting components and the respective composition of the melt. The composition of the melt is described by the mass fractions or by the volume fractions of the initially dosed components:

$$w_{S,A(B)} = \frac{m_{S,A(B)}}{m_{S,A(B)} + m_{S,B(B)}} \tag{3.17}$$

$$\varphi_{S,A(B)} = \frac{V_{S,A(B)}}{V_{S,A(B)} + V_{S,A(B)}} \tag{3.18}$$

Studies on the material behavior of two-substance systems have shown that not only the rheological but also the thermodynamic parameters of two-substance systems can only be determined approximately by a linear averaging of the data of the starting materials. Before discussing details of the rheological behavior, the thermodynamic characteristic melting enthalpy will be discussed. The enthalpy of a substance significantly influences the plasticizing behavior in the extrusion process. This applies to both filled polymers and polymer blends.

3.2.1.2 Thermodynamic Material Data of Two-Substance Mixtures

As an example of the thermodynamic behavior of two-substance systems, the behavior of the solid enthalpy and the melting enthalpy in a polypropylene filled with chalk and a polymer blend of polyamide and polyethylene is given here. The enthalpy curves for both material systems are shown as a function of the mass fractions in Figure 3.23 and Figure 3.24.

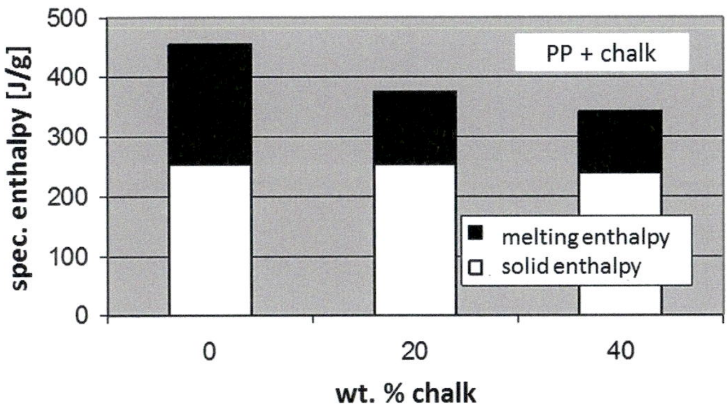

Figure 3.23 Example of melting enthalpy for a PP chalk compound (see [13])

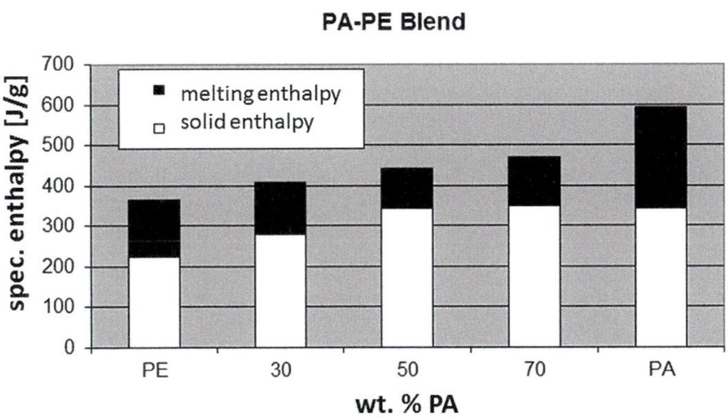

Figure 3.24 Example of melting enthalpy for a PA/HDPE polymer blend (see [13])

A closer analysis of the compound data reveals that the enthalpy of solids correlates with the mass fraction of the filler. However, the course of the melting enthalpy, i.e., the portion of enthalpy required for the actual plasticizing of the semi-crystalline plastic PP, is striking. This decreases disproportionately to the proportionate polymer mass. From this it can be concluded that the microstructure change caused by adding the chalk has changed the thermodynamic properties of the polymer phase. Also, with the mixture of two polymers, for which one could assume that the thermodynamic data can be calculated from the linear combination, the real behavior looks different (see Figure 3.24). Confirmations of these findings can be found in the literature, among other sources [14–20].

3.2.1.3 Viscosities of Two-Substance Mixtures

(a) Viscosity of Polymer–Filler Compounds

Filling and reinforcing polymers with a wide variety of substances – in most cases of mineral origin – not only affects the properties of the final product, but also significantly affects the processing properties. As already shown earlier, this results in a change in both the thermodynamic properties and the rheological properties explained later. Here, too, different approaches based on theoretical considerations or experimental investigations have been described in the literature [15, 21–25], some of which are to be analyzed in more detail.

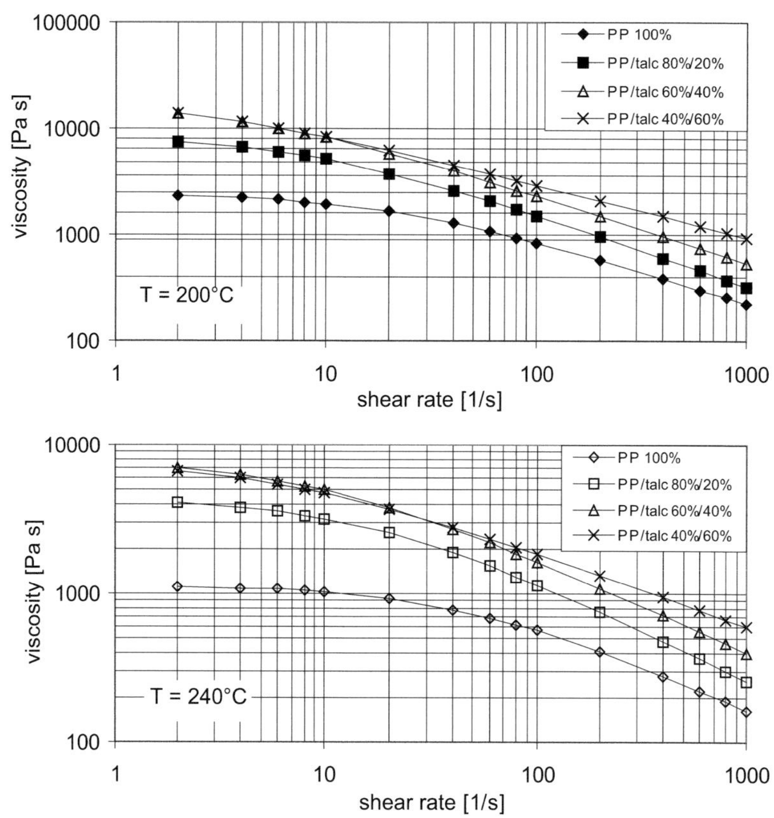

Figure 3.25 Viscosity curves of a polypropylene filled with talcum (see [13])

Figure 3.25 shows an example of the rheological behavior of a filled polymer depending on the mass fraction of the mineral filler at two different temperatures. The shift of the viscosity curve to higher viscosity values with increasing filler content is clearly visible.

Table 3.4 Equations for Calculating the Viscosity of Filled Polymers

Einstein	$\eta_S = \eta_F (1 + 2.5ø)$
Hashin	$\eta_S = \eta_F \left(1 + \frac{2ø}{1 - ø}\right)$
Maron and Pierce	$\eta_S = \eta_F \left(1 - \frac{ø}{ø_0}\right)^{-2}$
Mooney	$\eta_S = \eta_F \exp\left(\frac{2.5ø}{1 - kø}\right)$

Figure 3.26 shows an example of a comparison between a measured viscosity curve and the curves calculated with the equations given in Table 3.4 for different polymer-filler combinations. The descriptions of the viscosities of the filled polymers in the examples shown differ widely from the measured results. A more precise analysis reveals that the existing description equations do not sufficiently take into account the influence of temperature, shear rate, and type of filler.

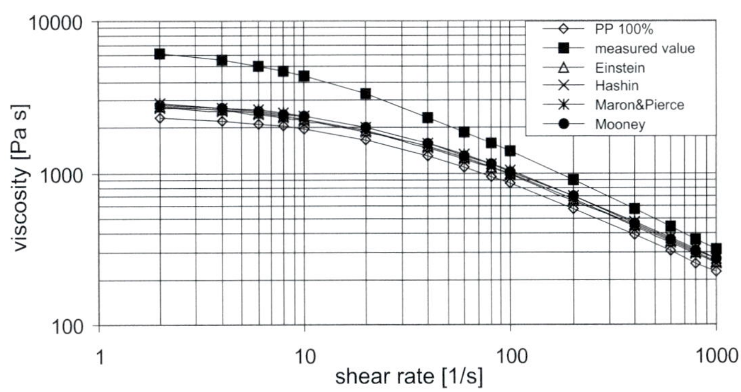

Figure 3.26 Comparison between calculated and measured viscosity values (PP-talcum 40/60 wt.-%, T = 240 °C) (see [13])

(b) Viscosity of Polymer-Polymer Blends

When describing the viscosity of polymer mixtures, the compatibility of the polymers used plays a decisive role. To simplify matters, one can differentiate between homogeneous mixtures of compatible polymers, partially and partially compatible, and heterogeneous mixtures of incompatible polymers [6-11].

Also, in the existing and published descriptive approaches, most of which are based on experimental investigations of various kinds, mixing capability and viscosity ratio are also given as decision variables.

3.2.1.4 Compatible Polymer Blends

The simplest and most common approach is the so-called logarithmic mixing rule according to Arrhenius (see Table 3.5). For compatible polymers with a ratio of zero viscosities less than 10, this approach for low shear rates leads to good results [17, 18].

3.2.1.5 Immiscible (Incompatible) Polymer Blends

For an evaluation of the approaches to describe the viscosity of incompatible polymers, various blends of a mixture of a high-density polyethylene with different contents by weight with low density polyethylene, a polypropylene, and a polyamide have been prepared on a co-rotating twin-screw extruder without a compatibilizer. The results of the following rheological investigation are exemplarily documented in Figure 3.27. It is clear that a uniform description is only possible to a limited extent due to the differences in the gradients.

Due to the complexity of polymer–polymer mixing systems (cf. [26]), only a few systems will be presented here without claiming to be complete.

For an analysis of the quality of the description, an incompatible polymer blend of 20/80% by weight HDPE/PA was used. Approaches according to Arrhenius [9, 11, 17] and Bingham [19, 20] (cf. Table 3.5) are shown exemplarily for comparison in Figure 3.28.

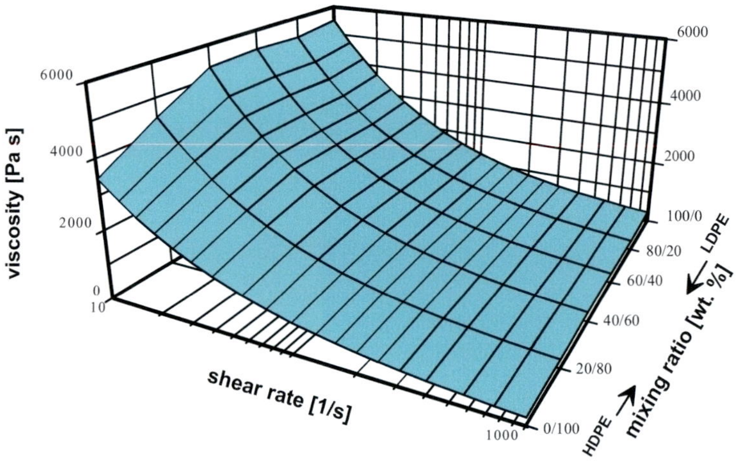

Figure 3.27 Viscosity curves for different polymer blends at standard operating temperatures (see [13])

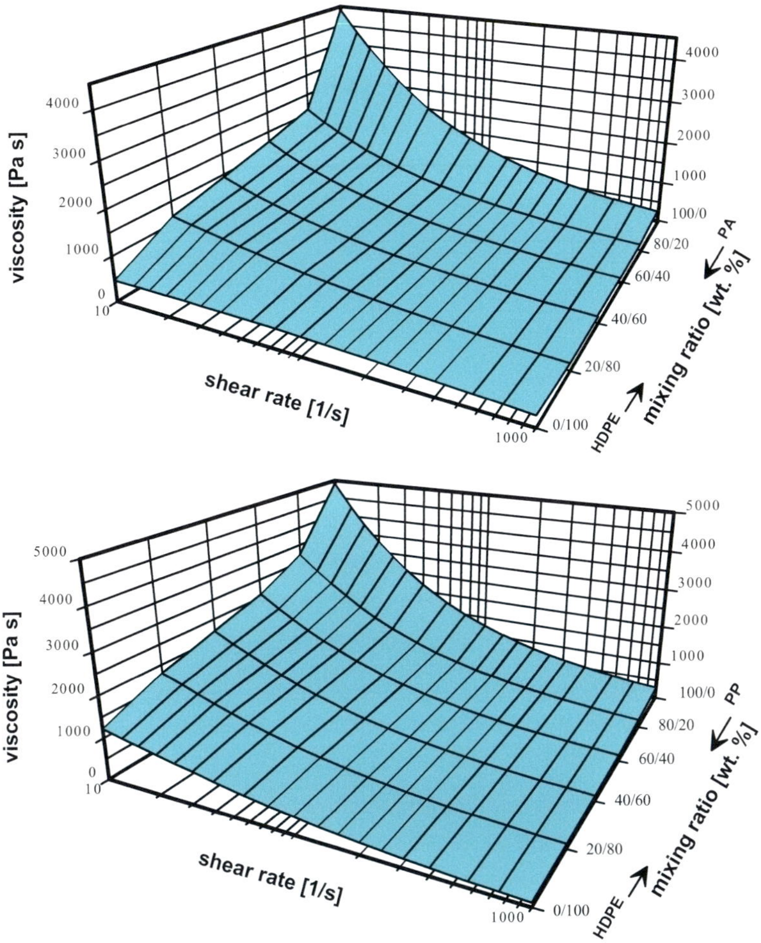

Figure 3.27 Viscosity curves for different polymer blends at standard operating temperatures (see [13]) *(continued)*

Table 3.5 Equations for Calculating the Viscosities of Two-Phase Miscible Polymer Blends

Arrhenius	$\log \eta_b = w_1 \log \eta_1 + w_2 \log \eta_2$
Bingham	$\frac{1}{\eta_b} = w_1 \frac{1}{\eta_1} + w_2 \frac{1}{\eta_2}$
Utracki	$\eta_b^{1/3.4} = w_1 \eta_1^{1/3.4} + w_2 \eta_2^{1/3.4}$
Grunberg	$\ln \eta_b = w_1 \ln \eta_1 + w_2 \ln \eta_2 + 2 w_1 w_2 G$

It can be shown that all approaches strictly related to the calculation of the zero viscosities do not adequately reflect the viscosity of the polymer blends. However, some of the more complex approaches given in the literature [19, 20] were not included in the diagrams for reasons of clarity but led to poorer results in the analysis.

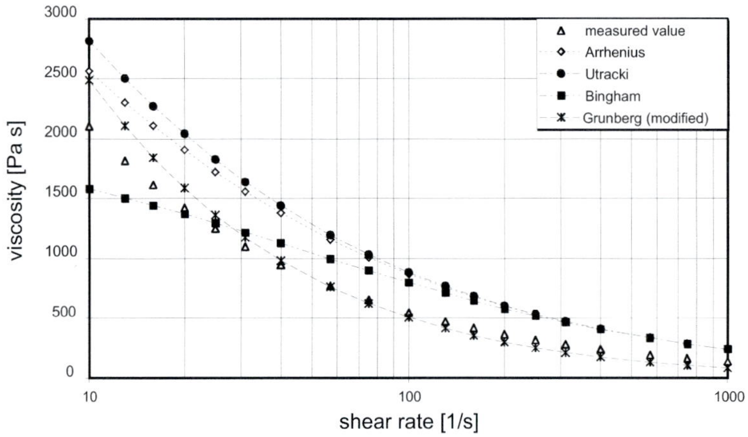

Figure 3.28 Comparison between measured and calculated viscosity values (HDPE/PA wt.-% 20/80, T = 240 °C) (see [13])

Another interesting approach was published by Grunberg in [20]. This approach, developed on the basis of the Arrhenius approach, essentially contains a constant G as an extension, which causes a parallel shift of the Arrhenius approach. In Figure 3.28, this shifting constant was modified as a function of the shear rate so that an exact description becomes possible. The disadvantage of this approach is the need to know selected measured viscosity curves for adaptation (cf. [27]).

In summary, it can be stated that despite the large number of approaches, the material data of two-substance systems must be checked in individual cases for the quality of the prediction of the rheological material behavior. Only in exceptional cases can the behavior be calculated exactly in advance using general approaches. Nevertheless, the calculation of material data from the combination of additives is a requirement for describing the process behavior of two-component systems. In this context, it may be necessary in practice to develop special approximation equations – but in any case, to check the quality of the description equations used.

3.2.2 Process Behavior during Plasticizing of Two-Substance Polymer Systems

Most mathematical methods and simulation calculations are related to the process behavior of the melt phase only or the transport of solids in an extruder. Plastification is complicated and difficult to describe using mathematical methods. Also, there are few reliable experimental results as a basis for the development of a description, so the extent to which the description can be generally transferred and remain valid must be checked in each individual case.

But, especially with multi-component systems, the melting zone and the melting process are particularly interesting and have a significant influence on the behavior of the entire extrusion or processing line.

The quality of the dispersion through the melting zone, in which high shear stresses act, is also decisively determined [28, 29]. The plasticizing behavior in a co-rotating twin-screw extruder has been analyzed and described using physical-mathematical modeling [30–32].

In the case of plasticizing in the presence of a filler, a masterbatch, or different polymers, the existing description must be extended and adapted for a calculation. The plasticizing of mixtures of two substances is a complex process, since the material properties of the melt already formed change constantly along the plasticizing process depending on the melting process and thus the boundary conditions for the calculation. Mixing and dispersing in the zone is also very intensive and changes the material properties [33, 34].

For a more detailed understanding, it must be clear that with filled polymers, the resulting melt portion is first wetted and highly filled with the entire filler. As the melting process progresses, the filler content in the melt changes up to the defined percentage by weight. Rheological and thermodynamic properties are constantly changing and must be described iteratively in sections for a design.

The same applies to polymer blends where at first the low-melting temperature component starts with plasticizing and enters into a mixture with the high-melting component. Rheology and thermodynamic properties are influenced by the melting and must be taken into account, e.g., mixing gaps between the two components [4, 5]. For extended modeling for disperse melting in the processing of compounds and polymer blends on co-rotating twin-screw extruders, fundamental agreements must first be defined [35].

The necessary modification step compared to the melting model for single-component systems is that, in the case of processing two polymer components into a blend in the screw channel, three phases must be considered at all times. Figure 3.29 shows the assumed macromorphology in the case of blending for a volume element in the screw channel.

It is assumed that the screw section in this case is fully filled.

The melt is understood as a homogeneous mixture of the already melted material of both components. Gaps in the mixture due to thermodynamic incompatibility are not taken into account. Due to the different melting enthalpy and mass proportions, the composition of the melt phase changes in the direction of flow as a result of the melting process of both polymers. In order to determine the properties of the melt phase from the properties of the components involved, simple mixing rules can be used in the case when both phases can be mixed with each other and evaluated step by step in the course of the melting calculation (Table 3.5).

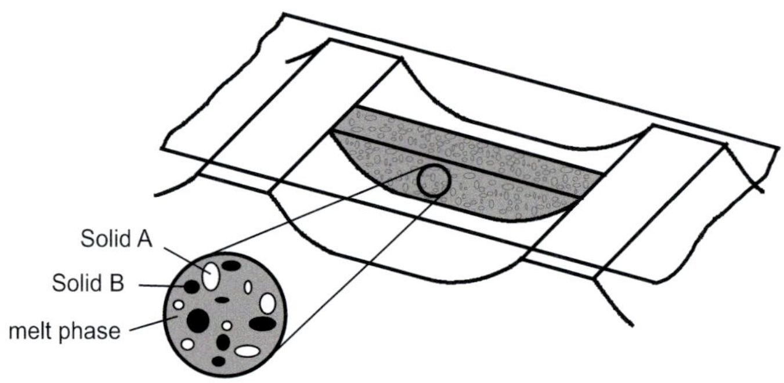

Figure 3.29 Phases considered in the calculation of the melting process (polymer blend or compound)

Analogous to the single-phase melting model, the calculation of multi-phase systems is carried out stepwise along the melting zone. The following quantities must be known at the start of melting or as input quantities for each step:

1. solid content of both components,
2. solid temperature,
3. average melt temperature.

The determination of the solid content at the start of melting is subject to the assumption that the solid particles will be arranged with a certain packing density. This model assumption is to be used both for considerations in the feed zone of an extruder and for the entry into a melting zone.

To simplify matters, the ideal spherical shape should be assumed for the solid particle shape (for both granules and fillers). The dimensions of commercially available granulates are in the range 2–5 mm. The particle sizes of light fillers, such as precipitated silicas, aluminum silicates, chalks, or similar, lie in a range of 0.1–5 µm [36] during processing. Due to the two monodisperse starting materials, a bidisperse particle distribution must be assumed. Any statistical particle size distributions of the individual fractions are integrated via averages.

The solid content in a screw channel is generally defined as the ratio of solid volume to channel volume:

$$F = \frac{V_F}{V_{Channel}} \tag{3.19}$$

This leads to the melt portion in fully filled screw sections:

$$S = 1 - F \tag{3.20}$$

For the determination of solid content, the number of particles must be considered. This can be determined for a monodisperse particle size distribution taking into account the channel geometry of a co-rotating twin-screw extruder in a volume section [32]. Due to the shape of the channel, not all particles in the elementary cells are used to determine the number of particles in the volume element under consideration (see Figure 3.30). For other channel geometries, e.g., rectangular channels in single-screw extruders, this model must be adapted accordingly. Since particles are partly outside the channel, the number is corrected by an area factor [31]. For a cubic face-centered arrangement, this results in:

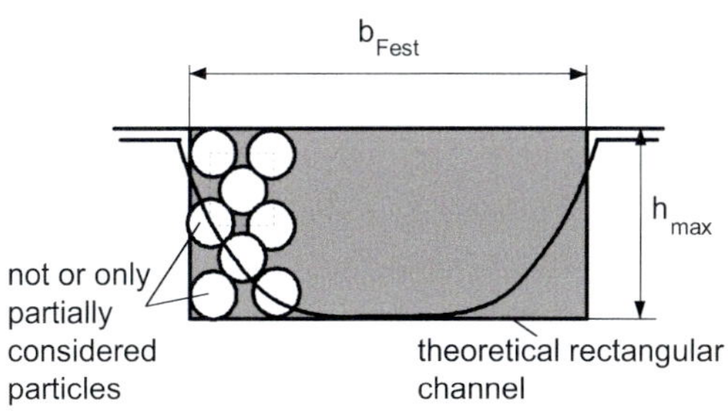

Figure 3.30
Cubic face-centered particle arrangement in the screw channel [31]

Hereby, the approximate particle number can be described as shown in [37, 38], assuming a cubic face-centered spherical packing for the channel section of constant geometry being considered:

$$F = \frac{n_{gr} \frac{4}{3} \pi \left(\frac{d_{gr}}{2} \right)^3}{A_{Channel} a_{cfc}} \tag{3.21}$$

For a bidisperse fill, as is assumed in a two-substance system, this solid content must be corrected. The first step is to consider porosity.

According to Jeschar, the description of porosity is based on bidisperse fills [39]. For the following consideration of a melting zone, the porosity of a fill is equal to the melt fraction. It is assumed that the remaining channel volume of the spaces between the solid particles is completely filled with melt. The example of a compound (melt + filler + granulate) results in:

$$\psi = \frac{V_{\mathrm{Tot}} - V_{\mathrm{Fi}} - V_{\mathrm{Gr}}}{V_{\mathrm{Tot}}} \tag{3.22}$$

Furthermore, a fundamental distinction must be made between two systems:

(a) The bidisperse fill is produced by adding fine particles to the monodisperse coarse fill. The fine particles arrange themselves in the cavities of the coarse particles without disturbing the structure. For particles with very different dimensions, as is the case with a system of plastic granulate and filler powder, these requirements are met.

(b) In the other case, the larger particles are dispersed in the unaffected powder matrix.

Figure 3.31 shows both limit cases.

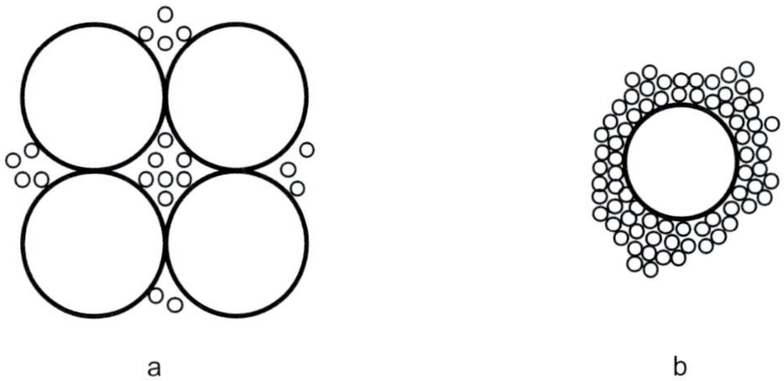

Figure 3.31
Limit cases of bidisperse spherical packings [39]

Figure 3.32 shows the two limit curves of cases (a) and (b) as well as data measured by Jeschar. In both cases, the porosity of the monodisperse fills was set at 0.38 [39].

The intersection of the two boundary curves represents the minimum possible total porosity. The volume fraction of fine material, $Q_{\mathrm{F,min}}$, at which this porosity occurs is calculated for equal porosities of the monodisperse packings by equating case (a) and case (b) according to [39].

$$Q_{\mathrm{F,min}} = \frac{\psi_{\mathrm{Gr,Fi}}}{1 + \psi_{\mathrm{Gr,Fi}}} \tag{3.23}$$

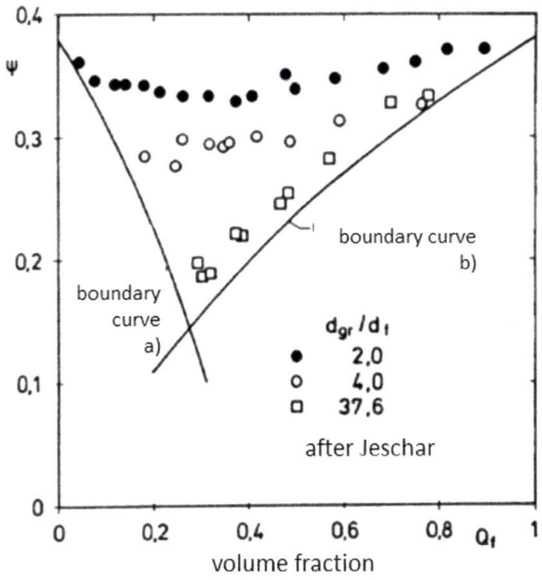

Figure 3.32
Porosity of bi-spherical spherical packings as a function of the solid content according to Jeschar [39]

For the granules/filler system, however, it must be assumed that the porosities of the monodisperse packs are different. The filler will settle in the gaps between the granules and reduce the overall porosity. In this case, the volume fraction of fine material at which the lowest total porosity occurs is calculated as follows:

$$Q_{F,min} = \frac{\psi_{Gr} - \psi_{Gr}\psi_{Fi}}{1 - \psi_{Gr}\psi_{Fi}} \tag{3.24}$$

It is taken into account here that the filler is also available in a porous package. For the porosity of the monodisperse packing, the maximum (possible) tightest sphere packing is assumed to give ψ = 0.26 (porosity of an undisturbed face-centered cubic packing). The proportion by volume of the filler is calculated on the basis of the dosed operating mode:

$$Q_F = \frac{V_{Fi,dos}}{V_{Gr,ASB} + V_{Fi,dos}} \tag{3.25}$$

The validity limit of Equation (3.25) can be calculated in accordance with [40] over a degree of saturation in case (a), resulting from the bulk densities of both components.

3.2.2.1 Calculation of the Melting Behavior of Two-Substance Systems

In the following, the calculation of the melting process of two-substance systems is discussed according to the principles of solids determination. In analogy to the disperse melting model [31], which forms the basis of the description for two-substance systems, only the essential modeling steps are explained here to show the general approach. The model can be seen in detail in [13] and [31].

The following conventions must be made:

1. The particle shape is assumed to be the ideal sphere.
2. The particles are dispersed in the melt.
3. One single particle or one particle per component is being considered.
4. The reference particles are not melted off by neighboring particles.
5. Between the solid particles and the surrounding melt, there is no relative speed.

Existing melting models can be modified on the basis of these considerations. In each step of the calculation along the plasticizing process, the material properties must be determined anew using mixing rules. As an example, in [13] on the basis of [30], a model for calculating the plasticizing process for two-substance systems is worked out in detail. In addition to a large number of other publications on investigations of plasticizing behavior and mathematical models, [31, 41–44] should be mentioned here as an example.

Based on these considerations, it is not only plasticizing behavior but also feeding behavior as well as dispersion and mixing processes that can be modeled in an extruder. Mathematical models and simulation calculations can thus take real material behavior into account – but the complexity of the models and simulations increases disproportionately compared to single-material systems.

3.2.3 Final Remarks for Use in Practice

The main intention of the explanations given earlier and the demonstration of the complexity of multi-component systems is that, often, wrong conclusions are drawn on the basis of simple assumptions (single-component systems) in design/simulation calculations. However, since not all material and process parameters can usually be determined in practice, it is possible to describe approximations for certain cases involving extruder screws with fully filled zones.

But also in practice, when operating extruders or especially compounding processes, the deviating process behavior due to the modification of polymers is often underestimated.

In summary, practical examples can be mentioned here:

- Collapse of the feeding behavior of grooved bush extruders.
- Unexpected performance increase in the extruder during the addition of mixing systems.
- Insufficient plasticizing performance even with a minor modification of polymers.
- Inclusions and gels when adding minor additives to the extrusion product.
- “Flushing” of degassing openings – modification of the polymers can lead to changes in the filling degree and conveying behavior.
- “Plugging” (insufficiently balanced channel volume) of melting zones in the barrier screws of single-screw extruders.

3.2.4 Conclusion

Despite the large number of publications and papers on polymer blends and polymer–solids mixing systems and their process behavior, the mathematical description and simulation of extruders is not comprehensively described. Each model presentation contains a number of assumptions and a number of simplifications. The approaches shown in this section help to understand the process characteristics of polymer mixing systems, but an application-specific adaptation must be checked in each individual case.

References for Section 3.2

[1] Paul, D. R., Newman, S., *Polymer Blends,* Academic Press Inc., New York (1978)

[2] Sok, K., *Polymer Compatibility and Incompatibility Principles and Practices,* Harwood, New York (1982)

[3] Kaiser, W., *Kunststoffchemie für Ingenieure,* Carl Hanser Verlag, Munich, Vienna (2006)

[4] Wendorff, J. H., *Kunststoff-Legierungen – Theoretische Grundlagen,* oral presentation, DKI Darmstadt (1992)

[5] Jungnickel, B. J., *Physik und Werkstoffkunde von Polymerlegierungen und -mischungen,* oral presentation, DKI Darmstadt (1992)

[6] *Handbuch der Kunststoffextrusion: Extrusionsanlagen; Band II,* Carl Hanser Verlag, Munich, Vienna (1986)

[7] Gächter, R., Müller, H., *Kunststoff-Additive,* Carl Hanser Verlag, Munich, Vienna, New York (1989)

[8] Flory, P. J., *Discuss. Faraday Soc.* (1970) 49

[9] Utracki, L. A., *Polymer Alloys and Blends,* Carl Hanser Verlag, Munich, Vienna, New York (1983)

[10] Olabisi, O., Robeson, M., Shaw, T., *Polymer-Miscibility,* Academic Press (1979)

[11] Utracki, L. A., Kamal, M. R., Melt Rheology of Polymer Blends. *Polym. Eng. Sci.* (1982) 22, pp. 96–114

[12] Kretschmer, K., *Untersuchung und Beschreibung des Prozess- und Mischverhaltens von Mischelementen für Gleichdrall-Doppleschneckenextruder,* Uni-GH Paderborn, Dissertation (2004)

[13] Flecke, J., *Grundlagen zur Simulation des Prozeßverhaltens eines Gleichdrall-Doppelschneckenextruders bei der Aufbereitung mineralisch gefüllter Polymere,* Uni-GH Paderborn, Dissertation (2000)

[14] Christov, S.F., Skorokhodov, I.I., Shuraleva, Z.V., The concentration dependence of viscosity: mixtures of homogenous polymers. *Vysokomol-Soed.* (1978) A20, pp. 1705–1911

[15] Kim, Y.W., Han, C.D., Rheological Properties and Microstructure of Multiphase Polymeric Systems. *J. Appl. Polym. Sci.* (1976) 20, pp. 2905–2912

[16] Montfort, J.P., Marin, G., Arman, J., Monge, Ph., Viscoelastic properties of high molecular weight polymers in the molten state, II. influence of the molecular weight distribution on linear viscoelastic properties. *Rheol. Acta* (1979) 18, pp. 623–628

[17] Montfort, J.P., Marin, G., Arman, J., Monge, P., Viscoelastic properties of high molecular weight polymers in the molten state. *Rheol. Acta* (1979) 18, pp. 623–628

[18] Schuch, H., Application of a blending rule for the complex of polymer melts. *Rheol. Acta* (1988) 4(27), pp. 384–396

[19] Irving, J.B., *Viscosity of Binary Liquid Mixtures: A Survey of Mixture Equations* (1977) National Engineering Laboratory, Report No. 630

[20] Irving, J.B., *Viscosity of Binary Liquid Mixtures: The Effectiveness of Mixture Equations* (1977) National Engineering Laboratory, Report No. 631

[21] Sun, L., Aklonis, J.J., Salovey, R., Effect of Modification of Filler Composition on Rheology. *Polym. Eng. Sci.* (1993) 33, pp. 1308–1319

[22] Maron, S.H., Pierce, P.E., Application of Ree-Eyring generalized flow-theory to suspensions of spherical particles. *J. Colloid Sci.* (1956) 11, pp. 80–95

[23] Mooney, M., The viscosity of concentrated suspensions of spherical particles. *J. Colloid Sci.* (1951) 6, pp. 162–170

[24] Kataoka, T., Kitano, T., Oyanagi, Y., Sasahara, M., Viscous properties of calcium carbonate filled polymer melts. *Rheol. Acta* (1979) 18, pp. 635–639

[25] Einstein, A., Eine neue Bestimmung der Moleküldimension. *Ann. Phys.* (1906) 19, p. 289

[26] Heindl, M., *Einfluss von Dehnströmungen auf die Morphologieausbildung in Polymerblends,* Universität Erlangen-Nürnberg, Dissertation (2005)

[27] Potente, H., Flecke, J., Simulation und Auslegung von Aufbereitungsmaschinen, Materialverhaltens von Zweischneckensystemen. *Kunststoffberater 5* (1997)

[28] Raasch, J., *Beanspruchung und Verhalten suspendierter Feststoffteilchen in Scherströmungen hoher Zähigkeit,* TH Karlsruhe, Dissertation (1961)

[29] Chen, H., Sundararaj, U., Nandakumar, K., Modeling of Polymer Melting, Drop Deformation, and Breakup under Shear Flow, *Polym. Eng. Sci.* (2004), 44(7), p. 1258–1266

[30] Ansahl, J., *Grundlagen für die Auslegung dichtkämmender Gleichdrall-Doppelschneckenextruder,* Uni-GH Paderborn, Dissertation (1993)

[31] Melisch, U., *Grundlagen zur Simulation des Plastifizier- und Förderprozesses eines dichtkämmenden Gleichdrall-Doppelschneckenextruders,* Uni-GH Paderborn, Dissertation (1998)

[32] Armstroff, O., Zettler, H.D., Ansätze zur Berechnung von Zweischneckenextrudern. *Kunststofftechnik* (1973) 9(12), pp. 240–243

[33] Reichert, H., Desagglomeration organischer Farbpigmente in Scherströmungen hochzäher Flüssigkeiten. *Chem. Ing. Tech.* (1973) 45(6), pp. 391–395

[34] N.N., *Optimierung des Compoundierprozesses durch Rezeptur- und Verfahrensverständnis,* VDI-Verlag (1997)

[35] Potente, H., Melisch, U., Flecke, J., SIGMA-Rechnergestütze Prozeßsimulation dichtkämmender Gleichdrall-Doppelschneckenextruder, Parts 1 und 2, Plastics No. 1, Issue November/December 1995 and January/February 1996

[36] Limper, A., Kautschukverarbeitung, lecture manuscript, Uni-GH Paderborn (1997)

[37] Potente, H., Melisch, U., Aufschmelzprozeß in Gleichdrall-Doppelschneckenextrudern. Final report on on the DFG research project, PO 171/31-1 (December 1995) Uni-GH Paderborn

[38] Potente, H., Flecke, J., *Analysis and Modelling of the Residence Time Distribution in Intermeshing Co-Rotating Twin Screw Extruders based on Finite Element Simulations* (1997) 55. Antec 97, Session M21, Extrusion, Toronto

[39] Schlünder, E. U., Tsotsas, E., *Wärmeübertragung in Festbetten, durchmischten Schüttgütern und Wirbelschichten,* Georg Thieme Verlag, Stuttgart (1988)

[40] Schöppner, V., *Simulation der Plastifiziereinheit von Einschneckenextrudern,* Uni-GH Paderborn, Dissertation (1995)

[41] White, J. L., Szydlowski, W., Min, K., Kim, M.-H., Twin Screw Extruders; Development of Technology and Analysis of Flow. *Adv. Polym. Technol.* (1987) 7(3), pp. 295–332

[42] White, J. L., *Twin Screw Extrusion, Technology and Principles,* Carl Hanser Verlag, Munich, Vienna, New York (1991)

[43] White, J. L., Chen, Z., Simulation of Non-Isothermal Flow in Modular Co-rotating Twin Screw Extrusion. *Polym. Eng. Sci.* (1994) 34(3), p. 229

[44] Lacey, A. A., Nikolopulous, C. V., A 1D model for polymer melting during modulated temperature differential scanning calorimetry. *J. Appl. Math.* (2006) 71(2), pp. 186–209

3.3 Diffusive Mass Transport in Polymers

Heino Thiele

In single- and multiphase systems with more than one species, differences in local concentrations are equalized by diffusion. In multiphase systems, the mass transport by diffusion also takes place between the phases. This process is of technical importance in sorption processes, like the removal of solvents and residual monomers from polymers. In continuously operated screw machines, this is a quasi-stationary process. For this process important basic relationships are discussed below. This includes information about the time- and space-dependent concentration profiles near the phase boundaries and relationships to calculate the material flux caused by diffusion between the phases and the associated change in the mean concentration of the liquid phase. The explanations to the equilibrium concentration and the diffusivity complete the basics of diffusive mass transport in polymers. The explanations are limited to polymers at temperatures above the glass transition temperature. They are existing as melts and contain low-molecular, dissolved substances which can be removed from the polymer melt by desorption.

3.3.1 Mechanisms of Mass Transport

Theoretically based relationships to estimate the diffusive mass transport are often based on Fick's first law and the mass balance established for a volume element of the fluid system [1, 2]. According to Fick's first law, the mass flux by diffusion is proportional to the local concentration gradient. With the mass flow rate per unit area $\dot{m}_j$ as a characteristic value of the mass flux, the mass density ρ_j, representing the mass of species j per unit volume and the local coordinate x, the following relationship results in the one-dimensional case:

$$\dot{m}_j = -D_f \frac{\partial \rho_j}{\partial x} \tag{3.26}$$

The proportionality factor is the diffusivity D_f. As is shown later in this chapter, the concentration gradient $\partial \rho_j / \partial x$ itself is a function of the diffusivity D_f. The mass flux therefore is not directly proportional to diffusivity, as postulated by some mass transfer models according to the two-layer hypothesis [1, 2].

In polymers, mechanisms of diffusive mass transport that deviate from Fick's first law are also observed. For polymer melts above the glass transition temperature, these deviations are in most cases negligible [3]. Therefore, in the further discussions, the validity of Fick's first law is assumed.

3.3.1.1 Concentration Distribution Near the Phase Interface

In general, the mass transport in fluid systems is described by a partial differential equation and the associated initial and boundary conditions for the time- and space-dependent concentration derived from the mass balance. If changes in density ρ and diffusivity D_f can be neglected in the fluid, analytical solutions of the differential equations exist for simple geometries (plate, sphere, cylinder) and specific initial and boundary conditions [1, 2, 4]. Despite the simplifying assumptions, these solutions in combination with other model specifications are useful to describe the diffusive mass transport in polymers as a function of significant process parameters.

An important model for the diffusive mass transfer between a liquid and a gaseous phase is the penetration hypothesis first established by Higbie [5]. In its basic assumptions it corresponds to the surface renewal theory later postulated by Danckwerts [6]. The main features of the hypothesis are shown schematically in Figure 3.33. Higbie assumes that a liquid element at time $t = 0$ comes from deeper layers of the liquid to the interface. In the time interval $0 \leq t \leq t_E$ the concentration in the element changes nonstationary according to Fick's second law. At the end of the time interval, the liquid element returns to the deeper liquid layers and is mixed into the pool.

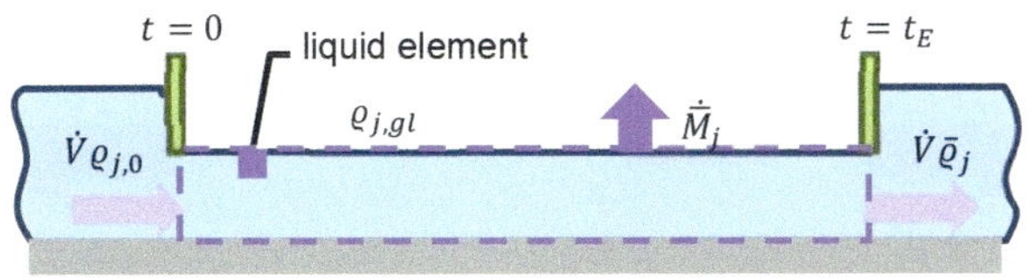

Figure 3.33 Schematic illustration for interfacial mass transport using the penetration hypothesis as an example

During the residence time at the interface, the liquid element moves at a constant velocity. Within the element no velocity gradient exists. Convective influences on mass transfer in the liquid element are neglected. The diffusive mass transport only takes place perpendicular to the interface. Any curvature of the interface is neglected. This simplification is allowed as long as the radius of curvature at the interface r_K is large compared to the thickness of the concentration layer δ_c.

At the interface, the concentrations on the liquid side and the gas side correlate to the individual equilibrium conditions. For continuity reasons, the mass flux of species *j* is identical on both sides of the phase boundary. According to Fick's first law the ratio of concentration gradients at the phase boundary therefore is inversely proportional to the ratio of the diffusivities in the liquid and the gaseous phase. In polymer melts the diffusivities are several orders of magnitude lower than those in gases [7]. A significant concentration gradient therefore only exists on the polymer side during mass transfer between polymer melt and gas. It is therefore sufficient to consider only the diffusive mass transport in the polymer melt.

Under these conditions the following partial differential equation is obtained to calculate the concentration $\rho_j(x,t)$ on the liquid side of the phase boundary as a function of the local coordinate *x* and time *t* (Fick's second law):

$$\frac{\partial \rho_j}{\partial t} = D_f \frac{\partial^2 \rho_j}{\partial x^2} \tag{3.27}$$

The initial and boundary conditions corresponding to the penetration hypothesis are:

Initial condition:	$\rho_j = \rho_{j,0}$	for $x > 0$ and $t = 0$
1st boundary condition:	$\rho_j = \rho_{j,gl}$	for $x = 0$ and $t > 0$
2nd boundary condition:	$\rho_j = \rho_{j,0}$	for $x \to \infty$ and $t > 0$

$\rho_{j,0}$ refers to the concentration at which the liquid element reaches the interface at time $t = 0$. $\rho_{j,gl}$ is the equilibrium concentration of the liquid phase at the boundary. The coordinate *x* denotes the vertical distance from the phase boundary. With the initial and boundary conditions given above, the solution of the differential equation is:

$$\rho_j = \rho_{j,gl} + \left(\rho_{j,o} - \rho_{j,gl}\right) \operatorname{erf}\left(\frac{x}{2(D_f t)^{1/2}}\right) \tag{3.28}$$

The concentration distribution in the liquid near the interface depends on the equilibrium concentration at the interface $\rho_{j,gl}$, the initial concentration $\rho_{j,o}$, the diffusivity D_f, the vertical distance from the interface x and the time t. The function $\operatorname{erf}\left(X^*\right)$ is the Gaussian error function, where the quantities D_f, x, and t are grouped as the dimensionless parameter X^*:

$$X^* \equiv \frac{x}{2(D_f t)^{1/2}} \quad \text{Dimensionless parameter} \tag{3.29}$$

The Gaussian error function

$$\operatorname{erf}\left(X^*\right) = \frac{2}{\sqrt{\pi}} \int_{I_u}^{I_0} e^{-X^{*2}} dX^* \tag{3.30}$$

is available in mathematic function libraries for spreadsheets and common programming languages. The values of I_u and I_o for the lower and upper integration limits have to be considered as parameters. In the case considered here, the lower integration limit has the value $I_u = 0$. The upper integration limit is $I_o = X^*$. With these integration limits it is $\operatorname{erf}(0) = 0$ and $\operatorname{erf}(\infty) = 1$. In the range $0 \le X^* \le \infty$ the values of the error function therefore change within the limits $0 \le \operatorname{erf}\left(X^*\right) \le 1$.

Equation (3.28) illustrates immediately that for any values of the dimensionless parameter $X^* > 0$ changes in the concentration of the fluid element are only possible if the condition

$$\left(\rho_{j,o} - \rho_{j,gl}\right) \neq 0 \tag{3.31}$$

is fulfilled. For values $\left(\rho_{j,o} - \rho_{j,gl}\right) > 0$ the concentration ρ_j in the element decreases. This is the case of desorption. In contrast, in the case of absorption, $\left(\rho_{j,o} - \rho_{j,gl}\right) < 0$, the concentration ρ_j increases in the element. For $\operatorname{erf}(0) = 0$, $\rho_j = \rho_{j,gl}$, whereby the first boundary condition of the differential equation (3.27) is fulfilled. For $\operatorname{erf}(\infty) = 1$ it is $\rho_j = \rho_{j,0}$. This satisfies both the initial condition and the second boundary condition.

By Equation (3.28) it is obvious to define a dimensionless concentration c_j^* as a ratio of local concentration differences by the following:

$$c_j^* \equiv \frac{\rho_j - \rho_{j,gl}}{\rho_{j,0} - \rho_{j,gl}} \text{ Related local concentration difference} \tag{3.32}$$

The definition of c_j^* considers that it is not the concentration itself but the difference between the local concentration and the concentration at the phase boundary $\left(\rho_j - \rho_{j,gl}\right)$ that is relevant for the diffusive mass transport. The maximum possible change in the concentration $\left(\rho_{j,0} - \rho_{j,gl}\right)$ is chosen as the reference value. With this definition, changes of the related local concentration difference c_j^* take place within the following limits:

$$1 \geq c_j^* \geq 0 \tag{3.33}$$

The use of dimensionless groups reduces the number of independent variables. In the case considered here the dimensionless concentration c_j^* is only a function of the dimensionless parameter X^*. It has the further advantage that the results are independent of the consistent units of the dimensional quantities. Thus, in the preceding results in place of the mass density ρ_j used so far, other definitions for the content of species j such as weight fraction w_j, molar concentration c_j, or mole fraction x_j can be used. The definition of the different content quantities is given in ISO 80000-9, so this is not discussed in detail here.

From Equation (3.28) follows for the related local concentration difference c_j^*:

$$c_j^* = \operatorname{erf}\left(X^*\right) = \operatorname{erf}\left(\frac{x}{2(D_f t)^{1/2}}\right) \tag{3.34}$$

For the dimensionless gradient $\partial c_j^* / \partial X^*$ the following equation results:

$$\frac{\partial c_j^*}{\partial X^*} = \frac{2}{\sqrt{\pi}} \exp\left(-X^{*2}\right) \tag{3.35}$$

As mentioned above the related local concentration difference c_j^* is only a function of the dimensionless coordinate X^*. The functional relationship is plotted in the left part of Figure 3.34. The most significant change of c_j^* is in the range of $0 \leq X^* \lesssim 2.5$. At the phase boundary at $X^* = 0$ it follows $\partial c_j^* / \partial X^* = 2 / \sqrt{\pi}$. For values of $X^* \gtrsim 2.5$ it is $c_j^* \approx 1$ and $\partial c_j^* / \partial X^* \approx 0$.

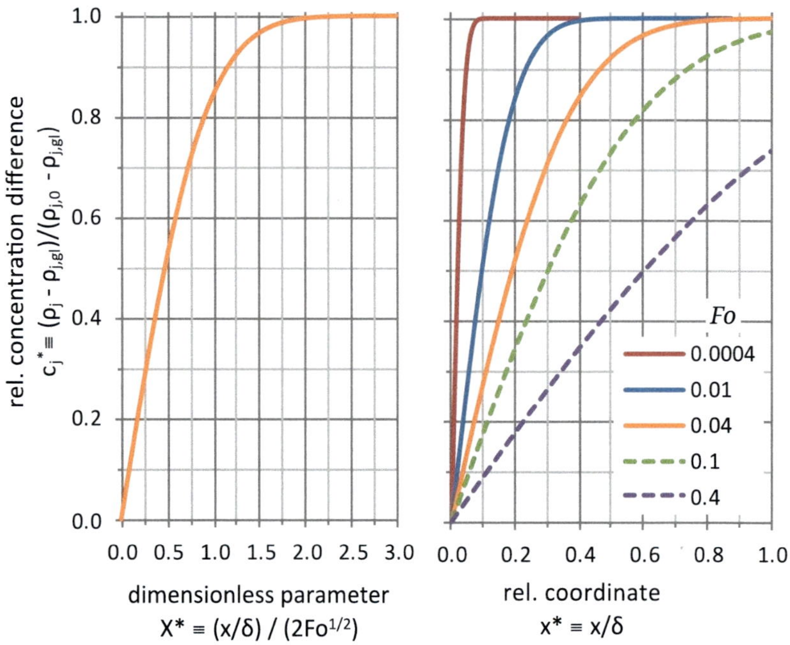

Figure 3.34 Related concentration difference c_j^*;
left: as a function of the dimensionless parameter X^*;
right: As a function of the related coordinate x^* with the Fourier number *Fo* as a parameter.

Due to the 2nd boundary condition, the solution of the differential equation (3.27) without further restrictions is only valid for a semi-infinite medium extended to the *x*-direction. The validity for liquid layers of finite thickness δ is limited. A statement about the validity range is obtained with the requirement $\partial \rho_j / \partial x \approx 0$ at $x = \delta$. This corresponds to the physically reasonable boundary condition of a wall that is impermeable to diffusive mass transport. As can be seen from the plot of c_j^* in the left part of Figure 3.34 in dimensionless groups this condition is very well fulfilled for all values of $X^* \gtrsim 2.5$.

In order to illustrate the limited validity as a function of the layer thickness δ in a dimensionless form, the dimensionless parameter X^* is substituted by the relative distance from the phase interface x^*

$$x^* \equiv \frac{x}{\delta} \quad \text{Relative distance from the phase interface} \tag{3.36}$$

and the Fourier number *Fo*

$$Fo \equiv \frac{D_f t}{\delta^2} \quad \text{Fourier number} \tag{3.37}$$

Formally, this is achieved by extending the dimensionless coordinate X^* in the numerator and denominator by a factor of $1/\delta$. With this extension the following relationship results:

$$X^* \equiv \frac{x/\delta}{2(D_f t/\delta^2)^{1/2}} = \frac{x^*}{2Fo^{1/2}} \tag{3.38}$$

With $x/\delta = 1$ and $X^* \lesssim 2.5$ for liquid layers of finite thickness the range of validity of Equation (3.35) is limited by the condition:

$$Fo \lesssim 0.04 \tag{3.39}$$

This limit is illustrated in the right part of Figure 3.34. The diagram shows c_j^* as a function of the relative distance from the phase interface $x^* \equiv x/\delta$ with the Fourier number *Fo* as parameter. One can recognize how the range covered by the change in concentration grows with increasing values of the Fourier number *Fo*. For values $Fo \lesssim 0.04$ the gradient $\partial c_j^*/\partial x^* \approx 0$ at $x^* = 1$. As can be noticed from the dashed curves, this condition is no longer fulfilled for values of the Fourier number $Fo \gtrsim 0.04$.

From Equation (3.35) follows for the gradient defined with $x^* = x/\delta$ as coordinate:

$$\frac{\partial c_j^*}{\partial x^*} = \frac{1}{\sqrt{\pi Fo}} \exp\left(-\frac{x^{*2}}{4Fo}\right) \tag{3.40}$$

With $x^* = 1$ and $Fo = 0.04$ it follows for $\partial c_j^*/\partial x^* = 5.44 \cdot 10^{-3} \approx 0$. For $x^* = 1$ and $Fo = 0.05$ it is $\partial c_j^*/\partial x^* = 1.70 \cdot 10^{-2}$. From the values it can be seen that $\partial c_j^*/\partial x^*$ for $Fo > 0.04$ increases significantly at the position $x^* = 1$. The calculated values confirm that the selected limit for the validity range is appropriate.

The minimum layer thickness δ_{min} of a liquid element at the phase interface is a function of the residence time *t* and the diffusivity D_f. With $Fo = 0.04$ it follows:

$$\delta_{min} = 5\left(D_f t\right)^{1/2} \tag{3.41}$$

With the assumption $D_f = 10^{-8}$ m^2/s and $t = 1$s the minimum layer thickness is $\delta_{min} = 0.5$ mm. When processing polymer melts in screw machines, the product $D_f t$ is generally lower than the value assumed here. The results can then also be applied to layer thicknesses $\delta_{min} < 0.5$ mm. Usually in screw machines the upper limit of the Fourier number $Fo = 0.04$ therefore does not limit the range of validity for the relationships derived with the penetration hypothesis for the diffusive mass transport in polymer melts.

Instantaneous and Average Mass Flow Rates

For the design of machines like extruders or other devices for sorption processes the knowledge of the concentration profiles at the phase boundary alone is not sufficient. However, it forms the basis to derive relationships to estimate the mass fluxes caused by diffusion between the phases. Suitable reference values for the dimensionless representation of mass flow rates result from the individual application. For this reason, the results for instantaneous and average mass flow rates are not presented here in dimensionless groups. Only the relationships for the mean liquid concentration are presented again as a dimensionless group.

From Equation (3.26) for the local concentration gradient at the phase boundary at $x=0$ the following relationship results:

$$\left.\frac{\partial \rho_j}{\partial x}\right|_{x=0} = \left(\rho_{j,0} - \rho_{j,gl}\right)\frac{1}{\sqrt{\pi D_f t}} \tag{3.42}$$

With this expression and Fick's first law, Equation (3.26), it follows for the mass flux diffusing from the liquid element at the phase boundary into the gas phase $\dot{m}_j$:

$$\dot{m}_j = -\frac{1}{\sqrt{\pi}}\left(\rho_{j,0} - \rho_{j,gl}\right)\left(\frac{D_f}{t}\right)^{1/2} \tag{3.43}$$

The relationship given by Equation (3.43) describes the diffusive mass transport on the liquid side of the phase interface. Both the proportionality $\dot{m}_j \sim \left(\rho_{j,0} - \rho_{j,gl}\right)$ as well as $\dot{m}_j \sim \sqrt{D_f / t}$ are typical of this kind of molecular mass transport.

For many practical tasks, instead of the instantaneous value of the mass flux $\dot{m}_j$ information about the time average value $\bar{\dot{m}}_j$ in the time interval $t_0 \leq t \leq t_E$ is required. With the mathematical definition of the time average value and Equation (3.43) for $\dot{m}_j$ it follows:

$$\bar{\dot{m}}_j \equiv \frac{1}{\left(t_E - t_0\right)}\int_{t_0}^{t_E} \dot{m}_j dt = -\frac{2}{\sqrt{\pi}}\left(\rho_{j,0} - \rho_{j,gl}\right) D_f^{1/2} \frac{\left(t_E^{1/2} - t_0^{1/2}\right)}{\left(t_E - t_0\right)} \tag{3.44}$$

With this relation the time average of the mass flux $\bar{\dot{m}}_j$ for any time interval $t_0 \leq t \leq t_E$ can be calculated. With regard to the limit

$$\lim_{t_0 \to t_E}\left(\frac{t_E^{1/2} - t_0^{1/2}}{t_E - t_0}\right) = \frac{1}{2t^{1/2}} \tag{3.45}$$

for $t_0 \to t_E$ then $\bar{\dot{m}}_j = \dot{m}_j$ applies, as shown in the comparison of Equation (3.44) with Equation (3.43).

For the time interval $0 \leq t \leq t_E$ the relationship is simplified because of $t_0 = 0$:

$$\bar{\dot{m}}_j = -\frac{2}{\sqrt{\pi}}\left(\rho_{j,0} - \rho_{j,gl}\right)\left(\frac{D_f}{t_E}\right)^{1/2} \tag{3.46}$$

The time average value $\bar{\dot{m}}_j$ for the interval $0 \leq t \leq t_E$ differs in this case from the instantaneous value $\dot{m}_j$ at time $t = t_E$ by a factor of 2.

By integration of the mass flux $\dot{m}_j$ or $\bar{\dot{m}}_j$ over the phase interface S the instantaneous and average mass flows diffusing over the phase interface are obtained. For the average mass flow $\bar{\dot{M}}_j$ in the time interval $0 \leq t \leq t_E$ it follows:

$$\bar{\dot{M}}_j = -\frac{2}{\sqrt{\pi}}\left(\rho_{j,0} - \rho_{j,gl}\right) D_f^{1/2} \iint t_E^{-1/2} dS \tag{3.47}$$

The integral expression considers that the phase interface S can vary with time and that different values are possible for the time interval $0 \leq t \leq t_E$ in which a liquid element is present at the phase interface.

Multiplying the average mass flow $\bar{\dot{M}}_j$ with time t_E gives the mass $\Delta\bar{M}_j$ exchanged by diffusion between the phases in the time interval $0 \leq t \leq t_E$:

$$\Delta\bar{M}_j = \bar{\dot{M}}_j t_E = -\frac{2}{\sqrt{\pi}}\left(\rho_{j,0} - \rho_{j,gl}\right) D_f^{1/2} \iint t_E^{1/2} dS \tag{3.48}$$

Equations (3.47) and (3.48) indirectly represent the basic relationship between diffusive mass transport over a phase interface and the relevant parameters of the liquid system, the operating conditions, and the geometry. The direct dependency can only be derived from the specific application by determining the individual values for liquid properties, operating conditions, and geometry for each of the variables on the right side of Equations (3.47) and (3.48).

With the relationships reported here for the instantaneous and average mass flow rates, it is common practice in process engineering to define mass transfer coefficients. This historically caused procedure is also used to describe the mass transport in screw machines [8–10]. However, it does not provide any additional information in the theoretically deduced relationships for mass transport, but it requires additional calculation effort in the application [1]. Therefore, no mass transfer coefficients are introduced here but direct relationships to determine the mean concentration of the volatile species in the polymer melt are derived from the mass balance.

Change of Mean Concentration

The mean value of the concentration of a species *j* may be defined for a volume or an area of the system. The suitable definition depends on the specific application. For machines or other devices with quasi-stationary throughput, as considered here, the mean (transport-) concentration $\bar{\rho}_j$ in an area perpendicular to the direction of flow is commonly used:

$$\bar{\rho}_j \equiv \dot{M}_j / \dot{V} = \left(\iint \dot{m}_j dA \right) / \dot{V} \tag{3.49}$$

In this case the mean concentration $\bar{\rho}_j$ is defined by the ratio of mass flow rate $\dot{M}_j$ of species *j* to volume flow rate $\dot{V}$ passing the cross section A. In the quasi-stationary case this is identical to the mean (volume-) concentration in the volume $V = \dot{V}\Delta t$ which results from the flow rate $\dot{V}$ in the time interval Δt through the cross-section A. This agrees with the average concentration usually determined by taking a sample from the bulk flow.

The change of mean concentration $\bar{\rho}_j$ is correlated with the mass transport caused by diffusion via a balance equation. Within the balanced region, conversions of the diffusing species are excluded. With this assumption, the balanced section is free of sources or sinks of species *j*. For the mass balance, therefore, only the mass flow rates caused by diffusion and convection across the boundaries of the balanced section are considered. Depending on the kind of application, different boundary conditions must be considered at the boundaries. The phase interface always forms a boundary of the balanced section. The material flow perpendicular to the interface is caused solely by diffusion of species *j*. At the other boundaries, the diffusive mass transport is zero (fixed wall) or negligible compared to the convective part of mass transport (quasi-stationary flow through the boundary).

In the case shown schematically in Figure 3.33 the direction of motion of the liquid element at the phase interface corresponds to the direction of flow of the volume flow rate $\dot{V}$. All liquid elements have the identical residence time t_E at the interface. In the quasi-stationary case, the mass in the balanced volume remains constant. All liquid elements that reach the interface at time $t = 0$ have the initial concentration $\rho_{j,0}$. In the volume shown by the dashed lines the convective mass flow rate $\dot{M} = \dot{V}\rho_{j,0}$ enters. In the time interval $0 \leq t \leq t_E$ caused by diffusion the mass flow rate $\bar{M}_j$ of species *j* is removed via the phase interface. As a result, at $t = t_E$ the mean concentration is reduced to the value $\bar{\rho}_j$. The corresponding mass flow rate is $\dot{V}\bar{\rho}_j$. With Equation (3.47) for the average mass flow rate $\bar{\dot{M}}_j$ caused by diffusion, the equation below from the mass balance at $t_E = const.$ follows:

$$\left(\bar{\rho}_j - \rho_{j,0}\right)\dot{V} = -\frac{2}{\sqrt{\pi}}\left(\rho_{j,0} - \rho_{j,gl}\right)\left(D_f / t_E\right)^{1/2} S \tag{3.50}$$

If the volume flow rate $\dot{V}$ is replaced by the mass flow rate $\dot{M}$ and the density ρ of the liquid according to $\dot{V} = \dot{M} / \rho$, after some rearrangement the following relationship is obtained:

$$\frac{\rho_{j,0} - \overline{\rho}_j}{\rho_{j,0} - \rho_{j,gl}} = \frac{2}{\sqrt{\pi}} \rho \left(\frac{D_f^{1/2} S}{t_E^{1/2} \dot{M}} \right) \tag{3.51}$$

The relationship given by Equation (3.51) is used by several authors [9, 11, 12] to characterize the devolatilization characteristics in co-rotating screw machines. The expression on the left is the ratio of the decrease in concentration $\left(\rho_{j,0} - \overline{\rho}_j\right)$ to the maximum possible change $\left(\rho_{j,0} - \rho_{j,gl}\right)$. This corresponds to the definition of an efficiency η_j^* for the removed species *j*:

$$\eta_j^* \equiv \frac{\rho_{j,0} - \overline{\rho}_j}{\rho_{j,0} - \rho_{j,gl}} \tag{3.52}$$

If the mean value of the related concentration difference $\overline{c}_j^*$ is defined in analogy to the definition of the related local concentration difference c_j^* according to Equation (3.32), then it follows:

$$\overline{c}_j^* \equiv \frac{\overline{\rho}_j - \rho_{j,gl}}{\rho_{j,0} - \rho_{j,gl}} \tag{3.53}$$

Between the dimensionless mean concentration $\overline{c}_j^*$ and the efficiency η_j^* the following relationship exists:

$$\overline{c}_j^* = 1 - \eta_j^* \tag{3.54}$$

In the following $\overline{c}_j^*$ is preferably used instead of η_j^*.

Equation (3.50) describes the dependency of the mean concentration $\overline{\rho}_j$ on the relevant parameters of diffusive mass transport and the mass flow rate $\dot{M}$. It is important to note that under the conditions on which Equation (3.50) was derived, there is a direct relationship between the volume flow rate $\dot{V}$ and the exposure time t_E of a liquid element at the interface. This relationship is generally dependent on the velocity profile in the flow cross-section. The volume flow rate $\dot{V}$ and the time t_E therefore cannot be treated as independent variables.

If one assumes for the case shown in Figure 3.33 that the horizontally arranged plate moves at constant velocity, no velocity gradient occurs in the cross-section of flow. The residence time t_E at the interface is identical for all elements. In this case, $t_E = V / \dot{V}$. With the volume V and the layer thickness of the liquid $\delta = V / S$ follows $S / \dot{V} = t_E / \delta$. With this substitution for $S / \dot{V}$ results from Equation (3.50), for the related average concentration difference $\overline{c}_j^*$ the relationship:

$$\overline{c}_j^* = 1 - \frac{2}{\sqrt{\pi}} Fo^{1/2} \tag{3.55}$$

The Fourier number *Fo* is defined by Equation (3.37), where δ is given by the height of the liquid layer $\delta = V / S$ and the time by $t = t_E$. The dimensionless concentration difference $\overline{c}_j^*$ is only a function of the Fourier number *Fo*. The validity of Equation (3.55) is limited to values of the Fourier number $Fo \lesssim 0.04$. This corresponds to changes of $\overline{c}_j^*$ within the range $1 \geq \overline{c}_j^* \gtrsim 0.77$. This demonstrates again that in Equation (3.51) the parameters $\dot{V}$ and t_E cannot be regarded as independent variables. For these reasons, Equation (3.51) is not adequate to calculate quantitative values of the change of the mean concentration in screw machines.

In other relationships derived from the penetration hypothesis, the case is considered in which the direction of mean flow is perpendicular to the direction of motion of the liquid elements at the interface [8, 10, 13, 14]. For these conditions in [14] a model for the mass transport in co-rotating multiple-shaft screw machines is proposed, which is shown schematically in Figure 3.35.

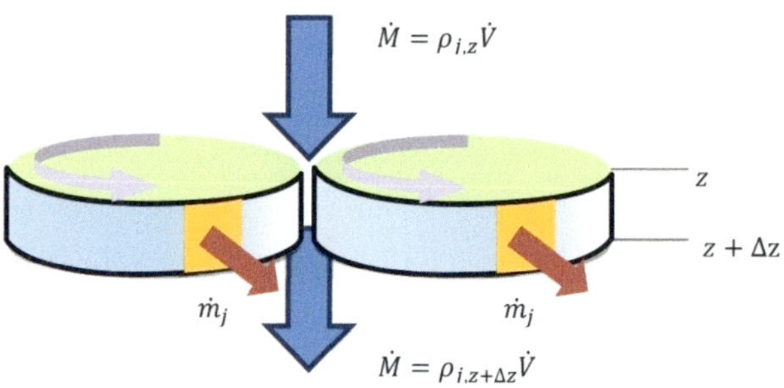

Figure 3.35
Model for the diffusive mass transport in co-rotating multiple-screw machines [14].

An element of length Δz from a shaft rotating about the *z*-axis is considered in a device that is passed by a liquid in the *z*-direction. At time $t = 0$ from the liquid pool a surface element of a liquid film is produced on the rotating shaft. In the time interval $0 \leq t \leq t_E$ the surface element moves with constant circumferential speed perpendicular to the *z*-axis and is mixed back into the pool at the end of the interval. During its time at the interface, species *j* diffuses with the mass flux $\dot{m}_j$ from the surface element into the adjacent gas phase. This process is consistent with the penetration hypothesis as long as an element of differential length is considered in the *z*-direction.

By mixing the liquid film into the axial liquid flow, both the mean concentration of the liquid as well as the initial concentration of the liquid elements reaching the interface change in the *z*-direction. If ideal mixing is assumed in the liquid pool across to the direction of flow, the initial concentration $\overline{\rho}_{j,0}(z)$ at which a liquid element reaches the interface at position *z* is equal to the mean liquid concentration at this position $\overline{\rho}_{j,z}$. Due to the mass flow rate $\dot{M}_j$ diffusing in the length Δz

from the liquid film into the gas phase the mean concentration changes at the position $z+\Delta z$ to the value $\overline{\rho}_{j,z+\Delta z}$. With the concentration values $\overline{\rho}_{j,z}$ and $\overline{\rho}_{j,z+\Delta z}$ and Equation (3.47) for the time average of the mass flow rate $\dot{\overline{M}}_j$ the relationship is obtained from the mass balance:

$$\left(\overline{\rho}_{j,z+\Delta z}-\overline{\rho}_{j,z}\right)\dot{V}=-\frac{2}{\sqrt{\pi}}\left(\overline{\rho}_{j,z}-\rho_{j,gl}\right)D_f^{1/2}\iint t_E^{-1/2}dS \tag{3.56}$$

With the limits $\lim\limits_{\Delta z\to 0}\left(\Delta z\right)=dz,\ \lim\limits_{\Delta z\to 0}\left(\overline{\rho}_{j,z+\Delta z}-\overline{\rho}_{j,z}\right)=d\overline{\rho}_j$ and the surface element $dS=dsdz$ with ds as line element of the interface, the differential equation results from the balance equation (3.56) to determine the related mean concentration difference $\overline{c}_j^*$:

$$\frac{d\overline{c}_j^*}{\overline{c}_j^*}=-\frac{2}{\sqrt{\pi}}\frac{D_f^{1/2}}{\dot{V}}\iint t_E^{-1/2}ds\,dz \tag{3.57}$$

For an analytical solution of the differential equation, it is assumed that only the related mean concentration difference $\overline{c}_j^*$ is a function of the axial coordinate z and the other quantities are independent from z. This generally applies to the mass transfer in bubble-free liquid layers, where it can be assumed that both the line element dS and thus the shape of the interface $\int dS$ and the time interval $0\le t\le t_E$ in which an element of the interface is used for the diffusive mass transport are independent of the axial coordinate z. The size of the interface $S=z\int ds$ thus increases linearly with the axial coordinate z. Under these conditions, with the dimensionless group Z^*:

$$Z^*\equiv\frac{2}{\sqrt{\pi}}z\frac{D_f^{1/2}}{\dot{V}}\int t_E^{-1/2}ds \quad \text{Devolatilization number} \tag{3.58}$$

and the initial condition $\overline{c}_j^*=1$ for $Z^*=0$ results in the solution of the differential equation:

$$\overline{c}_j^*=\exp\left(-Z^*\right) \tag{3.59}$$

The argument Z^* implicitly includes all essential quantities from the material properties, the operating conditions, and the geometric dimensions on each of which the related mean concentration difference $\overline{c}_j^*$ depends. Z^* is therefore referred to as the devolatilization number. In this correlation, $\dot{V}$ and t_E are two independent variables.

The relevant thickness δ is given by the thickness of the different liquid layers forming the interface in the specific device. It is mainly determined by design variables and is independent of the volume flow rate $\dot{V}$. The validity limit of the

Fourier number $Fo \lesssim 0.04$ is generally not reached for the reasons given in connection with Equation (3.41). In contrast to Equations (3.51) and (3.55), Equation (3.59) can be used to calculate the dimensionless concentration difference in the entire range $1 \geq \overline{c}_j^* \geq 0$. Within the range of validity, $\overline{c}_j^*$ is independent of the thickness δ. For identical degassing times t_E for all line elements ds of the interface S with $S = z \int ds$ and $\dot{V} = \dot{M} / \rho$ for Z^* the expression results:

$$Z^* = \frac{2}{\sqrt{\pi}} \rho \frac{D_f^{1/2} S}{t_E^{1/2} \dot{M}} \tag{3.60}$$

A comparison with Equation (3.51) shows that in this case Z^* formally matches the term to the right in Equation (3.51).

The differences between the related mean concentration difference $\overline{c}_j^*$ in the two cases are plotted in Figure 3.36. In both cases the related mean concentration difference $\overline{c}_j^*$ is only a function of the devolatilization number Z^*. For the values of $\overline{c}_j^*$ calculated with Equation (3.51) (curve B) the part plotted with the full line represents the function within the range $Fo \lesssim 0.04$ or $Z^* \lesssim 0.22$ and the part characterized by the dashed line is out of that range. Within the range of validity of Equation (3.51) the deviations of the related mean concentration difference $\overline{c}_j^*$ (curve A) calculated with Equation (3.59) are small despite the different flow scenarios according to Figure 3.33 and Figure 3.35. Mathematically this can be explained by the fact that the change in the inlet concentration of the liquid elements reaching the interface is not yet significantly apparent in the case shown in Figure 3.35. This influence increases with increasing values of Z^*. This has a dampening effect on the concentration characteristic according to curve A calculated with Equation (3.59) as curve A shows, for $Z^* \to \infty$ the value for $\overline{c}_j^*$ approaches the value of zero asymptotically according to the reciprocal exponential function $\overline{c}_j^* = 1 / e^{Z^*}$.

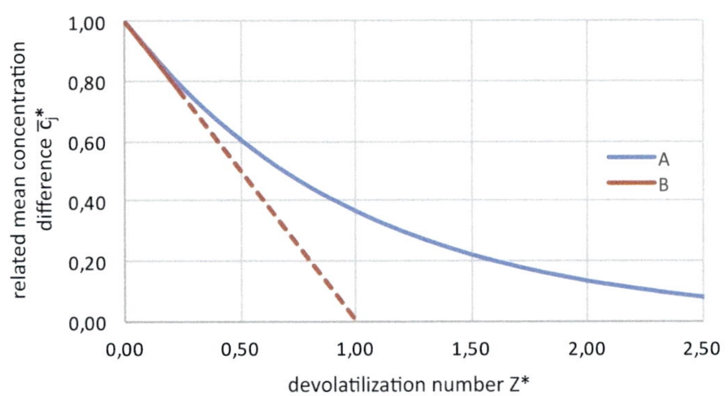

Figure 3.36 Related average concentration difference $\overline{c}_j^*$ as function of the dimensionless axial coordinate Z^*; curve A by Equation (3.59); curve B by Equation (3.51)

The devolatilization number Z^* defined by Equation (3.58) can be interpreted by various power products of independent variables, depending on the kind of application. For applications in screw machines, in [14] it is proposed to replace the time t_E by the rotational frequency of the shafts n and the angle of rotation θ_E at which the shafts have rotated in the time interval $0 \leq t \leq t_E$ according to the following relationship:

$$t_E = \theta_E / (2\pi n) \tag{3.61}$$

As a reference value for the phase interface S the surface of a cylinder with the length L and the circumference πD is selected. With these geometric quantities the following dimensionless groups are formed:

$$S^* \equiv S / (\pi DL) \quad \text{Related phase interface} \tag{3.62}$$

$$L^* \equiv L / D \quad \text{Related length of devolatilization section} \tag{3.63}$$

$$z^* \equiv z / L \quad \text{Related axial coordinate} \tag{3.64}$$

$$ds^* \equiv ds / (\pi D) \quad \text{Related line element} \tag{3.65}$$

$$\theta_E^* \equiv \theta_E / (2\pi) \quad \text{Related angle of rotation} \tag{3.66}$$

With these quantities as well as the dimensionless groups

$$\dot{V}^* \equiv \dot{V} / (nD^3) \quad \text{Volume Flow number} \tag{3.67}$$

$$Pe \equiv nD^2 / D_f \quad \text{Péclet number} \tag{3.68}$$

and the abbreviation:

$$k_S^* \equiv 2\sqrt{\pi} \int \theta_E^{*-1/2} ds^* \quad \text{Interface number} \tag{3.69}$$

from Equation (3.58) for the dimensionless group Z^* the expression results:

$$Z^* = z^* \frac{L^*}{\dot{V}^* Pe^{1/2}} k_S^* = \left(zDn^{1/2} D_f^{1/2} / \dot{V}\right) k_S^* \tag{3.70}$$

As the expression in brackets shows, Z^* is proportional to the axial coordinate z. However, since the related length $L^* \equiv L / D$ is a geometric ratio common for screw machines, z / D is replaced by $z / D \equiv z^* L^*$. With this substitution, in the range $0 \leq z \leq L$ the related axial coordinate z^* changes in the range $0 \leq z^* \leq 1$. Another common dimensionless group for screw machines is the Volume Flow number $\dot{V}^*$ (dimensionless volume flow rate). The Péclet number Pe can be interpreted as the ratio of convective to diffusive mass transport. The Interface number k_S^* defined by Equation (3.69) considers in dimensionless form for bubble-free liquid layers the influence of the phase interface on the diffusive mass transport from the liquid to the gas phase. For each element of the interface, both the shape and the residence time necessary for the diffusive mass transport are considered. k_S^* itself is a function of the Volume Flow number $\dot{V}^*$ and geometrical groups:

$$k_S^* = f(\dot{V}^*, \text{geometrical groups}) \tag{3.71}$$

Since k_S^* depends of and thus is a function of n and $\dot{V}$, the proportionality $\ln\left(\overline{c}_j^*\right) \sim \sqrt{n} / \dot{V}$ cannot be derived from Equation (3.70) as it results for the cases in which k_S^* is independent of n and $\dot{V}$ [14].

For the solution of the differential equation (3.57) it is assumed that with exception of $\overline{c}_j^*$ all other variables are independent of z. As can be seen from the expression in parentheses in Equation (3.70) this is the case when the diameter D, the rotation frequency n, the diffusivity D_f, the volume flow rate $\dot{V}$ and the Interface number k_S^* do not depend on the axial coordinate z. As already explained, this applies to bubble-free fluid layers at constant values for the diffusivity D_f and the density ρ of the liquid. The related mean concentration difference $\overline{c}_j^*$ calculated with Equation (3.59) and plotted in Figure 3.36 by curve A, is valid for any material systems, operating conditions and geometric dimensions as long as the above conditions are fulfilled. If the Devolatilization number Z^* is known $\overline{c}_j^*$ can be calculated. On the other hand, for a prescribed value of $\overline{c}_j^*$ the corresponding value of Z^* can be determined from the relationship:

$$Z^* = -\ln\left(\overline{c}_j^*\right) \tag{3.72}$$

For a specified system with known values of the diffusivity D_f and the equilibrium concentration $\rho_{j,gl}$ the operating conditions (rotation frequency n, volume flow rate $\dot{V}$) and geometric dimensions (diameter D, length L) can be selected for a given initial concentration $\rho_{j,0}$ so that the required value of Z^* is achieved. The structure of Z^* shows that this is possible with different combinations of values from operating conditions and geometric dimensions. Rules for scale up and scale down are not required. The key criteria for the choice of operating conditions and geometric dimensions result from economic considerations.

If the related mean concentration difference $\overline{c}_j^*$ cannot be calculated due to missing values to estimate Z^* experimental results are required. The basic information on diffusive mass transport provided here can be used to evaluate the experimental results. If the experimental results correlate quantitatively or qualitatively with Equation (3.59) a plot of the logarithmic dimensionless concentration difference $\log\left(\overline{c}_j^*\right)$ is a linear function of the Devolatilization number Z^*. This linear relationship also applies to any of the quantities in Equation (3.72) if the other quantities or their combination of power products used to calculate Z^* are constant. If the experimental results expressed in this way deviate from the linear relationship, there are two main reasons for the deviations:

1. The fluid layers are not bubble-free.
2. The concentration dependency of the diffusivity cannot be neglected.

As an example, Figure 3.37 shows experimental results from Gestring [15] who investigated the mass transfer during the devolatilization of a solvent from a polymer solution in a co-rotating twin-screw machine. Gestring [15] uses the efficiency η_j^* defined by Equation (3.52).

For the representation selected here with Equation (3.54), η_j^* is converted into the related mean concentration difference $\overline{c}_j^*$. On the logarithmically scaled vertical axis and the arithmetically scaled horizontal axis, the converted values (•) of the related mean concentration difference $\overline{c}_j^*$ are plotted as a function of the related axial coordinate z^*.

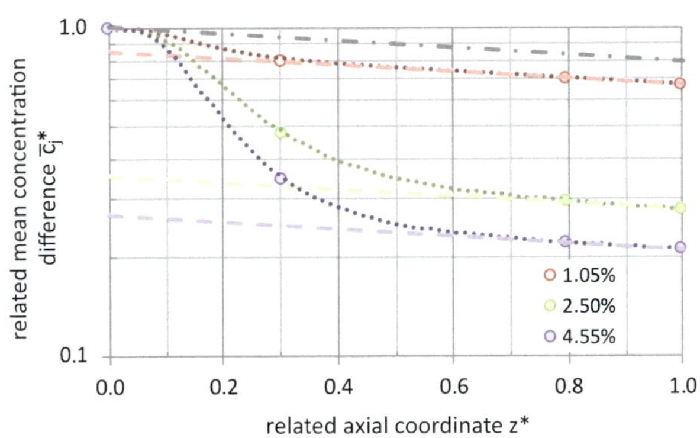

Figure 3.37 Related average concentration difference $\overline{c}_j^*$ as a function of the related axial coordinate z^* for initial mass fraction of 1.05%, 2.50%, and 4.55%. • Experimental results of Gestring [15] converted with $\overline{c}_j^* = 1 - \eta_j^*$.

In the experimental results shown, Gestring varied the initial concentration by three values of the mass fraction under otherwise constant operating conditions. The mean concentration was determined at three distances from the entry into the devolatilization section. With the dimensionless groups used here, the experimental conditions are represented by the values of the related length of the devolatilization section $L^* = 8.276$, the Volume Flow number $\dot{V}^* = 0.1289$, and the Péclet number $Pe = 4.098 \cdot 10^6$.

According to the relation predicted by the model, all experimental values should be represented in the semi-logarithmic plot by one straight line, independent of the initial concentration. The experimental results deviate significantly from this. The dashed lines indicate that the measured values approach parallel lines with increasing values of z^*. The approximation takes place at the mass fraction of 1.05% at significantly lower values of z^* than at the higher values of the mass fraction. The differences between the mass fractions of 1.05% and 2.50% are significantly greater than between 2.50% and 4.55%. Since all measured values approach to straight lines with identical gradients regardless of the inlet concentration in Figure 3.37, the concentration dependency of the diffusivity can be excluded as the reason for the deviations in the studied concentration range.

If one shifts the straight lines parallel until they cross the vertical axis at $\bar{c}_j^* = 1$, the dash dot straight line shows the expected dependence of $\bar{c}_j^*$ from z^* according to the model. With the value of $\bar{c}_{j,1}^*$ at $z^* = 1$ the Interface number k_S^* can be determined with the relationship following from the Equations (3.59) and (3.70):

$$k_S^* = -\ln\left(\bar{c}_j^*\right)\dot{V}^* Pe^{1/2} / L^* \tag{3.73}$$

The dimensionless expression $\dot{V}^* Pe^{1/2} / L^*$ is determined by the experimental conditions. In this example $\bar{c}_{j,1}^* \approx 1.79$ gives the value $k_S^* \approx 7.4$. The straight line corresponding to this value represents the change of the mean concentration in a bubble-free liquid flow.

Gestring [15] explains the different change of concentration in the devolatilization section by the different intensity of foaming of the polymer solution depending on the feed concentration. He notes that with decreasing mass fraction of the diffusing species at the inlet, the intensity of foam formation decreases. The maximum foam expansion is reached in the first third of the devolatilization section. After that the foam collapses. At the end there is a bubble-free sector. Figure 3.38 shows the different flow patterns in the devolatilization section [15].

Figure 3.38 Flow forms in the devolatilization section [15]

In order to consider the influence of bubble growth and bubble decay in the presented model for mass transport in screw machines, it is proposed in [14] to extend the devolatilization number Z^* by an additive term F_B which is dependent on the axial coordinate z^* and takes into account the influence of the bubbles on mass transport. If F_B is multiplied by the dimensionless group $\dot{V}^* Pe^{1/2} / L^*$ and this expression is denoted by K_B^*, from Equations (3.59) and (3.70) for $\bar{c}_j^*$ the relationship results:

$$\overline{c}_j^* = \exp\left[-\frac{L^*}{\dot{V}^* Pe^{1/2}}\left(z^* k_S^* + K_B^*\right)\right] \quad (3.74)$$

The term K_B^* is an integral quantity which considers the effect of the additional mass transport by bubbles on the mean related concentration difference $\overline{c}_j^*$. It is given by the integral of the function $k_B^*\left(z^*\right)$:

$$K_B^* = \int_0^{z^*} k_B^*\left(z^*\right) dz^* \quad \text{Integral bubble function} \quad (3.75)$$

where in [14] for $k_B^*\left(z^*\right) = k_{B0}\exp\left(-k_{B1}z^* - k_{B2} / z^*\right)$ is introduced. In order to allow adaptation to experimental results in a larger range, the relationship is extended by a further parameter:

$$k_B^* = \frac{k_{B0}^*\left[\left(1 - \exp\left(-k_{B1}^* z^{*k_{B2}^*}\right)\right)\right]}{\exp\left(k_{B3}^* z^{*k_{B4}^*}\right)} \quad \text{Bubble function} \quad (3.76)$$

The bubble function k_B^* determines in dimensionless form the additional mass transport from the polymer melt into the gas space caused by bubbles. Simplified this additional mass transport occurs in three steps. In the first step, bubbles are formed by local decompression. In the second step the bubbles grow and absorb the diffusing species. During this step, some of the bubbles are transported to the phase interface by convection. In the third step, some of the bubbles that have reached the interface collapse. The mass absorbed in this part of the bubbles is removed from the liquid. In the quasi-stationary state, this leads to a locally constant mass flow, which causes an additional reduction in the mean concentration. With the approach in Equation (3.74), this additional mass flow is formally considered as an additive term to the diffusive mass flow from the liquid layer depending on the related axial coordinate z^*. The Interfacial number k_S^* and the bubble function k_B^* or their integral values $K_S^* = z^* k_S^*$ and K_B^* can thus be used directly for a comparison of the parts of bubble-free and bubble-induced mass flow.

Figure 3.39 shows the bubble function k_B^* on the vertical primary axis and the integral bubble function K_B^* on the vertical secondary axis as a function of the related axial coordinate z^*. For the three pairs of curves plotted, the values for k_{B0}^* to k_{B4}^* to calculate k_B^* with Equation (3.76) are selected such that the values $\overline{c}_j^*$ calculated with Equation (3.74) reproduce the measured values of Gestring [15]. The result of the calculation is shown in Figure 3.37 by the dotted curves. The experimental values of Gestring [15] can be reproduced very well with the empirical approach for the bubble function k_B^*.

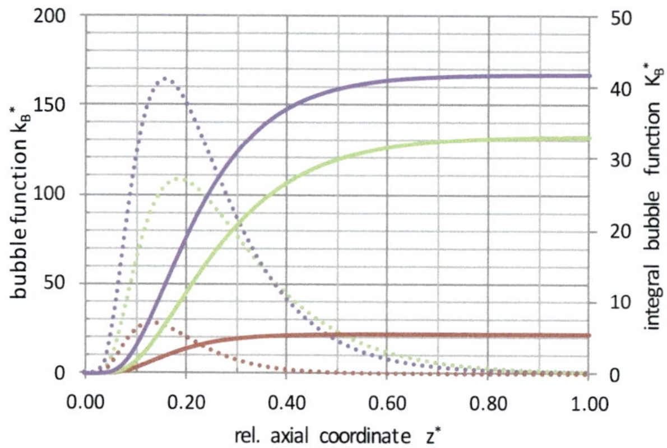

Figure 3.39 Bubble function k_B^* (dotted curves) and integral bubble function K_B^* (solid curves) as a function of the related axial coordinate z^*

The plot of the bubble function k_B^* represented by the dotted curves in Figure 3.39 shows the intensity of the additional mass transport caused by the bubbles. The greater the mass fraction of the diffusing species at the inlet to the devolatilization section, the greater is the intensity. This corresponds to the intensity of foaming observed by Gestring [15]. The maximum is reached for values of the related axial coordinate z^* within the range $0.1 \lesssim z^* \lesssim 0.2$. In the range $0.4 \lesssim z^* \lesssim 0.8$, k_B^* has dropped to the value $k_B^* \approx 0$. Up to this range the integral bubble function K_B^* increases. As soon as $k_B^* \approx 0$ and the integral bubble function K_B^* has reached a constant value, with increasing values of z^* no additional mass transport by bubbles occurs. A further decrease of the mean concentration $\overline{c}_j^*$ is caused only by the diffusive mass transport from bubble-free liquid layers, which is determined by the Interface number k_S^*.

A comparison of the value of $k_S^* \approx 7.4$ for bubble-free mass transport with the values of k_B^* in Figure 3.39 determined from the experimental results of Gestring [15] with a maximum value of $k_B^* \approx 165$ shows that much higher local mass flow rates are possible when bubbles occur. The integral value of the Interface number $K_S^* = z^* k_S^*$ increases linearly with z^* while the integral bubble function K_B^* in Figure 3.39 at the experimental conditions of Gestring [15] in the range $0.4 \lesssim z^* \lesssim 0.8$ reaches a maximum value dependent on the mass fraction at the inlet. Looking at the values at $z^* = 1$, the value of $K_S^* \approx 7.4$ is larger than the value of $K_B^* \approx 5$ at the lowest mass fraction at inlet, but significantly lower than $K_B^* \approx 42$ at the highest value of the mass fraction. To obtain the same value for $\overline{c}_j^*$ as for $K_B^* \approx 42$ under bubble-free conditions, L^* must be increased by a factor of about six. Considering that a value of $K_B^* \approx 40$ is already reached at $z^* = 0.5$, the ratio of the length of the degassing sections is $L_S^* / L_B^* \approx 11.4$ for the same change in $\overline{c}_j^*$.

Similar effects of foaming as obtained with a supersaturated polymer solution can be observed by adding a stripping agent [13, 16]. Therefore, it may be assumed that also for the use of stripping agents the change in concentration in the devolatilization section can be described with the semi-empirical approach according to Equation (3.74). However, there is a lack of suitable experimental results to proof the expected relevance of the semi-empirical approach in this case. This is due to incomplete information about the experimental conditions and material properties which are required to estimate the expression $\dot{V}^* Pe^{1/2} / L^*$ and the relative average concentration difference $\overline{c}_j^*$ as a function of the relative axial coordinate z^*. In addition, the mean concentration is rarely determined at different distances from the beginning of the devolatilization section under constant test conditions. A definite interpretation of the test results is therefore not possible. This can be deduced directly from the plot in Figure 3.37 assuming that the test results are not available for three different values, but only for a single value of z^*. The observation of the asymptotic approximation of the experimental values to parallel lines and the associated conclusions (no influence of the concentration dependence of the diffusivity, differentiation between the parts of bubble-free and bubble-containing sections of mass transport) would not be possible.

In order to reliably evaluate the bubble function k_B^* in consideration of material properties, operating conditions and geometric dimensions, further systematic experiments are required which meet the demands described above. Further development of suitable models for the generation, growth and destruction of bubble swarms in shear and extensional flow of polymer melts are needed. On these models it may be possible in future to derive theoretically founded relationships for the bubble-enhanced mass transport in screw machines as it is already possible today for the bubble-free case.

3.3.2 Influencing Quantities of the Material Properties

As the discussion of the mass transport mechanisms has demonstrated, the equilibrium concentration and the diffusivity are the most important material properties influencing the diffusive mass transport.

Equilibrium concentration

A diffusive mass transport between the polymer melt and the adjacent gas phase requires a difference between the mean liquid concentration in the polymer melt $\overline{\rho}_j$ and the local concentration of the polymer melt at the phase interface $\rho_{j,gl}$. The local concentration $\rho_{j,gl}$ at the phase interface on the liquid side is in equilibrium with the local concentration of the diffusing component on the gas side of the interface.

In gases it is common to use the partial pressure of the component under consideration p_j as a quantity for the concentration. With this convention, the equilibrium at the phase boundary consists of a relationship between the mass density $\rho_{j,gl}$ on the liquid side and the partial pressure $p_{j,gl}$ on the gas side of the phase boundary. If the concentration gradient on the gas side, as explained in Section 3.3.1, is negligible, the partial pressure at the phase interface is equal to the partial pressure in the gas. It is:

$$p_{j,gl} = p_j \tag{3.77}$$

Therefore, in the following p_j is also used for the reference of the equilibrium partial pressure at the phase interface.

Thermodynamically based equilibrium relationships are mainly derived as a function of the volume fraction ϕ of the polymer as a concentration quantity. In multi-component systems, the sum of the volume fractions of all components is equal to one. In a binary system, the relationship between the volume fraction ϕ_j of the diffusing component and the volume fraction ϕ of the polymer follows from this:

$$\phi_j = 1 - \phi \tag{3.78}$$

With the densities of the pure substances $\tilde{\rho}$ (density of the pure polymer) and $\tilde{\rho}_j$ (density of pure component *j*) the mass fraction and the mass concentration can be determined from the volume fraction. For the component *j* between the mass fraction w_j and the volume fraction ϕ_j the following relationship exists:

$$w_j = \frac{\phi_j \tilde{\rho}_j}{\phi_j \tilde{\rho}_j + (1 - \phi_j)\tilde{\rho}} \tag{3.79}$$

The mass concentration ρ_j is obtained by multiplying the mass fraction w_j with the density ρ of the melt at the corresponding temperature:

$$\rho_j = \rho w_j \tag{3.80}$$

At low concentrations of the diffusing component in the polymer melt, the relationship between the partial pressure p_j and the volume fraction ϕ_j of the volatile component at constant temperature is linear and can be described by Henry's law [1, 2] If the partial pressure p_j is related to the vapor pressure $\tilde{p}_j$ of the pure component *j*, with the Henry coefficient H^* the relationship is obtained:

$$\phi_j = H^* p_j / \tilde{p}_j \tag{3.81}$$

In this form of Henry's law, the Henry coefficient H^* is a dimensionless parameter and almost independent of temperature. The temperature dependence of the equilibrium concentration is considered by the dependence of the vapor pressure $\tilde{p}_j(T)$ of the pure component j at the temperature T. The vapor pressure $\tilde{p}_j(T)$ of pure liquids can generally be described in a wider temperature range by the Antoine equation:

$$\ln(\tilde{p}_j) = A - \frac{B}{C + T} \tag{3.82}$$

The parameters A, B, and C have been determined for a large number of liquids and can be found in material databases.

At higher concentrations of the diffusing component, the linear relationship between ϕ_j on the liquid side and p_j on the gas side of the phase interface is no more present. In practice, an approach based on the Flory-Huggins theory has proven successful for polymers for describing the equilibrium concentrations between the liquid and gas phases over the entire concentration range. The following equation is obtained with the designations used here:

$$\ln(p_j / \tilde{p}_j) = \ln(\phi_j) + (1 - \phi_j) + \chi(1 - \phi_j)^2 \tag{3.83}$$

χ refers to the Flory-Huggins's interaction parameter, which records interactions between the polymer chains and the low-molecular component. However, χ is not independent of ϕ or ϕ_j, as originally assumed by Flory and Huggins, and can therefore only be assumed to be constant within a limited range of concentration. Newer approaches include the concentration dependence of the interaction parameter n different forms [17]. An approach similar to the Antoine equation to determine χ as a function of ϕ is proposed in [18]:

$$\chi = \frac{A}{(1 - B\phi)^2} + C \tag{3.84}$$

The parameters A, B, and C are assumed as constants for a given material system. With this approach and the Flory-Huggins relationship, Equation (3.83), experimentally determined equilibrium values can be described in a wide concentration range.

Figure 3.40 shows the volume fraction ϕ_j as a function of the related partial pressure $p_j^* \equiv p_j / \tilde{p}_j$ at equilibrium for different values of the interaction parameter χ in double logarithmic scale.

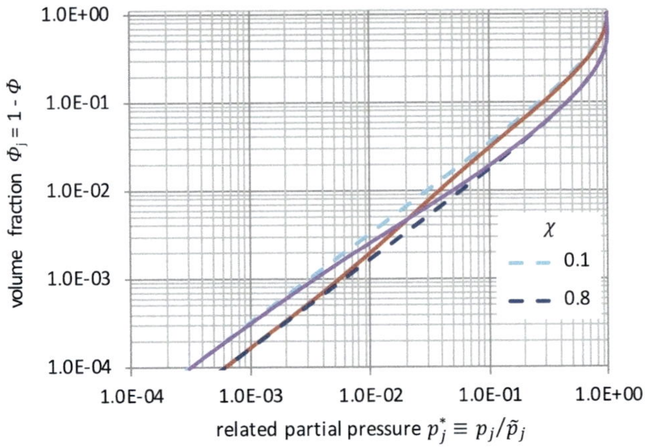

Figure 3.40 Volume ratio ϕ_j as a function of the related partial pressure $p_j^* \equiv p_j / \tilde{p}_j$ For different values of the interaction parameter χ at equilibrium

The dashed curves show the dependence for constant values of $\chi = 0.1$ and $\chi = 0.8$. Within these limits usually are the experimentally determined values of χ for polymer solutions containing organic solvents or monomers [17, 18]. The concentration dependence of χ according to Equation (3.84) applies to the solid curves. The values of the parameters A, B, and C are determined in such a way that the transition between the limiting curves is in the range $0.001 \lesssim p_j^* \lesssim 0.1$.

Figure 3.40 illustrates that the influence of the interaction parameter χ on the phase equilibrium is relatively small compared to the influence of the related partial pressure p_j^*. The volume fraction ϕ_j changes as a function of the related partial pressure p_j^* by powers of 10. In contrast, for a constant value of p_j^*, the value of ϕ_j in the range $0.1 \leq \chi \leq 0.8$ differs up to a factor of 2. The equilibrium concentration at the phase interface can be calculated at least approximately with the vapor pressure $\tilde{p}_j(T)$ of the pure component *j* at a specified partial pressure p_j even without exact knowledge of the interaction parameter χ.

Diffusivity

The diffusivity D_f depends on the molecular properties of the polymer solution as well as of pressure and temperature. Values for the diffusivity may differ by more than the factor [7]. For low-molecular substances, theoretically established relationships for the diffusivity have been applied to both gases and liquids. For polymers, on the other hand, the results of experimental investigations are generally used to determine the diffusivity. Either some of the parameters required in the models must be determined experimentally for specific systems or the models are not suitable for all conditions in order to describe the diffusive mass transport with sufficient accuracy [3].

References for Section 3.3

[1] Brauer, H., *Stoffaustausch einschließlich chemischer Reaktionen,* Sauerländer, Aarau and Frankfurt/Main (1971)

[2] Bird, R. B., Stewart, W. E., Lightfood, E. N., *Transport Phenomena,* 2nd ed., John Wiley, New York (2002)

[3] Masaro, L., Zhu, X. X., "Physical models of diffusion for polymer solutions, gels and solids", *Prog. Polym. Sci.* (1999) 24, pp. 731–775

[4] Crank, J., *The Mathematics of Diffusion,* Oxford University Press, London (1975)

[5] Higbie, R., "The rate of absorption of a pure gas into a still liquid during short periods of exposure", *Trans. Amer. Inst. Chem. Engrs.* (1935) 31, pp. 365–369

[6] Danckwerts, P. V., "Significance of liquid-film coefficients in gas absorption", *Ind. Eng. Chem.* (1951), 43(6), pp. 1460–1467

[7] Cussler, E. L., *Diffusion: mass transfer in fluid systems,* 3rd ed., Cambridge University Press, Cambridge (2009)

[8] Latinen, G. A., "Devolatilization of Viscous Polymer Systems", *Advances in Chemistry* (1962) 34(19), pp. 235–246

[9] Collins, G. P., Denson, C. D., Astarita, G., "Determination of Mass Transfer Coefficients for Bubble-Free Devolatilization of Polymeric Solutions in Twin-Screw Extruders", *AIChE J.* (1985) 31(8), pp. 1288–1296

[10] Biesenberger, J. A., „Devolatilization in Shear Flow Fields", in *Entgasen beim Herstellen und Aufbereiten von Kunststoffen,* Verein Deutscher Ingenieure, VDI-Verlag, Düsseldorf (1992), pp. 31–52

[11] Schuler, W., „Hochrechnung von Versuchs- auf Produktionsextruder", in *Der Doppelschneckenextruder: Grundlagen und Anwendungsgebiete,* VDI-Gesellschaft Kunststofftechnik, VDI-Verlag, Düsseldorf (1995), pp. 134–163

[12] Lechner, F., „Entgasen von Polymerschmelzen mit gleichläufigen Doppelschneckenextrudern", in *Der gleichläufige Doppelschneckenextruder: Grundlagen, Technologie, Anwendungen,* Carl Hanser Verlag, Munich (2007), pp. 191–212

[13] Padberg, G., „Grundlagen der Entgasung von Polymeren", in *Entgasen von Kunststoffen,* Verein Deutscher Ingenieure, VDI-Verlag, Düsseldorf (1980), pp. 1–23

[14] Thiele, H., Sapargaliyev, M., Pahl, M. H., „Entgasen strukturviskoser Medien im Mehrwellen-Hochleistungs-Reaktor TFR", in *Schriftenreihe der Verfahrenstechnik Universität Paderborn,* vol. 34, Pahl, M. H., Ed., Aachen, Shaker (2009), p. 127

[15] Gestring, I., „Entgasen von Polymeren", Universität Hannover, Hannover (2002)

[16] Barth, U., Schuler, W., „Restentgasen von Kunststoffen unter Berücksichtigung der Rückbildung von Monomeren", in *Entgasen beim Herstellen und Aufbereiten von Kunststoffen,* VDI-Gesellschaft Kunststofftechnik, VDI-Verlag, Düsseldorf (1992), pp. 199–236

[17] Wolf, B. A., "Making Flory-Huggins Practical: Thermodynamics of Polymer-Containing Mixtures", *Adv. Polym. Sci.* (2011) 238, pp. 1–66

[18] Konigsveld, R., Stockmayer, W. H., Nies, E., *Polymer Phase Diagrams,* Oxford University Press, Oxford (2001)

3.4 Influence Factors and Reduction of Degradation during Polymer Processing

Jens Hepperle

3.4.1 Introduction

Physical properties of polymers and polymer mixtures (polymer-polymer, polymer-solids) during compounding and processing are described in Sections 3.1 and 3.2. This section describes chemical reactions leading to polymer degradation typically occurring during extrusion or compounding. These degradation processes can be optimized in many cases.

During production and processing of blends and compounds, polymers are molten and mixed with other components, such as

- One or more other polymers (blending partners)
- Solid powders or masterbatches (like colors, fillers, stabilizers, functional additives)
- Liquids and suspensions (like color concentrates, rheological modifiers)

This section summarizes the main, usually undesired, chemical reactions causing polymer degradation during extrusion and compounding. Other favorable reactions used for reactive extrusion are not covered here. Many examples are also found in literature (e.g. [1–5]).

Chemical reactions occur not only during compounding and processing, but also during use of the final shaped parts (see Figure 3.41). During polymerization and the following work-up (like degassing, addition of additives) to the market-ready granules the polymer chemical stability might be influenced by remaining reaction partners (like solvents, catalyst residues, byproducts) and also depends on processing conditions (such as temperature, residence time). Remaining impurities (like metal ions or catalyst residues) can influence polymer stability during following processing steps or even during use of the final shaped products.

During processing of polymers, the following parameters play a role:

- Processing temperature/melt temperature and heat peaks within the extruder
- Intensity of shear and elongational flows within the extruder caused by screw design, processing parameters, and used polymer/solid properties
- Residence time from polymer melting of the granules until solidification
- Existence of other reaction partners, like impurities within the melt or water residues

- Reactions may also occur from contact of the melt with the environment, like with oxygen or the materials within the extruder, such as screws, housing, or nozzles; these reactions might influence product quality

Reactions may also occur by contact of the melt with the environment, like with oxygen or the materials used for screws or housing/nozzle materials. These reactions might influence product quality.

Final plastic parts are exposed with much longer exposition times compared to the polymer during polymerization, compounding, and processing. Reactions induced during use may be caused by light and UV radiation, by media (like water, acids or chemical bases), or other factors like microorganisms or radiation.

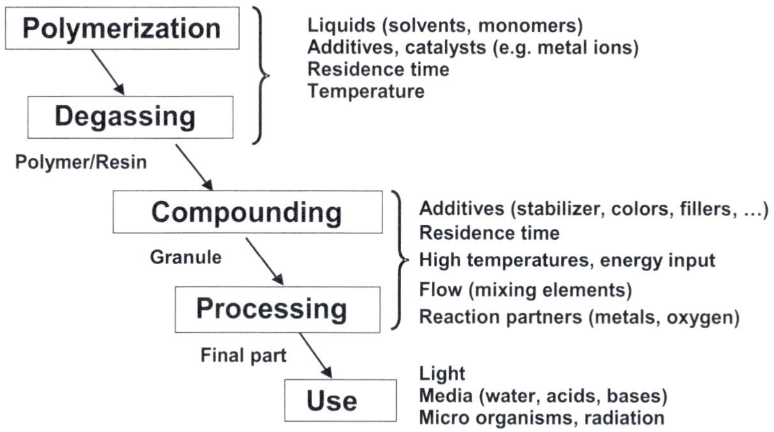

Figure 3.41 Influencing factors for polymer degradation during production, processing, and use

3.4.2 Chemical Reactions

An overview of reactions leading to polymer degradation is shown in Figure 3.42.

Mechanisms of degradation can be classified as thermal degradation, thermal-oxidative degradation (with oxygen being present), chemical degradation (without presence of oxygen), and mechanical degradation [1, 6, 7].

Different chemical reaction classes are often occurring in parallel. During polymer processing, usually a thermal degradation occurs caused by higher temperatures together with thermal-oxidative degradation via oxygen present in the environment. In addition, fillers (like titanium dioxide) might catalyze and speed up the degradation process.

The following sections show some examples of different degradation reactions.

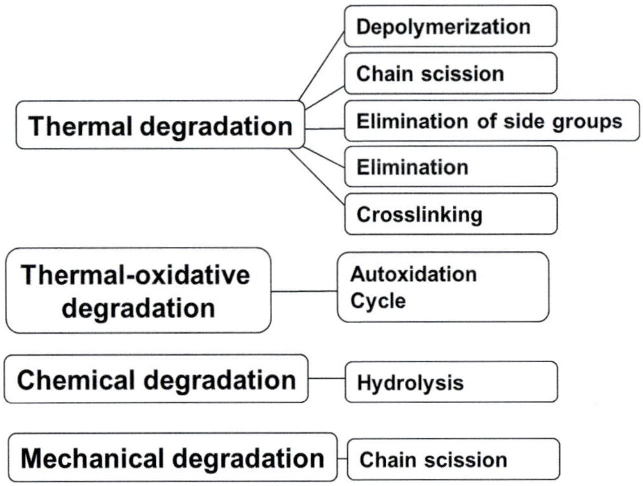

Figure 3.42 Overview of unfavorable chemical reactions during polymer processing

3.4.2.1 Damage through Thermal Degradation

Thermal degradation occurs at higher temperatures causing chemical changes in the polymer chains without another reaction partner. A purely pyrolytic reaction occurs only with complete exclusion of oxygen and without overlying oxidative degradation.

Reactions that might occur during thermal degradation are shown for vinyl polymers in Figure 3.43. Vinyl polymers comprise polystyrene (PS), polyethylene (PE), polypropylene (PP), polyvinyl chloride (PVC), and polymethyl methacrylate (PMMA).

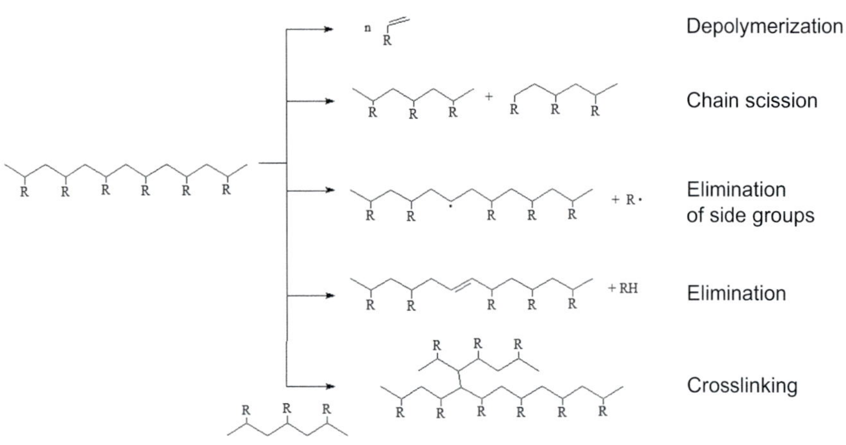

Figure 3.43 Overview of different degradation reactions of vinyl polymers [8]

Mechanisms of thermal degradation can be divided into chain-splitting reactions and depolymerization reactions [8]. Elimination of side groups happens more rarely. In contrast to reactions causing reduction of polymer molecular weight, molecular mass increase can occur via cyclization or crosslinking.

Division of pure thermal degradation and thermal-oxidative degradation is challenging as most polymers are not pure but include impurities (e.g. dissolved gases, impurities, solvent residues) that might react and lead as reaction partners to degradation processes.

(a) Depolymerization

Depolymerization is induced via thermal stress with exclusion of air or oxygen. In a kind of chain scission mechanism the reaction starts via splitting of monomers from the chain end. As a result, the mean degree of polymerization and the molecular mass decrease.

Typical polymers prone to depolymerization are polyoxymethylene (POM), polytetrafluorethylene (PTFE), polymethyl methacrylate (PMMA), and at higher temperatures also polystyrene (PS). In contrast, all polyolefins such as polyethylene and polypropylene are insensitive to depolymerization [8–10]. Split monomers caused by depolymerization can cause odor generation during extrusion and processing.

(b) Thermally Induced Chain Scission

Thermally induced chain scission from polymers occurs at much higher temperatures compared to usual mean processing temperatures. Therefore, this degradation type is only relevant at very high temperatures e.g. at $T > 300\,°C$ for polyolefins. This is caused by the high activation energy of thermal activated chain scissions in comparison to reactions with oxygen. As an example, thermally induced chain scission has an activation energy of 264 kJ/mol, whereas thermal-oxidative degradation requires an activation energy of only 59 kJ/mol. Therefore, for polyolefins thermal-oxidative degradation is dominant.

(c) Elimination of Side Groups

Side groups are chemical structures within polymers that can be eliminated during chemical degradation processes. This type of reaction preferably occurs when the bond energy between a carbon atom of the main chain and one in the side chain is *lower* compared to the bond energy between two carbon atoms in the main chain [8, 9, 11]. As the bond energy between carbon atoms is usually *lower* compared to that between carbon and other elements, side groups are usually not eliminated. An exception is chlorine, having a lower bond energy to carbon compared to a C-C bond. Therefore, the related polyvinyl chloride (PVC) is less stable compared to polyethylene. At higher temperatures polyvinyl chloride (PVC) splits off chlorine.

(d) Elimination and Crosslinking

Especially polyolefins show at low processing temperatures and sufficiently long reaction time a tendency toward crosslinking reactions. However, and at elevated temperatures a significant degradation caused by chain scissions may occur [12].

Formation of unsaturated molecules via chain scission is in competition with the consumption of unsaturated compounds via chain crosslinking reactions. Activation energy for crosslinking of polyethylene was determined to be 96 kJ/mol, whereas the chain scission requires 146 kJ/mol [12]. Crosslinking reactions are therefore favored at lower temperatures and chain scission reactions at elevated temperatures. Some inhomogeneities like gel particles can be formed at high degrees of crosslinking. This will in some cases lead to spots or nibs, a problem especially for blown films. Another crosslinking problem can be caused by unsaturated polymers (like rubber and rubber phases in blends) leading to visible spots, inhomegenities, or even crack formation influencing the mechanical properties of final parts.

3.4.2.2 Oxidative Degradation

Oxidative degradation is one of the most common degradation mechanisms during polymer processing. The presence of oxygen at very low concentrations has a catalytic effect on degradation of polymers. Oxidative degradation may lead to an increased chain scission, formation of volatile components, and discoloration.

Figure 3.44 shows the cycle of oxidative degradation, also called autoxidation. Reactions occurring in this cycle are similar for almost all polymers [13, 14]. Radical reactions at the polymer chain (R-R and R-H) are shown in Figure 3.45.

The initiation of the oxidation cycle during processing is caused by formation of alkyl radicals in the presence of oxygen via thermal energy or mechanical stress. Metal ion residues can also initiate formation of radicals, as well as small residues of peroxides or other oxygen-containing components or impurities.

The alkyl radicals react with oxygen to form peroxy radicals $ROO^{\bullet}$, practically without activation energy and therefore with similar reaction rate independent of temperature. Afterwards the peroxy radical reacts with a polymer chain (identified as RH) to form a hydroperoxide ROOH.

The activation energy for this specific reaction is pretty high and polymeric alkyl hydroperoxides have a lifetime of several tens of seconds, so therefore their decomposition is the rate-determining step [18]. Decomposition of hydroperoxides leads to formation of alkoxy radicals and hydroxy radicals. They can attack other polymer chains and recombine to form new alkyl radicals that can react with oxygen again (Figure 3.44). Termination reactions are the recombination of two alkyl radicals or one alkyl radical with another peroxy radical.

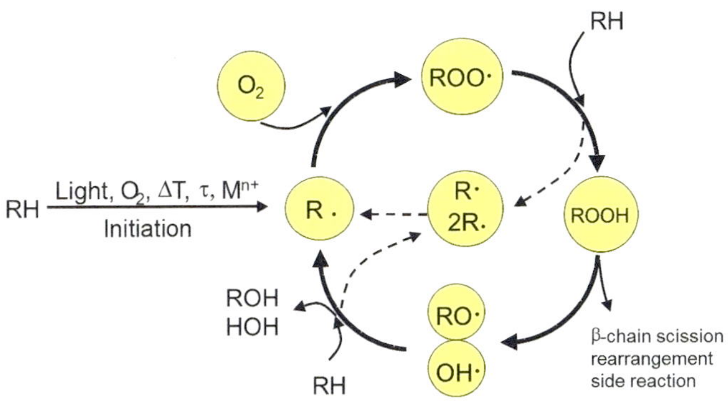

Figure 3.44 Oxidative polymer degradation via autoxidation [7, 13, 14]

Chains with lower molecular weight can be formed via recombination of two alkyl radicals (Figure 3.45). Main degradation products are alcohols, esters, carboxylic acids, as well as chains with carboxyl groups [15]. Figure 3.46 shows the chronological sequence of oxidation products from polyethylene. After an induction period the concentration of hydroperoxides is increasing significantly before these molecules react further. During degradation of polyethylene, carboxylic acids are formed, but also esters, lactones, and alcohols.

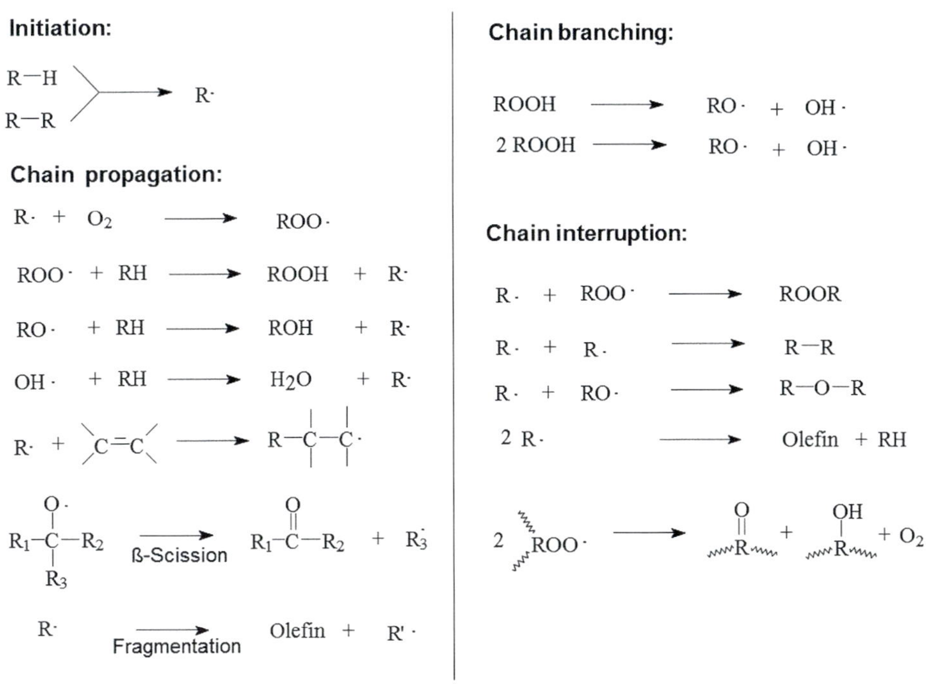

Figure 3.45 Reactions during thermal-oxidative degradation [13, 14]

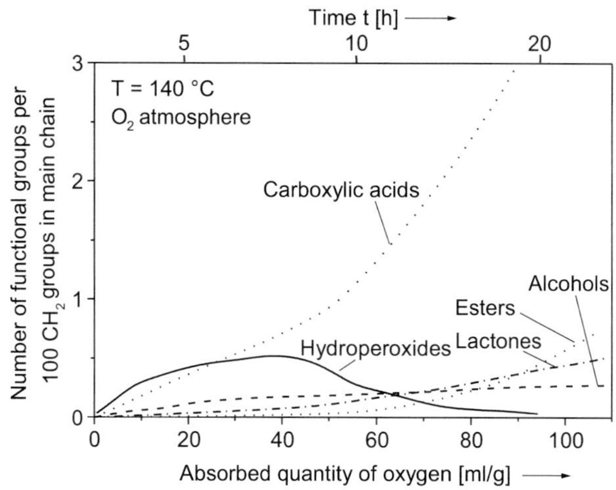

Figure 3.46 Oxidation products of polyethylene at 140 °C in oxygen atmosphere [15]

In summary, the central element of the oxidative polymer degradation during processing of polymers is formation and decomposition of hydroperoxides. The decomposition increases the more radicals are present within the reaction cycle. The cycle is autocatalytic, meaning that the formed reaction products are acting as a catalyst for further degradation reactions. This causes an increase in oxygen consumption after initiation of the first reaction. Some polymer properties caused by molecular weight loss can be changed as a consequence, like viscosity, mechanical properties, and others.

3.4.2.3 Chemical Degradation Reactions via Residual Water

Chemical degradation is practically relevant during processing when water from residual moisture content reacts in a hydrolysis reaction. Residual excess moisture can result from insufficient drying of the polymer granules prior to processing.

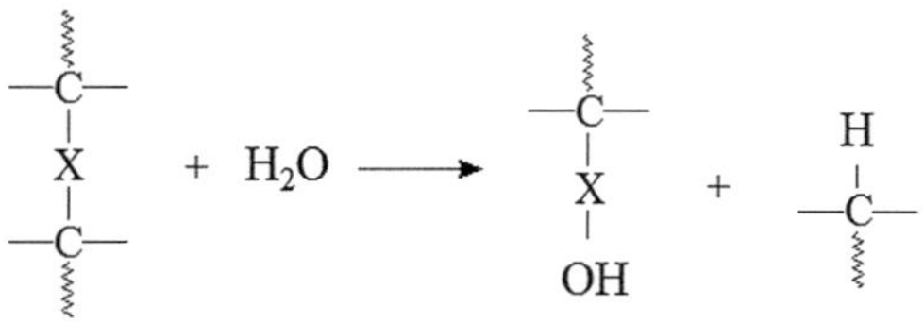

Figure 3.47 Scission of the polymer chain via hydrolysis [8]

Hydrolysis is a reaction of a suitable bond, e. g. an ester bond, with water. The reaction leads to scission of the polymer chain resulting in reduction of mean molecular weight. Technically meaningful polymers prone to hydrolysis are polyesters (PET, PC), polyamides (PA 6, PA 66), and polyurethanes (TPU).

Sufficient drying of the polymer granules prior to processing is recommended in order to avoid hydrolytic degradation. In addition, traces of alkaline or acidic reaction partners or metal ions should be avoided as they can catalyze the reaction. Water content can easily be detected with standard methods for low water contents (< 1 wt.-%), such as Karl-Fischer titration.

3.4.2.4 Degradation via Mechanical Stress

Mechanical stress during stretching of the entangled polymer chains can lead to mechanical degradation. As a consequence radicals might be formed leading to further degradation. As an example it was shown that during extrusion of polypropylene at a temperature of 230 °C, mechanical stress can form alkyl radicals [16].

However, in practice the effects of degradation caused by stresses during processing (via shear and elongational flow of the melt) cannot be segregated from thermal and thermal-oxidative degradation caused by temperature increase resulting from energy input during extrusion. A homogeneous temperature distribution within the extruder is more difficult due to the usually low heat conductivity of the polymer melt. Shear flow within the region of conveying and mixing elements leads to two effects [17]:

- Heating of the melt and increase of thermally induced degradation reactions (Arrhenius)
- Lowering of melt viscosity and quicker relaxation leading to lower effects of thermal stress on degradation

In practice, the effect of thermal degradation outweighs the one of degradation caused by mechanical stress.

3.4.2.5 Influence of Metals on Degradation

Metals and metal ions can be present in the melt via contamination during polymer reaction (e.g. catalyst residues), within silos used for polymer granule storage, or during processing via friction of the unmolten polymer or compounds at extruder walls. Residues of metal sparks or metal ions can operate as catalyst for the autoxidation reaction and therefore speed up degradation [13, 18].

Metal ions can provoke decomposition of hydroperoxides via formation of radicals. During examination of thermal decomposition of polypropylene it was found that the catalytic effect of metal ions for oxidative degradation increases according to the metal in the order Cu < Fe < Cr < Mn < Co, and with increasing redox potential [9]. These metal ions can react with hydroperoxides via formation of free radicals (alkoxy radicals $RO^•$ and peroxy radicals $RO_2^•$) [18].

The damaging effect of metals can be reduced with suitable stabilizers inducing formation of metal complexes [19].

In the case of influence of the used metal housing and/or screw materials on product quality, a modification of the materials should be considered. This can be done via use of other alloys for extruder screws and housings and in some cases also via specific coatings in the critical processing (e.g. mixing elements).

3.4.3 Relationship between Polymer Degradation and Properties

A summary of possible chemical reactions during processing and use as well as influencing factors is shown in Figure 3.48. More details can be found in several literature sources (e.g. [2, 7, 8]). The figure also shows the possible influence on some key polymer properties. As some of the properties might be linked, this is shown with an arrow.

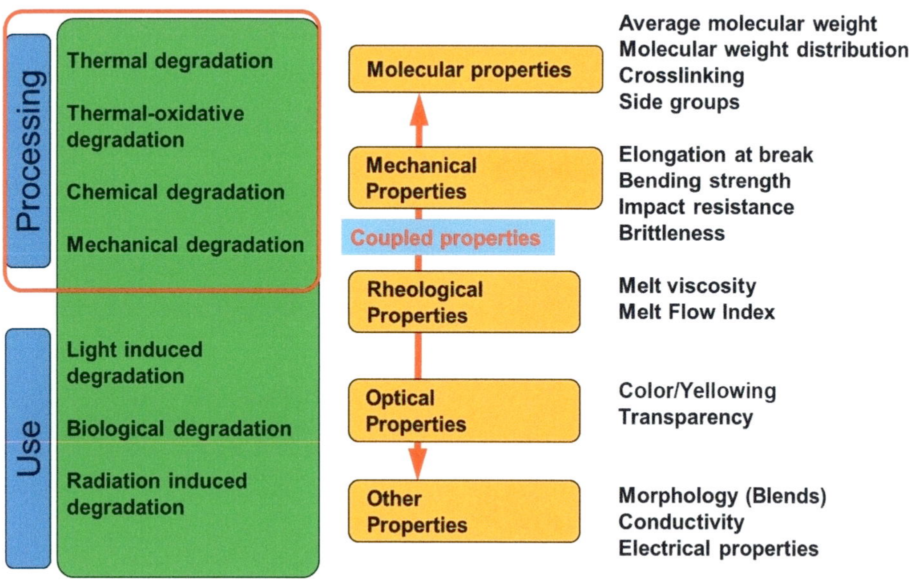

Figure 3.48 Examples of influencing factors of chemical degradation mechanisms on properties of polymers, blends, and compounds with related measurable/detectable parameters

a) Changes of the molecular structure of polymers

Degradation reactions can cause molecular weight reduction, a change of molar mass distribution or the formation of chain branches or crosslinks. These molecular parameters can influence rheological properties of the polymer melt (like viscosity) and the mechanical or optical properties of the processed granules or finished parts.

b) Changes of mechanical properties

The molecular composition and homogeneity of the mixture after compounding of polymers with fillers or after mixing of different polymers can have an influence on mechanical properties, like tensile strength or bending strength, as well as on brittleness or impact strength.

As an example Figure 3.49 shows the influence of the weight average molecular weight on the tensile strength at break of three different polymers.

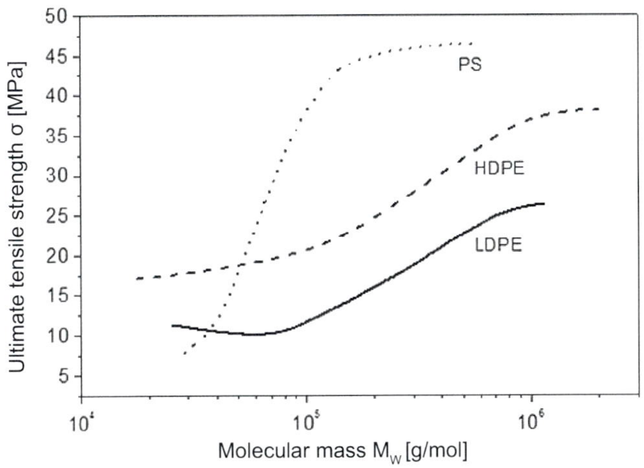

Figure 3.49 Influence of weight average molecular weight on tensile strength at break for different polyolefins (after [18])

Molecular weight degradation can influence mechanical properties significantly: below a critical molecular weight the tensile strength and impact resistance are drastically reduced. An example of the critical molecular weight is for polypropylene around 170,000 g/mol. For homopolymers, similar dependencies are found for elongation at break [14, 20, 21].

c) Change of flow behavior (rheological properties)

In addition to mechanical properties, molecular weight degradation may also change rheological properties of polymer melts. This changed flow behavior might result in changed processing behavior within the extruder.

The mean molecular weight and molecular weight distribution can be determined directly via gel permeation chromatography (GPC). In addition to changes of the mean molecular weight, the width of the molecular weight distribution can be narrowed through degradation or widened through crosslinking reactions [13]. Other parameters related to molecular weight distribution are the zero shear viscosity, solution viscosity or melt flow index (MFI) and melt volume index (MVI), respectively.

Already small changes of mean molecular weight via degradation can lead to a significant change of viscosity: with decrease of mean molecular weight by a factor of 2, the zero shear viscosity is diminished by a factor of 10 [20, 22, 23]. Rheological methods are therefore a very sensitive indicator for changes of the molecular structure of polymers.

d) Optical properties

Optical properties comprise color changes (e.g. yellowing, discoloration) that might be caused by oxygen-containing functional or chromophoric groups (conjugated double bonds). Color changes can be determined and quantified via optical or spectroscopic methods. Crosslinks might lead to gels or spots (punctiform impurities) that can influence optical properties in addition.

e) Other properties

Depolymerization leads to detachment of monomers, which can lead to odor formation, and related molecular weight changes can lead to mechanical property changes. Side group elimination can lead to emissions of undesired or toxic substances, e.g. emission of hydrochloric acid from polyvinyl chloride (PVC).

Aside from mechanical properties, polymer degradation may also cause changes of dielectric properties or electrical conductivity.

3.4.4 Reduction of Polymer Degradation during Processing

Degradation of polymers during processing can be reduced primarily via changes of the extruder screw design or processing parameters. In any case, attention should be paid to reduction of reaction partners (like water, metals, oxygen).

Additional additives can be used to prevent polymer degradation, e.g. physical or chemical modifiers. The following sections describe these and other measures.

3.4.4.1 Extruder Screw Design or Processing Parameters

a) Extruder length and cylinder/screw geometry

During processing of polymers a reduction of degradation is possible via suitable setup of the processing zones and screw design. The aim is to design the melting zone in a way that the polymer granules are processed to allow gentle fusion. In the following mixing zone processing length and energy input should be as high as required, but as low as possible.

A higher mass temperature leads to higher degradation compared to a longer residence time. Therefore, if technically possible, the choice of lower temperatures is favorable with the compromise of having possibly a longer processing zone and longer residence times. The dispersion result and granule homogeneity definitely need to be assessed as there is no decoupling possible of mixing efficacy, thermal stress, and polymer degradation.

b) Selection of mixing elements

With a careful selection of gentle mixing elements a compromise can usually be realized ensuring sufficient mixing efficacy and reducing mechanical as well as thermal stress during compounding. Although not necessarily accessible via usual temperature sensors, high shear rates can cause a local overheating of the melt, which can exceed the mean processing temperature by more than 50 °C [18]. Especially the area between mixing elements and their geometry influence this effect. The influence of these temperature spots on polymer degradation depends on thermal sensitivity and stability of the polymer or polymeric compound. In general, selection of gentle mixing elements can reduce damaging influence as long as the mixing result is reached.

c) Screw speed and throughput

Mechanical energy input in the melt is defined via specific energy input (see Chapter 4). Energy input can be minimized via low screw speed and high throughput resulting in lower energy input per mass and higher filling degree. This may result in higher residence time of the melt within the hot processing zones but allows lower mean mass temperatures in comparison to processing with high screw speed and lower throughput (see Chapter 4).

d) Recipe changes

As polymer degradation is influenced by residence time from melting up to solidifying as well as by possible impurities, a thorough cleaning of the extruder before or after changes to new recipes or compounds is recommended. Long times of heated equipment without screw movement (e.g. overnight) should also be avoided due to accelerated degradation at higher temperatures.

3.4.4.2 Changes of Melt Flow Behavior via Molecular Weight and Flow Modifiers

a) Flow modifiers

In order to improve incorporation of solids into melts their viscosity can be reduced via addition of low-molecular waxes. Due to their low melting point they can cause better wetting of added solids and they also lead to a reduction of melt viscosity. These additives allow in many cases a reduction of processing temperatures and as a result also a reduction of thermally induced degradation reactions.

b) Use of different polymer grades (molecular weight)

Reduction of melt viscosity during processing can also be achieved via use of polymers with lower average molecular weight (or higher melt flow index, respectively), e.g. via use of easy-flowing injection molding grades as mixing partner. This allows reduction of the temperature profile with similar processing conditions compared to more viscous polymer grades. The following mixing and compounding of lower-viscous grades of the same polymer is unproblematic due to miscibility of polymers with the same chemical structure but different molecular weight.

3.4.4.3 Minimization of Reaction Partners

a) Water

All polymers tending to split with presence of water (see Section 3.4.2.3) need to be dried thoroughly prior to processing. Especially all polyesters (PET, PC), polyamides (PA6, PA 66), and thermoplastic polyurethanes (TPU) are affected.

b) Oxygen

Reaction of the polymer with oxygen requires oxygen transported via diffusion from the surrounding air. Thermal-oxidative degradation during compounding can therefore in critical cases be reduced using an atmosphere of the inert gas nitrogen during processing.

c) Fillers

During compounding some solids and fillers used can be exchanged, especially for dyes (such as metal oxides) or other fillers that might harm stability of the polymer used.

3.4.4.4 Additives for Reduction of Polymer Degradation

Suitable stabilizers that are added homogeneously in small amounts (< 1 wt.-%) to the polymer prior to extrusion protect the polymer chains against influence of degradation caused by oxygen, heat, light, shear stress, or metal ions.

Figure 3.50 shows the effect of thermal-oxidative stabilizers used as degradation inhibitors within the autoxidation cycle [13, 19]. There are different stabilizer types that are effective toward different influencing factors within the autoxidation cycle.

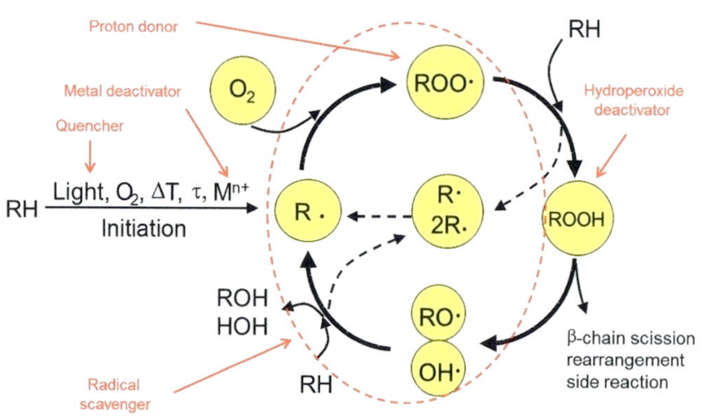

Figure 3.50 Influence of degradation inhibitors on the autoxidation cycle [13, 19]

These types of stabilizers block the ageing process and reduce thermo-oxidative damage during processing. In addition, they may enhance usage properties of the final part. In many cases stabilizers act synergistically as their effectiveness is enhanced through use of several stabilizers [8]. Commonly used stabilizer types are specified as follows [13, 14, 19].

a) Antioxidants (primary antioxidants)

These stabilizers consist of radical scavengers and proton donors leading to an interruption of the chain reactions during the autoxidation process (Figure 3.50). They consist of systems of organic phosphites, aromatic and sterically inhibited amines, and sterically inhibited phenols. Radical scavengers prevent extension of the reaction chain via inhibition of radicals ($R^{\bullet}$, $RO^{\bullet}$, $ROO^{\bullet}$). Proton donors prevent separation of H atoms from the polymer chain and block therefore formation of alkyl radicals that are initiating degradation reactions. They protect against damaging influences from oxygen and heat during processing and use and are consumed irreversibly. Their rate of consumption depends on intensity of external influences.

b) Costabilizers, hydroperoxide decomposers (secondary antioxidants)

The effectiveness of primary stabilizers can be improved significantly via use of costabilizers. An example is hydroperoxide degrading agents, such as thioesters or organic phosphorus compounds. They split hydroperoxides (ROOH) into non-radical stable products that are little reactive and thermally stable. They are used usually together with proton donors and are used up irreversibly. During compounding some additives used (e.g. carbon black or pigments) can also act synergistically and as costabilizers.

c) Metal deactivators

The harmful influence of active metal ions like transition metal ions (e.g. Cu^{2+}) via redox reactions can be reduced or avoided via complexation with chelate ligands.

In addition to their protecting functionality, stabilizers have to be distributed well within the polymeric matrix, must not lead to discoloration and need to be miscible and compatible with the polymer. Moreover, they need to be stable long-term and must not lead to blooming defects. Therefore, there are many different stabilizers depending on their purpose for processing and use.

Using stabilizers it has to be considered that a damage of polymer molecules can happen long before any property changes are visible or detectable. After an induction period, where no detectable changes can be measured, a rapid change of properties can occur. During induction period the stabilizers are used up and integration of oxygen leads to increase of the hydroperoxide concentration. The duration of this induction period depends on the system polymer/stabilizers/additives as well as the processing conditions (temperatures, oxygen concentration, residence time). It has been shown that for many stabilizers there is linear correlation between the stabilizer concentration and the duration of the induction period [13].

As soon as the stabilizer is used up, the autoxidation process is no longer stopped and molecular weight as well as molecular weight distribution can change rapidly – with corresponding changes of rheological and mechanical properties [8, 13]. Especially for plastic recycling processes this interrelationship is important as degradation reactions with added stabilizers are rarely following a simple mechanism that can be linearly extrapolated.

3.4.5 Summary

During extrusion, compounding, and forming operations (as injection molding, extrusion, blow molding) polymers are processed in the molten form. High temperatures used can lead to chemical reactions of fillers or other additives present within the melt. This can lead to significant damage within the extruded granules or the final part, as for example:

- Mechanical properties as reduced impact strength or elongation at break (increased brittleness, crack formation, etc.)
- Optical properties (transparency, gel particles, black dots, specks, etc.)
- Reduced quality for further processing or use (as reduced varnishability, delamination)

In addition, degradation reactions induced during processing can lead to a reduced life cycle time of final parts.

In order to reduce degradation processes, processing conditions during extrusion are recommended to be chosen allowing low thermal stress. Possible reaction partners (as excess water though insufficient drying) need to be reduced or avoided. During compounding of new recipes a systematic study of influencing factors of individual components is recommended.

If thermal degradation cannot be avoided with an optimized processing setup, additives can be used to improve polymer stability preventing undesired chemical reactions. These are antioxidants, costabilizers, or metal deactivators that reduce or avoid oxidative degradation of the polymers during processing.

References for Section 3.4

[1] Kircher, K., *Chemische Reaktionen bei der Kunststoff-Verarbeitung,* Hanser, Munich (1982)

[2] Kaiser, W., *Kunststoffchemie für Ingenieure,* 3rd ed., Carl Hanser Verlag, Munich (2011)

[3] Fedtke, M., *Reaktionen an Polymeren,* VEB Deutscher Verlag für Grundstoffindustrie, Leipzig (1985)

[4] Liu, N. C., Baker, W. E., "Reactive Polymers for Blend Compatibilization", *Adv. Polym. Technol.* (1992) 11 (4), pp. 249–262

[5] Xanthos, M., Dagli, S. S., "Compatibilzation of Polymer Blends by Reactive Processing", *Polym. Eng. Sci.* (1991), 31 (13), pp. 929–935

[6] Grassie, N., Scott, G., *Polymer Degradation and Stabilization,* Cambridge University Press (1985)

[7] Hepperle, J., "Schädigungsmechanismen bei Polymeren", in: *Polymeraufbereitung 2002: technischer Fortschritt zur Steigerung von Leistung und Produktqualität,* VDI-Verlag, Düsseldorf (2002), pp. 17–52

[8] Schnabel, W., *Polymer Degradation – Principles and Practical Applications,* Hanser International, Munich (1981)

[9] Hawkins, W. L., *Polymer Degradation and Stabilization,* Springer, Berlin (1984)

[10] Stivala, S. S., Kimura, J., Gabbay, S. M., "Thermal degradation and oxidative processes", in: *Degradation and Stabilization of Polyolefins,* Allen, N. S., Ed., Applied Science Publishers, New York (1983)

[11] Pauling, L., *The Nature of the Chemical Bond and the Structure of Molecules and Crystals,* 3rd ed., Cornell University Press, Ithaca, NY (1960)

[12] Clough, R. L., Billingham, N. C., Gillen, K. T., *Polymer Durability – Degradation, Stabilization and Lifetime Prediction,* Advances in Chemistry Series, Vol. 249, American Chemical Society, Chicago (1996)

[13] Zweifel, H., *Stabilization of Polymeric Materials,* Springer, Berlin (1997)

[14] Schwarzenbach, K., et al., "Antioxidants", in: *Plastics Additives Handbook,* Zweifel, H., Ed., Hanser, Munich (2001)

[15] Cheng, H. N., Schilling, F. C., Bovey, F. A., "^{13}C Nuclear Magnetic Resonance Observation of the Oxidation of Polyethylene", *Macromolecules* (1976) 9, pp. 363–365

[16] Sohma, J., "Mechano-radical formation in polypropylene by an extruder action and its after-effects", *Colloid Polym. Sci.* (1992) 270, pp. 1060–1065

[17] Casale, A., "Temperature Effect on Polymer Mechanochemistry", *J. Appl. Polym. Sci.* (1975) 19, pp. 1461–1473

[18] Allen, N. S., *Degradation and Stabilization of Polyolefins,* Applied Science Publishers Ltd., Barking (1983)

[19] Gugumus, F., "Antioxidantien", in: *Taschenbuch der Kunststoff-Additive,* 3rd ed., Gächter, R., Müller, H., Eds., Hanser, Munich (1990)

[20] Michler, G. H., *Kunststoff-Mikromechanik,* Hanser, Munich (1992)

[21] Si, L., Zheng, A., Yang, H., Guo, R., Zhu, Z., Zhang, Y., "A study on New Polymerization Technology of Styrene," *J. Appl. Polym. Sci.* (2002) 85, pp. 2130–2135

[22] Ferry, J. D., *Viscoelastic Properties of Polymers,* 3rd ed., John Wiley & Sons Inc, (1980)

[23] Macosko, C. W., *Rheology – Principles, Measurements, and Applications,* VCH Publishers, New York, Weinheim (1994)

■ 3.5 Calculation Basis for the Flow in Wedge Shaped Shear Gaps and Flow Properties of Filled Polymer Melts

Kalman Geiger, Gerhard A. Martin

3.5.1 Consideration of Pseudoplastic Flow Behavior of Plastic Melts in the Wedge Gap Flow and Key Numbers for the Evaluation of the Dispersion

3.5.1.1 Introduction – Deformation of Plastic Melts, Shear, and Elongation in the Wedge Gap Flow

The mixing homogeneity of melt templates is the most important criterion for melt quality in plastic compounding and processing. To achieve this homogeneity, the components must be uniformly distributed in the mixture. Miscible components that are compatible with each other can be uniformly distributed in mixing elements by shear at low shear stresses through enlargement of the boundary interface (stretching) and rearrangement of the resulting melt layers (wrinkling) [1–3]. Mixing of incompatible components requires before the distributive mixing process a division (dispersing) of the immiscible components to be incorporated into the melt. Dispersive mixing elements are intended to break down solid particles in the shear flow and should ensure a sufficient stress to break the particles. Here, the stress on the solid particles and the residence time of the particles in the shear flow are crucial [2]. The incorporation of liquid components (drops) in the melt requires the fragmentation of these drops in smaller ones. The dispersive mixing elements should ensure in a superimposed shear and elongational flow the enlargement of the drops by shearing and break up of these deformed drops through the elongational flow by stretching. The level of shear and elongational deformations and the residence time of the immiscible drops in the melt play an essential role for the results of the dispersive mixing processes [4]. Only a sensitive tuning of the stretching induced in the elongational flow and the residence time of the droplets in the mixing element can ensure the desired drop dispersing effect.

All kneading blocks in co-rotating twin-screw extruders include a wedge slit type flow channel, in which the macromolecules of the plastic melt are stretched. Figure 3.51 shows on the left a pair of kneading blocks of a co-rotating twin-screw extruder in radial section. The shear rates are uniform and high in the tip gap and intermeshing zone. In the entrance area on the other hand, they are low with superimposed elongational flow. Figure 3.51 on the right illustrates the deformation of molten drops in the tip area in detail. After the passage of the tip gap the melt drops are conveyed in the superimposed elongational flow and the shear flow, stretched and broken, and finally deformed and streaked in the tip gap.

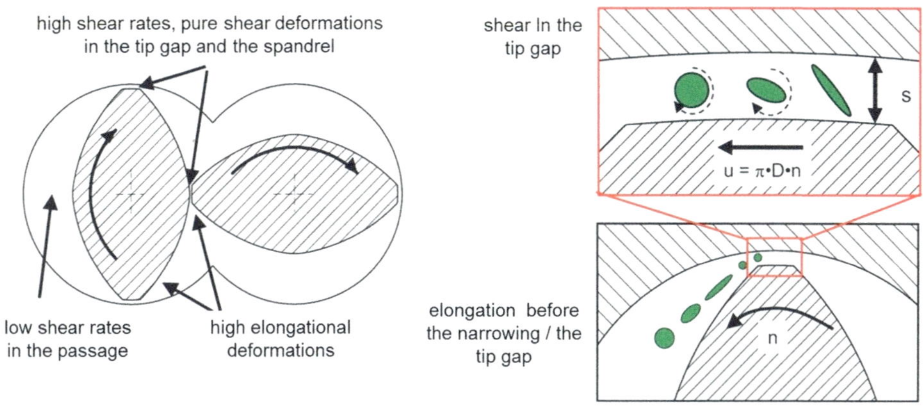

Figure 3.51 Kneading blocks in co-rotating screw elements and drop deformation (left) and break-up in a wedge shaped tip gap (right)

The combined wedge gap/flat gap system has been used for a long time in dynamic mixing elements, Figure 3.52. It provides elongational flow in the flow direction (convergent wedge shaped gap $\left(x=\left[0,l_K\right]\right)$) and subsequent shear flow in the flat gap $\left(x=\left[l_K,l_e\right]\right)$. The speed of the upper moving plate is constant.

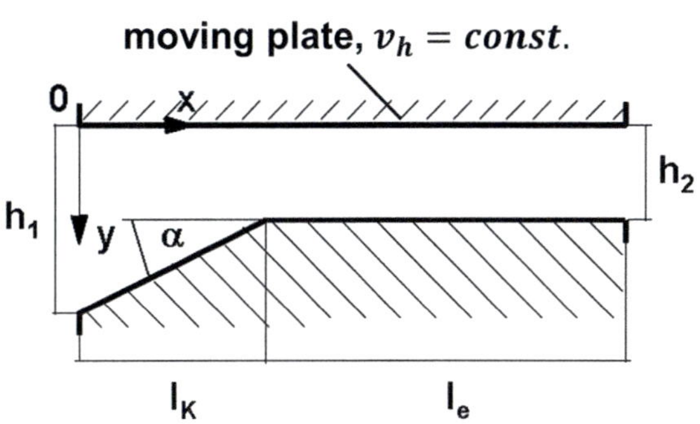

Figure 3.52
Combined wedge gap/flat gap system for elongational and shear flow

By cascading of several wedge gaps, the residence time of the drops in the dispersive mixing element must be adapted for the requirements of an efficient droplet dispersion. The polymeric entropy-elastic drop undergoes during the flow from one wedge gap to the next an elastic recovery deformation. In consequence the drop breaks in the mixing element.

Figure 3.53 shows a dynamic mixer consisting of a toothed mixing element and a shearing element followed by a second toothed mixing element. It is used successfully in high-power single-screw extruders for compounding and processing of polymer blends with additive modified plastic compounds. The dispersion of liquid components and solid agglomerates of fillers and color pigments takes place in the shear section of the dynamic mixer.

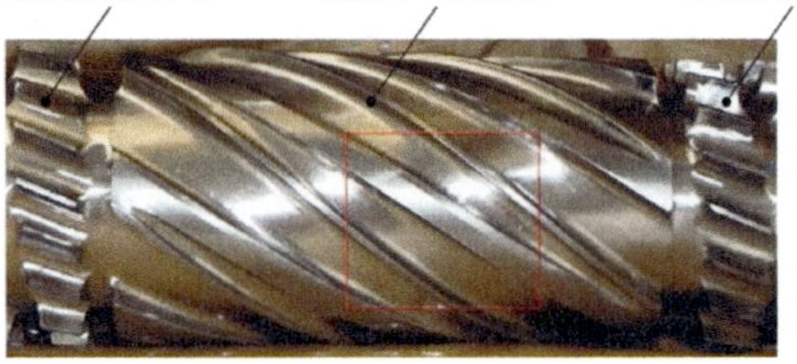

Figure 3.53 Dynamic mixer for single-screw extruders (Ematik [17, 18])

Figure 3.54 shows in detail the red marked area of Figure 3.53.

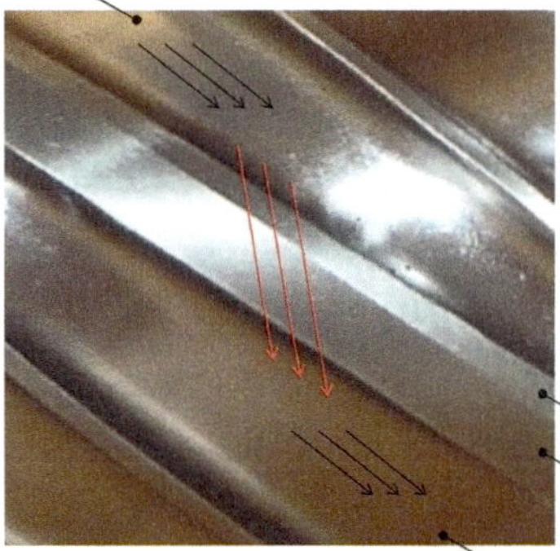

Figure 3.54 In the wedge gap (chamfer) elongational flow occurs, and in the parallel gap (shear flight) shear flow occurs

If unwound into the plane, the geometry of this gap corresponds exactly to the geometry of the wedge gap/flat gap system as shown schematically in Figure 3.52. The moving plate of this system is the cylinder wall of the extruder. The gap between the shear flight and the cylinder wall is 0.5 to 1.2 mm depending on the intensity of the required shear stress to break up the droplets and the agglomerates. The width of the shear flight is 4 to 24 mm depending on the required residence time. The incoming melt in the entry channel flows first through the wedge gap (chamfer) with large elongational deformation. The dispersive breakup of the droplets and agglomerates occurs in the shear flight.

The evaluation and the optimization of the mixing operation of the wedge shaped gap require a detailed analysis of the distribution of shear, elongation, and the resulting total deformation in flow and in gradient direction of the wedge gap flow. The goal of this contribution is to perform numerically the flow analysis for the wedge gap with flat boundary surfaces under consideration of a flow law for pseudoplastic liquids. In addition to this analysis, the flow is also calculated for the combined wedge gap/flat gap system. Shearing, planar stretching, and total deformation in the wedge gap exit are quantified by specific key numbers. The calculated mean residence time in the wedge gap allows the calculation of elongational strain rates and the Deborah number [12, 13] for characterization of the entropy-elastic state of the melt droplets in the flow.

3.5.1.2 Calculation of the Wedge Gap Flow for Highly Viscous Fluids

Figure 3.55 shows the wedge gap in a schematic representation with the most important geometry parameters. The boundary conditions are $v_x(x=0, y=0)=v_h$, $v_x(x=0, y=h_1)=0$, $v_x(x=l_K, y=0)=v_h$, $v_x(x=l_K, y=h_2)=0$, $\Delta p = p_1 - p_0 = 0$.

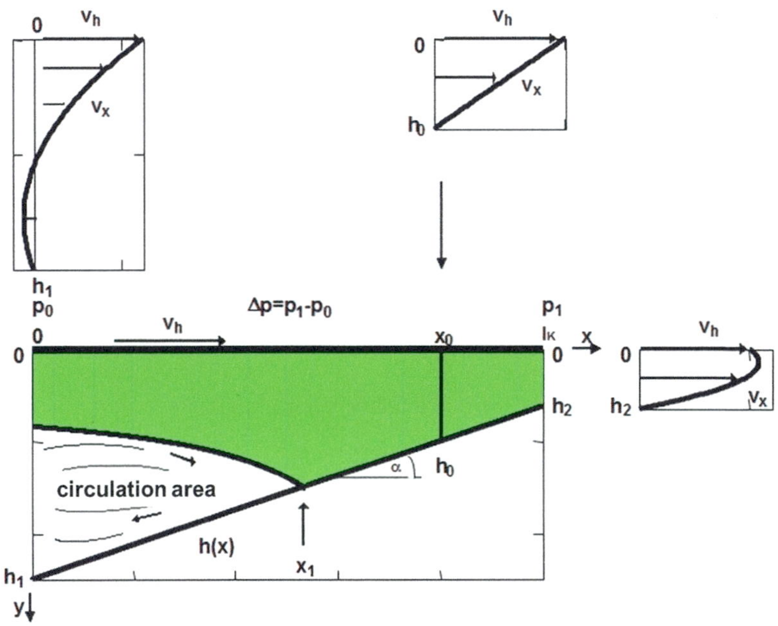

Figure 3.55 Schematic representation of the wedge gap flow

The natural coordinate system for the wedge gap is Cartesian. The geometry parameters of the wedge gap are: gap height at the entrance $h(x=0)=h_1$, gap height at the exit $h(x=l_K)=h_2$, wedge gap length l_K and wedge gap width b. The local gap height of the wedge gap with straight boundary surfaces is a function of the coordinate x and can be described with the following equation:

$$h(x)=ax+h_1 \quad \text{with} \quad a=\frac{h_2-h_1}{l_K} \tag{3.85}$$

The state variables of the wedge gap flow are the pressure at the moving plate $p(x)$, the velocity profile $v_x(x,y)$, and the therefrom derived distribution of the shear rate $\dot{\gamma}_{yx}(x,y)$, the shear $\gamma_{yx}(x,y)$ and the planar elongation $e_x(x,y)=\exp(\varepsilon_x)$ in the superimposed shear and elongational flow. The x-component of the momentum equation for highly viscous fluids with the common assumptions for laminar flow is:

$$\frac{\partial \tau_{yx}}{\partial y}=\frac{\partial p}{\partial x}=p_x \tag{3.86}$$

with the definition equation of the shear viscosity:

$$\frac{\partial}{\partial y}\left[\eta\left(\frac{\partial v_x}{\partial y}\right)\frac{\partial v_x}{\partial y}\right]=p_x \tag{3.87}$$

The viscosity function in Equation (3.87) is the Carreau approach [5]:

$$\eta\left(\frac{\partial v_x}{\partial y}\right)=\frac{A}{\left(1+B\left|\frac{\partial v_x}{\partial y}\right|\right)^C} \tag{3.88}$$

It should be noted that every other continuously differentiable flow approach can be applied for the numerical calculation of the wedge gap flow. For the solution of the non-linear differential equation (3.87), the boundary conditions on the moving plate and on the fixed inclined wall are formulated as follows:

$$v_x(x,y=0)=v_h \tag{3.89}$$

and

$$v_x(x,y=h(x))=0 \tag{3.90}$$

The solution of the system of equations consisting of Equations (3.86) to (3.90) can only be achieved numerically using the finite difference method (FDM) [8]. Only in the case of constant viscosity in Newtonian fluids is an analytical solution of this system of equations possible [7].

3.5.1.2.1 Definition of a Measure of Shear, Planar Elongation, and Total Deformation at the Wedge Gap Exit

The shear at $x=l_K$ is $\gamma_{yxtot}(x=h_2,y)$, the planar elongation $e_x(x=h_2,y)$, and the total deformation $\gamma_{tot}(x=h_2,y)$, achieved at the wedge gap exit, are characterized by suitably chosen integral key numbers. They will be here weighted with the local volumetric flow rate at the wedge gap exit in the interval $y = [0, h_2]$ according to the following equations:

$$\gamma_{yxtoth2} = 2 \int_{y=0}^{y=h_2} \frac{v_x(x = h_2, y)}{h(x = x_0) v_h} \gamma_{yxtot}(x = h_2, y) dy \tag{3.91}$$

$$e_{xtoth2} = 2 \int_{y=0}^{y=h_2} \frac{v_x(x = h_2, y)}{h(x = x_0) v_h} e_{xtot}(x = h_2, y) dy \tag{3.92}$$

$$\gamma_{toth2} = 2 \int_{y=0}^{y=h_2} \frac{v_x(x = h_2, y)}{h(x = x_0) v_h} \gamma_{tot}(x = h_2, y) dy \tag{3.93}$$

3.5.1.2.2 Calculation of the Means of the Residence Time, the Hencky Strain, the Strain Rate and the Tensile Stress at the Wedge Gap Exit

The average residence time t_K in the wedge gap is the quotient of the wedge gap volume V and the throughput G through the wedge gap according to the equation:

$$t_K = \frac{V}{G} = \frac{(h_1 + h_2) l_K}{v_h h(x = x_0)} \tag{3.94}$$

The Hencky strain – also known as natural strain – is the natural logarithm of the stretching. With the Equation (3.92) calculated key number for the planar stretching one obtains the mean Hencky strain at the wedge-gap exit as:

$$\varepsilon_{xtoth2} = \ln(e_{xtoth2}) \tag{3.95}$$

The average strain rate $\dot{\varepsilon}_{xm}$ then becomes with the average residence time t_K in Equation (3.94):

$$\dot{\varepsilon}_{xm} = \frac{\varepsilon_{xtoth2}}{t_K} \tag{3.96}$$

The elongational viscosity in the planar elongational flow for a Newtonian fluid is:

$$\mu_N = 4 \eta_N \tag{3.97}$$

The shear viscosity η_N at one for the wedge gap flow characteristic shear rate (at the position of the pure drag flow) [6] is calculated with Equation (3.88). With the Equations (3.96) and (3.97) one then obtains the mean tensile stress in the elongational flow component of the wedge gap flow:

$$\sigma = 4\eta_N \dot{\varepsilon}_{xm} \tag{3.98}$$

The representative value for the tensile stress σ is applied to the melt droplets. If this tensile stress reaches a characteristic critical value for the plastic melt, a brittle fracture [9] of the melt droplets occurs. For polymers with linear unbranched macromolecules this critical tensile stress lies in the range of 0.5 to 1.5 MPa [10, 11].

3.5.1.2.3 Calculation of the Deborah Number as a Measure of the Elasticity of the Viscoelastic Plastic Melt in the Wedge Gap

The viscoelastic deformation behavior of concentrated polymer solutions and melts can be judged, if the characteristic relaxation time of the material (material time) and the observation time during the deformation processes in the flow (process time) are compared. The relation of these two times is described by the dimensionless Deborah number *De*. The ratio of the *B* value is defined as the characteristic critical relaxation time in the Carreau approach (Equation (3.88)) [12]. The characteristic Deborah number [13] is determined as the quotient of the *B* value and the mean residence time of the melt in the wedge gap t_K [14]:

$$De = \frac{B}{t_K} \tag{3.99}$$

If the Deborah number is less than 1, the deformation behavior of the plastic melt is viscous. For Deborah numbers $De = 1$ one observes viscoelastic deformation behavior of the melt. For Deborah numbers $De > 1$ - in particular for $De \gg 1$ - the deformation behavior of the viscoelastic melt corresponds to that of a solid body.

3.5.1.3 Plastic Melts with Different Pseudoplastic Flow Behavior

The numerical calculation and analysis of the wedge gap flow was performed for six selected plastic melts taken from plastics compounding and plastics processing practice. In Table 3.6 the parameters of the Carreau approach (Equation (3.88)), the flow exponent of the power law approach by Ostwald/de Waele, $m = 1/(1-C)$, the practice-relevant processing temperature, and a rheological classification of these plastics are compiled.

The six plastic melts differ in terms of their zero shear viscosity level (described with the parameter *A*), of the expression of their pseudoplastic flow behavior (characterized with the parameter *C* or *m*), and of their elastic properties (quantified with the characteristic relaxation time of the melt, represented by the parameter *B*). Figure 3.56 shows the viscosity curves of these six selected plastics.

Table 3.6 Carreau Parameters of Six Different Plastic Melts [15, 16]

	A in Pas	$B \approx \lambda_{char}$ in s	C	m	T in °C	Rheological classification
PC	87	0.0013	0.096	1.01	340	Newtonian low viscous, inelastic
PA 6	245	0.001	0.51	2.02	250	weak pseudo plastic, low viscous, inelastic
PE-HD	34,550	2.20	0.59	2.46	160	pseudo plastic, highly viscous, highly elastic
PE-LLD	673	0.03	0.61	2.54	160	pseudo plastic, low viscous, inelastic
PP	3420	0.13	0.74	3.82	230	strong pseudo plastic, middle viscous, elastic
PS	21,610	0.59	0.80	4.90	160	strong pseudo plastic, highly viscous, highly elastic

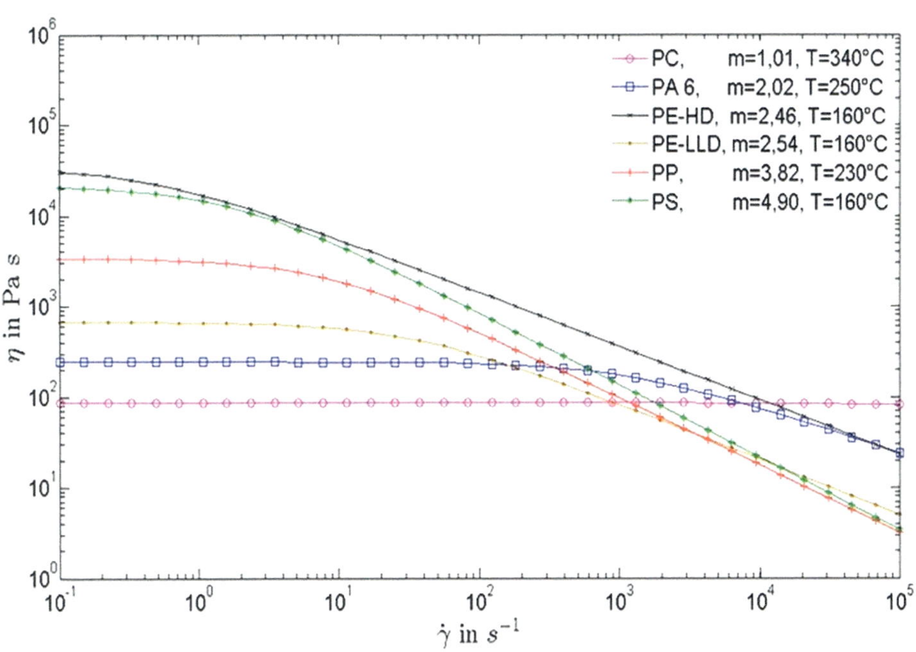

Figure 3.56 Viscosity curves of the six selected plastic melts [15, 16]

3.5.1.4 Results of the Simulation

Simulation results are presented and discussed for the six selected plastics, and thereby are shown the influence of the viscosity, the expression of the pseudoplastic flow behavior, and the viscoelastic properties of these plastic melts. The following geometry parameters were used for the wedge gap shown in Figure 3.52: wedge gap length $l_K = 4$ mm, wedge gap height at the wedge gap exit $h_2 = 0.8$ mm (practice-relevant values). The opening angle of the wedge gap α was varied in the interval [0°, 45°]. The moving plate speed v_h was varied between 1200 mm s^{-1} and 3600 mm s^{-1}. These values correspond to the circumferential speed of fast running extruders and are therefore practice relevant. The calculation of the flow took place under the assumption of constant melt temperature in the wedge gap, meaning isothermal conditions. Also in the case of the combined wedge gap/flat gap system the numerical calculations were performed isothermally first.

The flow properties of the plastic melt decisively influence the pressure profile, the velocity distribution in the flow and in the gradient direction, and consequently the formation of a circulation area in the wedge gap (see Figure 3.54). The shear rate, the shear and the planar elongational deformation, and the total deformation calculated from the shear and elongational deformations (see Equation (3.95)) allow a precise evaluation of the dispersive mixing process in the dynamic mixing elements. From the shear and elongational viscosity of the melt one obtains the level of applied tensile stresses on the melt drops.

3.5.1.4.1 Velocity Distribution in the Wedge Gap without and with Circulation Area

To demonstrate the influence of pseudoplastic flow behavior on the flow of the plastic melt in the wedge gap, calculation results are first shown for the plastic melts PC, PA 6, and PP (see Table 3.6 and Figure 3.56), respectively at the wedge gap angles $\alpha = 15°$ and $\alpha = 35°$, Figure 3.57.

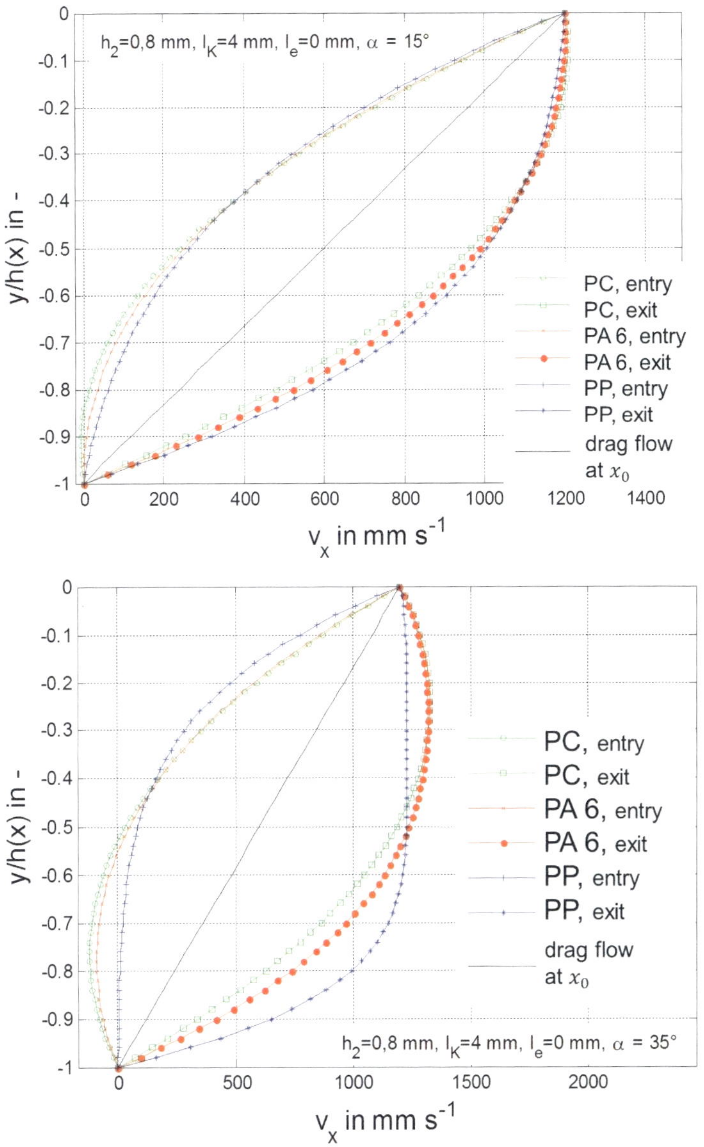

Figure 3.57 Velocity distributions in gradient direction in the wedge gap; above at $\alpha = 15°$ and below at $\alpha = 35°$ for PC, PA 6, and PP

3.5.1.4.2 Mean Value of the Total Deformation as a Function of the Wedge Gap Angle

Dispersive mixing elements of dynamic mixers in plastic compounding and plastic processing should be optimized in the geometry and in the operating mode for the respective application fields. The design of such mixing elements is currently purely empirical and the design criteria are based exclusively on experience from the compounding and the processing practice. The numerical analysis of the wedge gap flow for pseudoplastic fluids offers opportunities for the design of mixing elements tailored for the production of polymer blends. The deformation capacity of the wedge gap is calculated as a function of the wedge gap geometry.

The numerically calculated key number of the total deformation in the wedge gap is shown for opening angle α for the six plastic melts (see Figure 3.56 and Table 3.6) in Figure 3.58 in semi logarithmic representation.

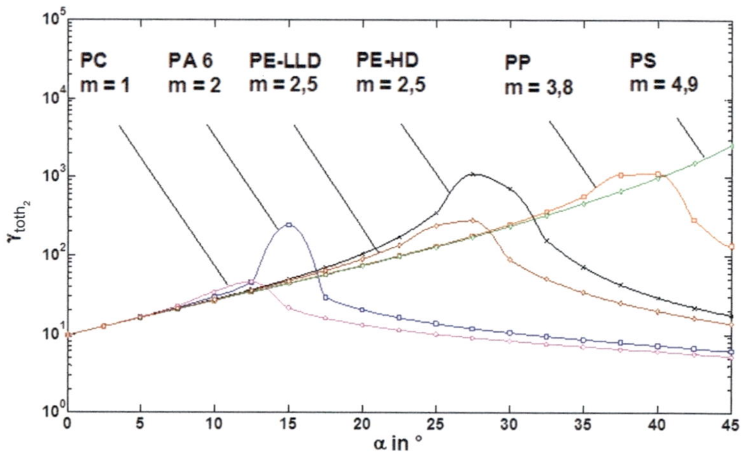

Figure 3.58 Total deformation as function of α for six different plastic melts, $v_h = 1200\ \text{mm s}^{-1}$

The maximal achievable total deformation in the wedge gap and the onset of a (usually undesirable) circulation area in the entry region of the wedge gap shift with increasing flow exponent m to significantly higher wedge gap angles α.

An increase in the total deformation of melt drops in dispersive mixing elements is achieved with the combined wedge gap/flat gap system as shown in Figure 3.52. Here the wedge gap brings the required elongation for the dispersive process, which is then stabilized in the subsequent flat gap. The melt drops in the shear flow are stretched by shear. The results of the numerical analysis of the deformation capacity of the combined wedge gap / flat gap system depending on the wedge gap opening angle α and the flat gap length l_e are summarized for all six plastic melts (rheological data see Figure 3.56 and Table 3.6) in the six diagrams in Figure 3.59 in semi logarithmic representation.

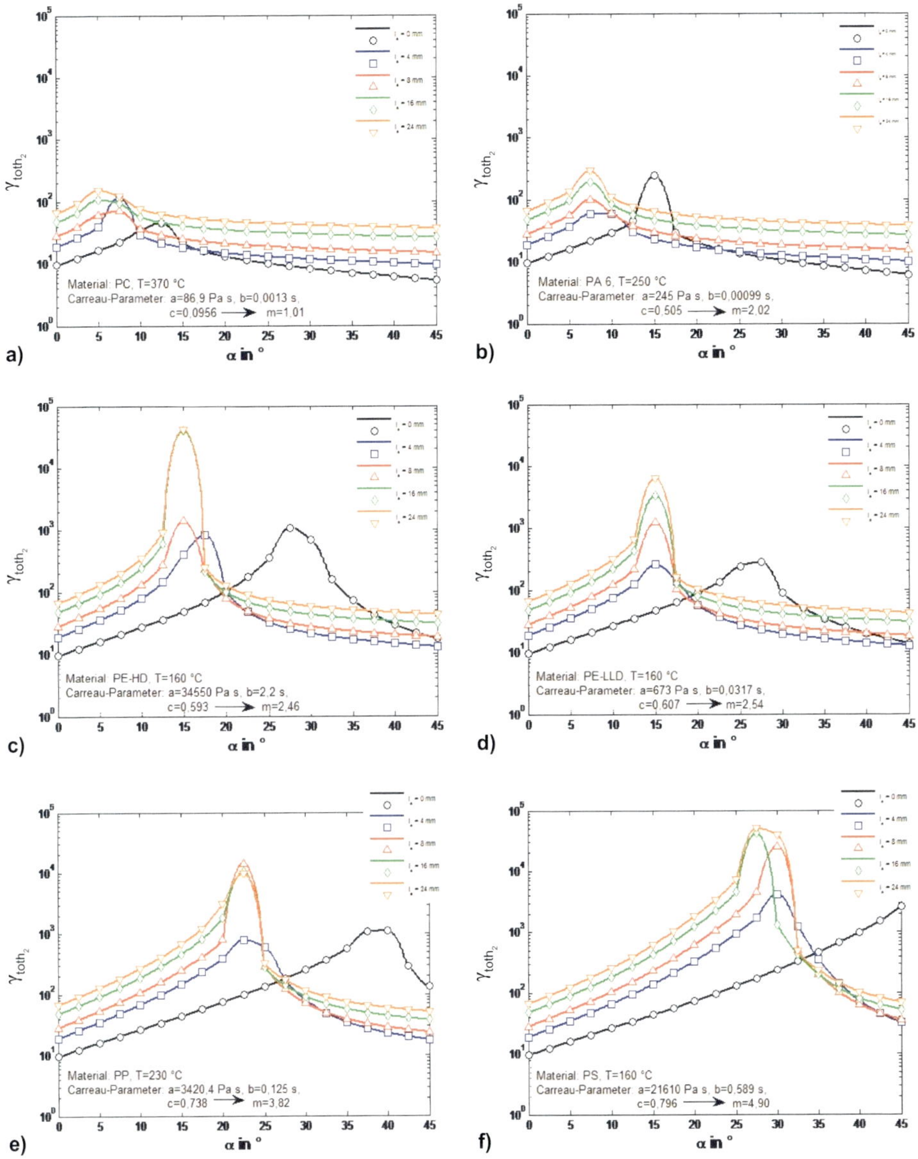

Figure 3.59 The total deformation in the wedge gap/flat gap system as function of the wedge gap opening angle α and the flat gap length l_e for the six plastic melts. Upper moving plate speed $v_h = 1200$ mm s^{-1}

Every wedge gap and wedge gap/flat gap system has a critical opening angle α for the maximal total deformation. At higher angles a circulation area appears and the total deformation is lower. With longer flat gap length l_e the total deformation increases because of the longer residence time of the melt. The (reasonably possible) length of the flat gap is limited by geometrical restrictions the by the shear heating of the melt. The circulation area should be avoided, because the residence time in this area is undefined long and unnecessary shear heating of the melt occurs.

3.5.1.4.3 Mean Residence Time in the Wedge Gap and in the Combined Wedge Gap/Flat Gap System

Figure 3.60 shows – for the plastic melt PP – the calculated mean residence time of the plastic melt in the wedge gap or in the combined wedge gap/flat gap system for $v_h = 1200\ \text{mm s}^{-1}$ depending on total length $l_K + l_e$ in semi logarithmic representation.

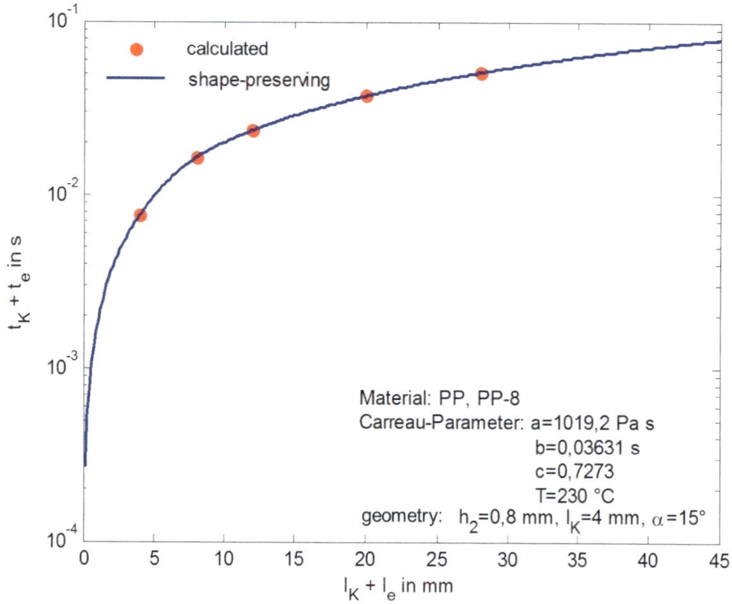

Figure 3.60 Mean residence time as a function of the total length of the wedge gap and the combined wedge gap/flat gap system exemplary for the plastic melt PP, $v_h = 1200\ \text{mm s}^{-1}$

3.5.1.4.4 Hencky Strain, Strain Rate and Tensile Stress Depending on the Moving Plate Speed v_h and the Wedge Gap Opening Angle α

The melt drops are exposed in the wedge gap high elongational strains at very high elongational strain rates. The mean Hencky strain at the wedge gap exit (according to Equation (3.95)), the mean elongational strain rate (according to Equation (3.96)), and the mean tensile stress (according to Equation (3.98)) are calculated for all six plastic melts. In each case the wedge gap opening angle α for the maximum mean planar stretching and the maximum total deformation has been reached. The results are shown for three moving plate speeds in Tables 3.7 to 3.12.

Table 3.7 Material: PC; Wedge Gap Opening Angle: $\alpha = 10°$

v_h in mm s^{-1}	$\gamma_{yxtoth2}$	γ_{toth2}	e_{xtoth2}	ε_{xtoth2}	$\dot{\varepsilon}_{xm}$ in s^{-1}	$\sigma \cdot 10^5$ in Pa
1200	8.3	34.4	2.5	0.9	123	0.4
2400	8.3	33.9	2.5	0.9	245	0.7
3600	8.3	33.7	2.5	0.9	367	1.1

Table 3.8 Material: PA 6; Wedge Gap Opening Angle: $\alpha = 15°$

v_h in mm s^{-1}	$\gamma_{yxtoth2}$	γ_{toth2}	e_{xtoth2}	ε_{xtoth2}	$\dot{\varepsilon}_{xm}$ in s^{-1}	$\sigma \cdot 10^5$ in Pa
1200	8.0	244.3	5.8	1.8	226	1.6
2400	7.9	68.5	4.1	1.4	366	2.0
3600	7.8	60.0	3.9	1.4	530	2.6

Table 3.9 Material: PE-HD; Wedge Gap Opening Angle: $\alpha = 28°$

v_h in mm s^{-1}	$\gamma_{yxtoth2}$	γ_{toth2}	e_{xtoth2}	ε_{xtoth2}	$\dot{\varepsilon}_{xm}$ in s^{-1}	$\sigma \cdot 10^5$ in Pa
1200	7.4	1537.4	44.7	3.8	409	6.3
2400	7.4	1751.1	48.1	3.9	834	8.5
3600	7.4	1868.7	49.9	3.9	1263	**10.1**

Table 3.10 Material: PE-LLD; Wedge Gap Opening Angle: $\alpha = 27°$

v_h in mm s^{-1}	$\gamma_{yxtoth2}$	γ_{toth2}	e_{xtoth2}	ε_{xtoth2}	$\dot{\varepsilon}_{xm}$ in s^{-1}	$\sigma \cdot 10^5$ in Pa
1200	7.3	892.7	24.5	3.2	351	1.2
2400	7.3	432.8	19.0	2.9	646	1.5
3600	7.3	421.7	19.0	2.9	968	1.7

Table 3.11 Material: PP; Wedge Gap Opening Angle: $\alpha = 35°$

v_h in mm s^{-1}	$\gamma_{yxtoth2}$	γ_{toth2}	e_{xtoth2}	ε_{xtoth2}	$\dot{\varepsilon}_{xm}$ in s^{-1}	$\sigma \cdot 10^5$ in Pa
1200	6.8	572.2	31.9	3.4	340	1.5
2400	6.8	651.0	35.5	3.6	700	1.9
3600	6.8	719.9	38.3	3.6	1072	2.1

Table 3.12 Material: PS; Wedge Gap Opening Angle: $\alpha = 40°$

v_h in mm s^{-1}	$\gamma_{yxtoth2}$	γ_{toth2}	e_{xtoth2}	ε_{xtoth2}	$\dot{\varepsilon}_{xm}$ in s^{-1}	$\sigma \cdot 10^5$ in Pa
1200	6.5	982.8	56.6	4.0	366	2.4
2400	6.6	1238.6	67.7	4.2	764	2.8
3600	6.6	1411.2	74.7	4.3	1173	3.1

In the six tables it can be seen that the total shear deformation $\gamma_{yxtoth2}$ at a given wedge gap geometry is independent of the moving plate speed v_h and the mean total deformation γ_{toth2} increases with the mean planar stretching e_{xtoth2} in all plastic melts. The achievable mean Hencky strain ε_{xtoth2} is high for the more viscous and more pronounced pseudoplastic melts. The strain induced deformation for each individual drop of the melt lies at a very high level. The high mean strain rates are increased with increasing moving plate speed v_h. The mean tensile stress reaches values of $\sigma = 10^6$ Pa for the highly viscous PE-HD melt. This was measured by Hürlimann [11] with a capillary rheometer. At the capillary exit melt fracture was observed. Bernnat [9] verified with in rheotens experiments comparable critical tensile stresses. Also at lower strain levels drop breakage is possible on a large number of melt drops when they are exposed to much higher local tensile stresses in the wedge gap. In general one can find that for more pronounced pseudoplastic flow behavior of the melt the wedge gap geometries with a large opening angle α are more effective. Furthermore, the increase in total shear deformation can be adjusted over the flat gap length. The total mean elongation, the mean strain rate, and the mean tensile stress are adjustable over the moving plate speed v_h.

3.5.1.4.5 Deborah Number Depending on the Moving Plate Speed

The Deborah number *De* (Equation (3.99)) was calculated for the six plastic melts in the wedge gap ($l_e = 0$ mm) for two moving plate speeds. The results are summarized in Tables 3.13 and 3.14.

Table 3.13 Deborah Number *De* for Six Plastic Melts in the Wedge Gap with the Moving Plate Speed $v_h = 1200$ mm s^{-1}, $\alpha = 15°$, $l_K = 4$ mm

Material	PC	PA 6	PE-HD	PE-LLD	PP	PS
t_K in ms	7.9	7.8	7.7	7.7	7.6	7.6
De	0.16	0.13	285	4.1	16.4	77.5

Table 3.14 Deborah Number *De* for Six Plastic Melts in the Wedge Gap with the Moving Plate Speed $v_h = 3600\ \mathrm{mm\ s^{-1}}$, $\alpha = 15°$, $l_K = 4$ mm

Material	PC	PA 6	PE-HD	PE-LLD	PP	PS
t_K in ms	2.6	2.6	2.6	2.6	2.5	2.5
De	0.49	0.38	858	12.4	49.2	232

Already at the moving plate speed $v_h = 1200\ \mathrm{mm\ s^{-1}}$ the Deborah numbers are significantly larger than 1 for the more pronounced entropy-elastic melts (with higher *B*-value of the Carreau approach, Equation (3.88)), so the viscoelastic melt drops show more the deformation behavior of a solid body. That means drop breakage due to the viscoelastic properties of the melt is favored when the characteristic relaxation time is sufficient large and the residence time of the melt in the wedge gap is sufficiently short (i. e., short wedge gap length with a few millimeters l_K and large wedge gap angle α). At lower disc speeds the Deborah numbers are smaller and the tensile stresses obtained are too low for a drop size reduction.

3.5.1.4.6 Local and Average Dissipative Temperature Increase in the Wedge Gap/Flat Gap System

The temperature increase in the melt by shear, planar elongation, and resulting total deformation at the exit of the wedge gap/flat gap system are plotted over the *y*-coordinate in Figure 3.61. The temperature of the melt rises very sharply on the fixed wall. The dispersive shear element can heat up and the dispersive mixing effect can be reduced. Therefore the geometry and the circumferential speed of the shear element must be adjusted due to the flow behavior and the thermal properties of the melt.

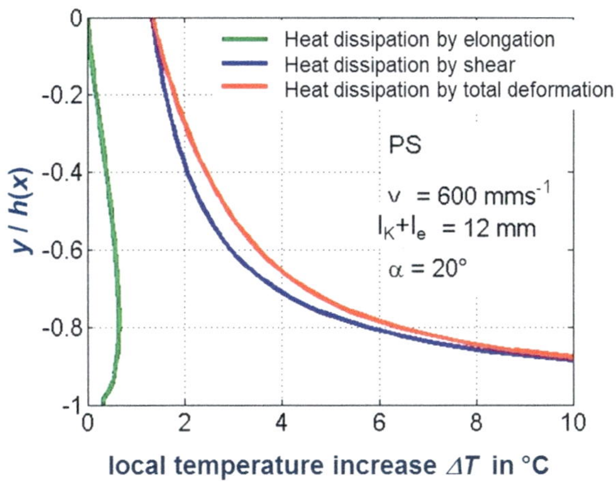

Figure 3.61 Local temperature increase for a PS melt at the wedge gap/flat gap exit

The next Figure 3.62 demonstrates the influence of the pseudoplastic flow behavior on the mean temperature increase of the plastic melt in the wedge gap/flat gap system for four melts with different pseudoplastic flow behavior. The viscosity data (Carreau parameter, flow exponent m) were taken from Table 3.6.

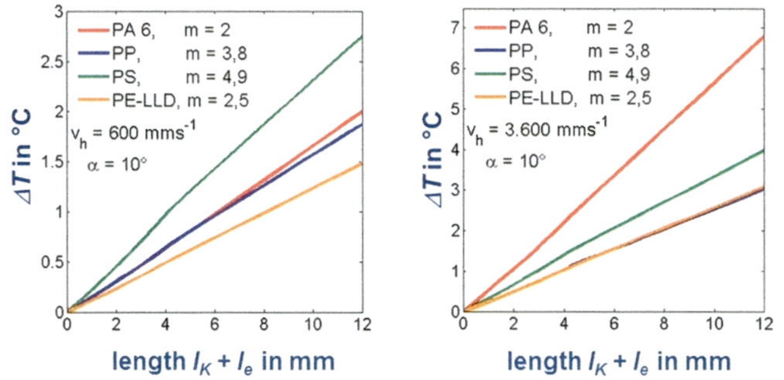

Figure 3.62 Mean temperature increase in the wedge gap/flat gap system for four plastic melts for two moving plate speed v_h

The flow behavior of the melt is described with the Carreau approach (Equation (3.88)). The parameters of this approach are compiled in Table 3.6 for a PA 6, a PE, a PP, and a PS melt. The two diagrams in Figure 3.61 and Figure 3.62 show the mean temperature increase over the length of the wedge gap/flat gap system. At lower moving plate speeds the temperature increase for the PS melt is higher than for the PA 6 melt. This results from the higher viscosity of this melt at lower shear rates. At high moving plate speed v_h the less pseudoplastic PA 6 melt dissipates more heat than the strongly pseudoplastic PS melt. At high shear rates the PA 6 melt is significantly more viscous than the PS melt.

3.5.1.4.7 Summary

The dispersion of liquid and solid components in plastic melts is an essential step in the plastics compounding and processing. This is achieved with wedge gap like narrow shear elements in single and twin screw extruders. Important key numbers were defined in such shear gaps with regard to the induced shear, elongational, and total deformations, and to the drop-dividing acting entropy-elastic tensile stresses. Examples were presented for different pronounced pseudoplastic and viscoelastic plastics. The analysis of the wedge gap flow offers the possibility to take into account the viscous and the viscoelastic properties of the plastic melts in dispersively acting mixing elements. The optimization of the operating parameters and the geometry and the improvement of the mixing result are possible in compounding and processing of plastics. This improves the quality of compounding and products.

The analysis of drop-dividing processes of liquid additives is the subject of research and should be performed for design of extrusion experiments. The drop comminution, the dissolution of agglomerates as well as the homogeneity of additivized plastics should be investigated experimentally and described using 3D simulation models. In particular, the influence of the viscosity difference between the matrix polymer and the incorporated liquid additives should be analyzed.

3.5.2 Modeling of the Flow Behavior of Highly Filled Plastics

3.5.2.1 Viscosity of Polymers with Different Filler Contents

The application potential of thermoplastics has always been expanded by targeted modification with additives and fillers. With fillers and reinforcement particles (GFRP and CFRP composite materials) highly filled composite materials are generated, with increased the strength and rigidity. To improve the extensibility and the impact resistance, thermoplastically processable partially crosslinked plastic compounds (TPV) are added. Compounds with minerals (zeolite molding compounds) and other complex systems for specific applications (e. g. composites highly filled with flame retardant for cable sheathing) would not be processable without the flowability of specific polymeric matrix materials. Highly filled melts need a special viscosity approach because their flow behavior is very different from that of other non- or less filled plastic melts. A sufficient physical description of such complex flow behavior is described by Musialek [19, 20].

Filled plastics are polymers with organic or non-organic fillers which have fibrous, platelet, or spherical geometry. Such melts consist of two or more components and two or more phases. Adding fillers into a polymer melt leads to a change of its rheological properties which affects the processing behavior in extrusion and injection molding.

The effect of fillers on the flow behavior of polymer melts can be considered and explained as illustrated in Figure 3.63 to Figure 3.66 [19]. Unfilled polymers show the typical pseudoplastic flow properties. With increasing shear rate the macromolecules orient in the flow direction and cause a viscosity reduction (Figure 3.63). If measured values are from oscillatory tests applying the Cox-Merz relation, the same is valid. If fillers are added in low concentration to the polymer melt pseudoplastic flow behavior is maintained.

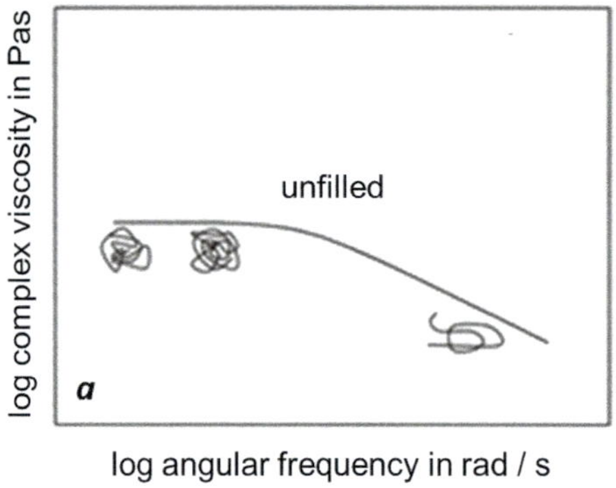

Figure 3.63
Viscosity curve of an unfilled or low filled polymer melt

Due to the so-called hydrodynamic effect the viscosity shifts over the entire shear rate range to higher values as shown in Figure 3.64 [21].

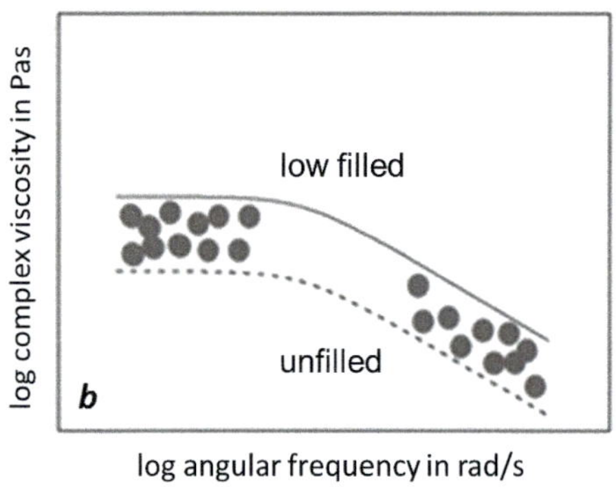

Figure 3.64
Viscosity curve of a filled polymer melt ($\phi < \phi_c$)

The flow behavior of the polymer melt with a low filler concentration ϕ and weak interparticular forces can be described with the Carreau approach (Equation (3.100)):

$$\eta_c(\omega) = \frac{A}{(1 + B\omega)^C} \tag{3.100}$$

A means the zero shear viscosity, B represents the transition between the Newtonian plateau and the pseudoplastic regime of the viscosity curve, and C characterizes the slope in the pseudoplastic flow regime. The dashed line in Figure 3.64 represents the flow behavior of the unfilled material. This flow behavior remains persistent as long as the filler concentration ϕ does not correspond to the rheological percolation threshold ϕ_c. When the rheological percolation threshold is reached, the fillers form a three-dimensional structure and lead to immobilization of the macromolecules in the matrix polymer melt. Consequently, the viscosity increases and a flow obstruction at low shear rates becomes apparent. In Figure 3.65

the dashed line indicates the flow behavior of the filled polymer with a concentration below the rheological percolation threshold ϕ_c [19].

In the filler network itself, the interparticular forces between the single particles are the cause of the cohesion of the modifier to a network. If the shear rate is increased, a higher shear stress is imposed on the filled polymer, which acts against the interparticular forces. An increase in the effective shear stress leads to a characteristic value to overcome the interparticular forces and initiates the destructuring of the fillers network and the pseudoplastic flow behavior of the compound occurs again.

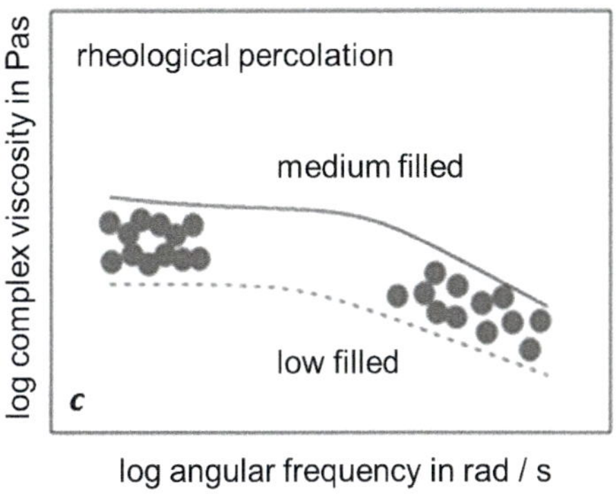

Figure 3.65
Viscosity curve of a filled polymer $(\phi = \phi_c)$

This destructuring point is marked by the turning point of the viscosity curve (Figure 3.65 and Figure 3.66). A further increase of the filler concentration leads to a pronounced flow obstruction, which is shown by a stronger viscosity increase in the low shear rate range. The viscosity curve in Figure 3.66 is characteristic for highly filled polymers [21].

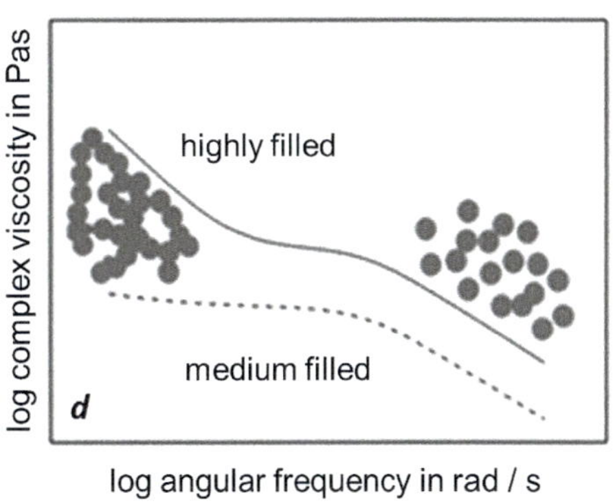

Figure 3.66
Viscosity curve of a highly filled polymer $(\phi > \phi_c)$

Fillers are for example SiO_2 particles [22–24], layer silicates [25, 26], glass fibers [27], or carbon nanotubes [28, 29]. Dispersed in different plastics such as PP, PE, or

PA 6, they lead to this typical flow behavior of the compound. The flow obstruction is explained in the literature with the building of particle networks. Flow restrictions are also observed for polymer blends based on polycaprolactone and thermoplastic starch [30]. The same is observed for PE/TPS blends reported by Ning et al. in [31].

Usually, plastics are in a molten state during processing. Consequently, the flow behavior and the viscosity play an essential role in the processing. To calculate the pressure loss exactly a model is needed to describe as accurately as possible the flow behavior of the plastic melt with the additives and the modifiers.

3.5.2.2 CARPOW Approach for the Viscosity Function of Highly Filled Polymers

The flow behavior is described for highly filled polymers with the CARPOW approach by Geiger [32]:

$$\eta = \frac{d}{\omega^n} + \frac{\eta_p}{(1+b\omega)^c} \tag{3.101}$$

The two flow exponents n and c are dimensionless and characterize the slope in the area of the flow obstruction and the slope in the area of the shear thinning respectively. Parameter η_p has the dimension of the shear viscosity (Pas) and b the dimension of time (s). The dimension of the parameter d (Pas^{1-n}) depends on the flow exponent n.

By normalizing the parameter d (Equation (3.101)) a physical material parameter is defined:

$$\eta_{tough} = d\omega_{TP}^{-n} \tag{3.102}$$

This η_{tough} can be interpreted as a consistency parameter for highly filled polymer melts and has the unit Pas. ω_T is the angular frequency or the shear rate in the turning point with the unit s^{-1} and indicates the breakup of the particle network. For an unfilled polymer or for polymers with a modifier concentration below the percolation threshold, $d = 0$. The CARPOW approach (Equation (3.101)) reduces to the Carreau approach (Equation (3.100)).

Figure 3.67 shows the viscosity curves of a long glass fiber reinforced polypropylene melt with different LGF weight concentrations. Table 3.15 includes the most important parameters of the CARPOW approach, the complex shear viscosity η^*_{TP} and angular frequency ω_{TP} at the turning point and the consistency parameter η_{tough} of PP-LGF compounds with different long glass fiber content.

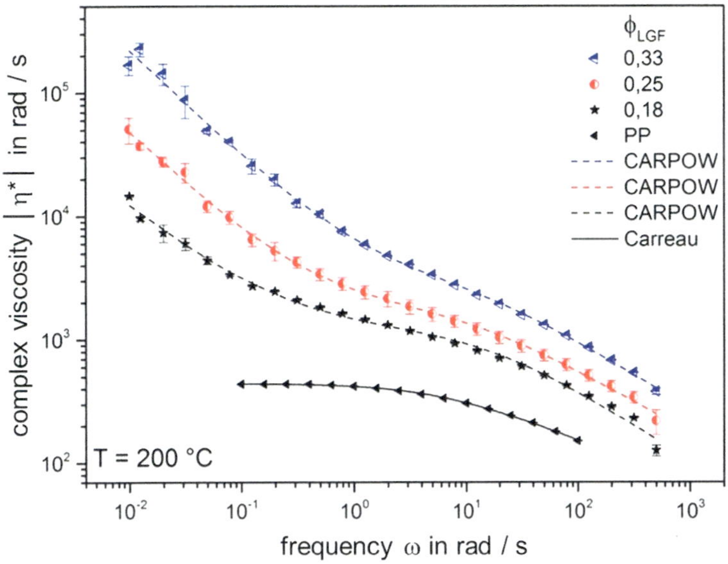

Figure 3.67 Viscosity curve of PP-LGF composite with different weight concentrations

With increasing LGF concentration, the viscosity of the composite increases in the range of low shear rates. It forms a so-called flow obstruction. The strong viscosity increase may be explained with the building of a complex three-dimensional structure of the glass fibers.

Table 3.15 Parameters of the CARPOW Approach, the Complex Shear Viscosity η^*_{TP} and the Angular Frequency ω_{TP} at the Turning Point and the Consistency Parameter η_{tough} of PP-LGF Compounds with Different Long Glass Fiber Content, T = 200 °C

PP/LGF Long glass fiber content	Parameters of the CARPOW formula					Turning point of the $\eta^*(\omega)$ function		Consistency parameter
	d	n	$\eta_p = a$	b	c	η_{TP}^*	ω_{TP}	η_{tough}
Vol. [%]	in [Pa s^{n-1}]	–	in [Pa s]	in [s]	–	in [Pa s]	in [rad s^{-1}]	in [Pa s]
0	0	0	452.2	0.23	0.33	0	0	0
18	2401.4	0.74	1138.3	0.06	0.59	1237.3	2.45	190.8
25	7318.7	0.75	1128.0	0.01	0.95	1259.2	10.45	233.2
33	19,856.3	0.81	1775.0	0.01	0.94	2043.9	16.56	506.4

3.5.2.3 Summary

Using the CARPOW model reflects the measured complex viscosity curves for highly filled polymers in a very good approximation. The unfilled polypropylene on the other hand has the typical pseudoplastic flow behavior which can be described with the Carreau approach. The flow behavior of filled polypropylenes can be modeled very well with the CARPOW approach.

References for Section 3.5

[1] Pfeiffer, A., VDI Kunststofftechnik, VDI-Verlag, Düsseldorf (1997), 119

[2] Martin, G. A., VDI Kunststofftechnik, VDI-Verlag, Düsseldorf (1997), 267

[3] Michaeli, W., Wolff, T., VDI Kunststofftechnik, VDI-Verlag Düsseldorf (1997), 279

[4] Martin, G. A., VDI Kunststofftechnik, VDI-Verlag Düsseldorf (1997), 443

[5] Carreau P. J., Ph.D. Thesis, University of Wisconsin (1968)

[6] Geiger, K., Grünschloß, E., Martin, G. A., Platz, E., *Chem. Ing. Tech.* (2013), 85(12), pp. 1904–1913

[7] Geiger, K., Grünschloß, E., 19th Stuttgarter Kunststoff-Kolloquium (2005), 4/V4, 1

[8] Winter, H. H., personal communication (1977)

[9] Bernnat, A., Ph. D. Thesis, Universität Stuttgart, Germany (2001)

[10] Kim, S., Dealy, J. M., *Polym. Eng. Sci.* (2002), 42(3), p. 495

[11] Hürlimann, H. P., Knappe, W., *Rheol. Acta* (1972), 11, p. 292

[12] Geiger, K., Martin, G. A., Sobotta, A., *Kunststoffe* (2011), 6, p. 44

[13] Reiner, M., *Phys. Today* (1964), 17(1), p. 62

[14] Vrentas, J. S., Jarzebski, C. M., Dudda, J. L., *AIChE J.* (1975), 21(5), p. 894

[15] *Kenndaten für die Verarbeitung thermoplastischer Kunststoffe,* Teil 2, Carl Hanser Verlag, Munich, Vienna (1982)

[16] *Kenndaten für die Verarbeitung thermoplastischer Kunststoffe,* Teil 4, Carl Hanser Verlag, Munich, Vienne (1986)

[17] Werkbild, Ematik GmbH, Magdeburg (2012)

[18] Martin, G. A., unpublished results (2012)

[19] Musialek, M., Geiger, K., Bonten, C., 24th Stuttgarter Kunststoff-Kolloquium (2015)

[20] Bonten, C., *Kunststofftechnik,* Hanser, Munich (2014)

[21] Musialek, M., Geiger, K., Poindl, M., Bonten, C., 23rd Stuttgarter Kunststoff-Kolloquium (2013)

[22] Chassagnau, P., *Polymer* (2008), 49, p. 2183

[23] Dorigato, A., Pegoretti, A., Penati, A., *Express Polymer Letters* (2010), 4, p. 115

[24] Ray, S., Okamoto, K., Okamoto, M., *Macromolecules* (2003), 36, p. 2355

[25] Ray, S., Yamad, K., Okamoto, K., Ueda, K., *Polymer* (2003), 44, p. 857

[26] Thomasset, J., Carreau, P. J., Sanschagrin, B., Ausias, G., *Non-Newtonian Fluid Mechanics* (2005), 125, p. 25

[27] Prashantha, K., Soulestin, J., Lacrampe, M., Krawczak, P., Dupin, G., Claes, M., *Compos. Sci. Technol.* (2009), 69, p. 1756

[28] Song, Y., Youn, J., *Carbon* (2005), 43, p. 1378

[29] Shin, B., Narayan, R., Lee, S., Lee, T., *Polym. Eng. Sci.* (2008), 48, p. 2126

[30] Sarazin, P., Orts, G., Li, W., Favis, B., *Polymer* (2008), 49, p. 599

[31] Ning, W., Jiugao, Y., Xiaofei, M., Ying, W., *Carbohydr. Polym.* (2007), 67, p. 446

[32] Geiger, K., 21st Stuttgarter Kunststoffkolloquium (2009)

4 Conveying Behavior, Pressure and Performance Behavior

4.1 Introduction of Conveying and Pressure Behavior of Highly Viscous Liquids in Extruders

Klemens Kohlgrüber

4.1.1 Throughput and Pressure Behavior, Dimensionless Key Figures

4.1.1.1 Shear Rate and Viscosity

Viscosity plays a dominant role in the process technology with highly viscous liquids. It can change considerably, for example with differences in speed in the liquid (shear), temperature, and material blends (fillers, lubricants, solvents, etc.). Changes by orders of magnitude are possible.

The basics of the conveying behavior of highly viscous liquids will be covered in this chapter. For the understanding of the coherences it will be helpful not to include initially the numerous dependencies of the viscosity.

The definitions of viscosity and shear rate are very important for the understanding of the basics, for which reason a short description is given in addition to the preceding chapters.

Figure 4.1 shows in the upper area the development of velocity of a plane flow.

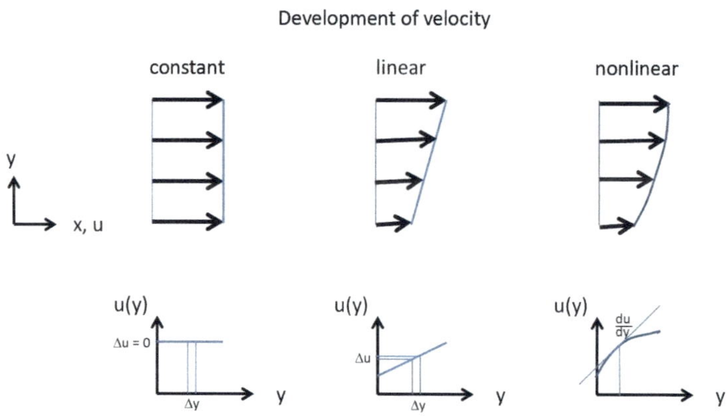

Figure 4.1 Demonstration and definition of the shear rate of a plane flow

Newton observed that without movement and without speed differences in a liquid no friction takes place. The case without speed difference is shown in Figure 4.1 left. If speed differences occur in a plane flow, then the local speed difference is described by the gradient of the speed. The gradient is in the middle of Figure 4.1 the difference quotient. This is called shear rate $\dot{\gamma}$. If the flow is more complex and there is no linear connection between the speed u and the spatial coordinate y, then the local gradient is described by the pitch of the tangent, in the figure right.

$$\dot{\gamma} = \frac{du}{dy} \tag{4.1}$$

In experiments Newton found out that the force F to the movement of the liquid is proportional to this gradient and the area A in direction of the movement:

$$F \sim A \cdot \dot{\gamma} \tag{4.2}$$

For the proportionality factor Newton has introduced the viscosity η:

$$F = \eta \cdot A \cdot \dot{\gamma} \tag{4.3}$$

The force applied to the area is called shear stress:

$$\tau = \frac{F}{A} = \eta \cdot \dot{\gamma} \tag{4.4}$$

The shear stress is in the unit (N/m^2), the shear rate ([m/s]/m = 1/s), in which the unit of the viscosity is given by Equation (4.4): ($N \cdot s/m^2$), which is mostly stated in (Pa·s). The conversion (Pa = N/m^2) is applied.

As mentioned, the viscosity may depend on many values. A constant viscosity is called "Newtonian viscosity". To understand the basics, a consideration of a constant viscosity makes sense. In a few cases this chapter describes also the shear thinning. Liquids whose viscosity decreases with increasing shear are called shear-thinning liquids. For further information on viscosity, please refer to Section 3.1.

4.1.1.2 Simple Qualitative Consideration on Simple Plane Flow

4.1.1.2.1 Shear Flow

In the literature relating to extruders, particularly in the past, the so-called "processing models", also named "pan models", have been considered for single-screw extruders (see Figure 4.2).

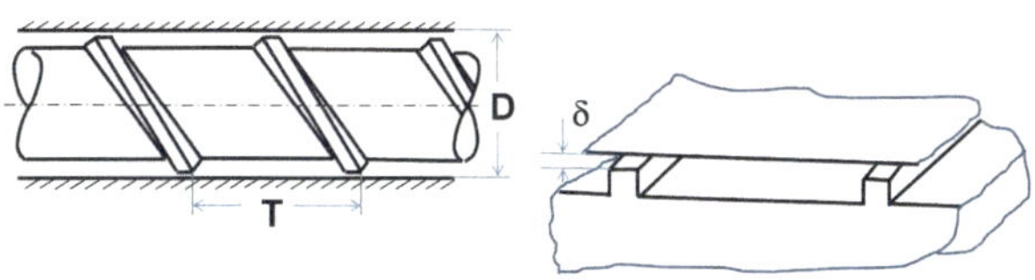

Figure 4.2 Processing of a single-screw extruder

Also in more recent publications, procedures for an easier modeling and for quantitative calculation of the flow are used for the co-rotating twin-screw extruder [1].

In Section 4.1.1.2.2 a very simple plane flow will be considered, not to execute quantitative calculations, but in order to develop the essential basics for the understanding of the product conveying and to clearly introduce the dimensionless key figures.

A "processing" with an extreme simplification for a co-rotating twin-screw extruder is depicted in Figure 4.3.

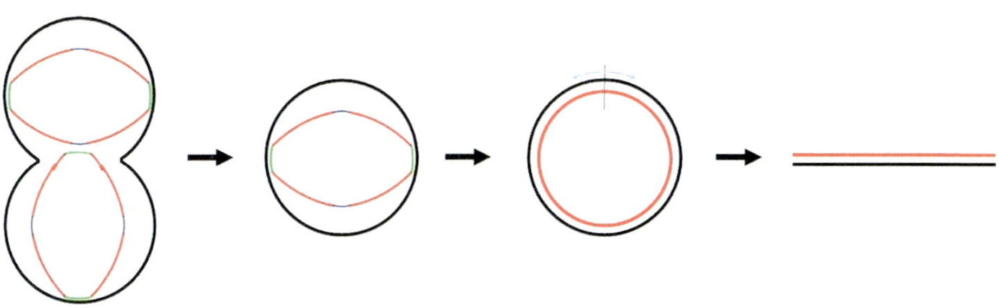

Figure 4.3 Simplification of the complex geometry for "plate-plate"

The cross-section of the extruder is shown on the left in the figure. The product conveying in longitudinal direction takes place by "conveying active elements", thus, for example, by screw connection of the shown geometry in the longitudinal direction of the extruder. The shown annular gap only conveys in the circumferential direction and also the processing for "plate-plate" shows initially no axial conveying. However, it will become apparent that this simple model can deliver all essential basics for the understanding of the conveying behavior in extruders.

The "plate-plate processing", also named two-plate model, becomes more concrete in Figure 4.4.

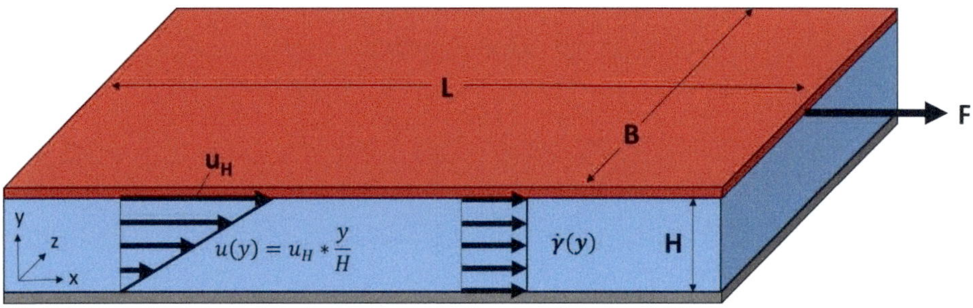

Figure 4.4 Flow between two plates, one of which is moving

The speed profile in the liquid is linear.

$$u(y) = u_H \cdot \frac{y}{H} \tag{4.5}$$

According to the definition of the shear rate this is constant and by Equation (4.4) the shear stress as well.

$$\dot{\gamma} = \frac{u_H}{H} = constant \tag{4.6}$$

The average speed is thus half the maximum speed.

$$\bar{u} = \frac{1}{2} u_H \tag{4.7}$$

The depicted flow is called drag flow. The drag flow factor 1/2 of this example is an important characteristic of the drag flow. It shows in the following sections the conveying capacity of a screw element.

The throughput as volume flow is calculated according to:

$$\dot{V} = \bar{u} \cdot cross\text{-}sectional\ area \tag{4.8}$$

in which the cross-sectional area is $H \cdot B$. If the throughput is related to the cross-sectional area and the speed u_H then a dimensionless key figure can be formed.

$$\dot{V}^* = \frac{\dot{V}}{H \cdot B \cdot u_H} = 0.5 \tag{4.9}$$

Thus, the dimensionless throughput can be interpreted as follows: Independent of the size and speed of the upper plate, the dimensionless throughput amounts to 0.5 (with the unit 1) and corresponds to the drag flow factor. This throughput is independent of the viscosity! Thus, the result is also valid for shear-thinning materials. Because of the constant shear rate, the viscosity of shear-thinning materials is constant above the cross-section although the absolute value may be lower. Speed and throughput remain the same; force F and shear stress τ are lower.

In Section 3.1 (Figures 3.18 and 3.19), the cases Newtonian, shear thinning, and shear thickening are compared.

Can the result be qualitatively transferred to extruders? For methods operating predominantly with drag flow, the result is mostly independent of the viscosity. At double- or multi-flighted geometry this is normally well fulfilled in the point of "maximum self-throughput" (explanation in Section 4.1.1.3). Here it is only pointed out to one example, matching this very well, Section 4.7, Figure 4.75.

4.1.1.2.2 Plate-Plate and Annular Gap

The primary geometric size of an extruder is the diameter D and, as independent process value, the speed n. Therefore, the Equation (4.9) shall be formulated with these values. With the circumferential speed and a so-called geometric similarity of all linear measures (they then change proportionally to each other at the scale-up/-down):

$$u_H = \pi \cdot n \cdot D \qquad B = L \qquad \frac{L}{D} = constant \qquad \frac{H}{D} = constant \tag{4.10}$$

results for the dimensionless volume flow rate in the annular gap:

$$\dot{V}^* = \frac{\dot{V}}{n \cdot D^3} = 0.5 \cdot \pi \tag{4.11}$$

The numerical value of the volume flow rate is not that interesting as it is referred to the circumferential direction. Here it is interesting how the key figure will be formed. In the following it will turn out that the given key figure with another relevance of the volume flow rate is an important key figure for extruders.

The question arises whether the development of speed in the annular gap is deviating strongly from that between two plates. At a very narrow annular gap this deviation will be low. Figure 4.5 shows the difference for smaller and bigger annular gaps.

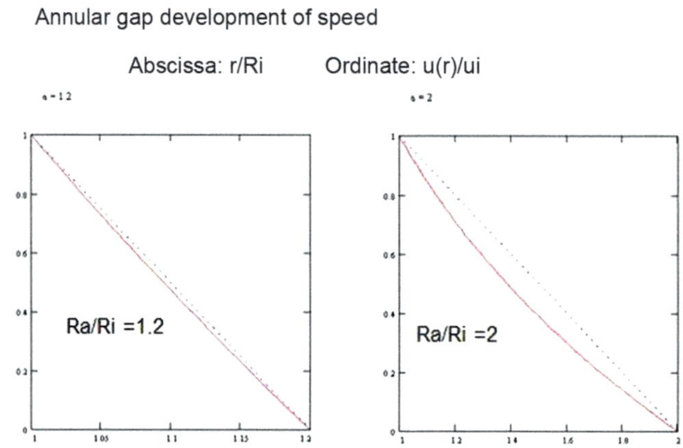

Figure 4.5 Comparison development of speed annular gap (continuous line) to plate-plate (dashed line)

The figure refers to an annular gap with rotating inner cylinder. At increasing diameter or radii ratio the deviation will increase as expected. The simplified plate-plate consideration is fully sufficient for the qualitative results shown.

4.1.1.2.3 Pressure Flow

So far, the pure drag flow was considered. In filled screw sections, pressures usually occur, a distribution of pressure over the cross-section and a pressure gradient along the screw. The pressure gradient can be carried out by a pump "upstream", i.e. in front of the screw section, and "downstream" by a pressure consumer or by the melt conveying in the screw section. The following Figure 4.6 illustrates this by an example plate-plate.

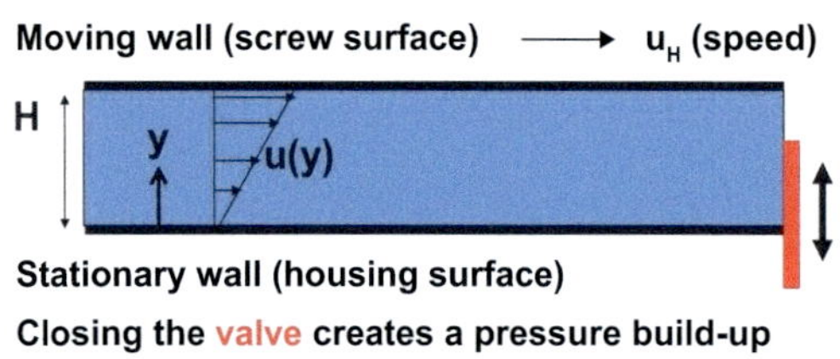

Figure 4.6 Simple flow with pressure build-up

First, a plane flowed channel with stationary walls shall be considered, Figure 4.7.

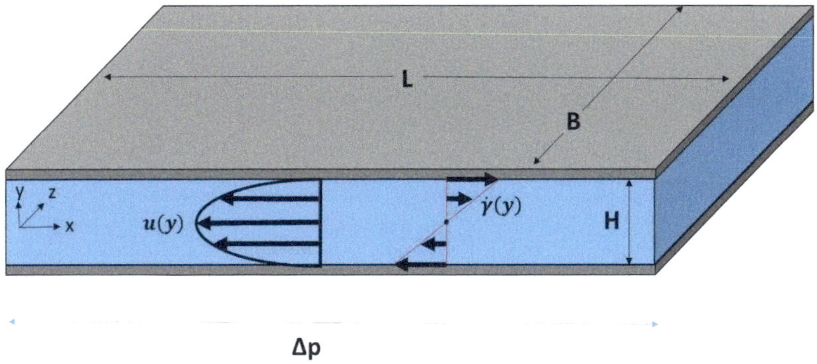

Figure 4.7 Flow of a liquid between two plates by a pressure gradient

For reasons of symmetry, the zero point of the height coordinate y is arranged in the middle. With $A_{longitudinal\ section} = L \cdot B$ and a balance of forces:

$$\Delta p \cdot A_{cross\text{-}section} = F_{liquid} = \tau \cdot A_{longitudinal\ section} \tag{4.12}$$

resulting after some calculations in:

$$u(y) = -\frac{|\Delta p|}{2L} \cdot \left(\left(\frac{H}{2} \right)^2 - y^2 \right) \tag{4.13}$$

In this case, the shear rate equation (4.1) is not constant.

$$\dot{\gamma} \sim y \tag{4.14}$$

After integration of the speed equation (4.13) the following throughput results:

$$\dot{V}_{pressure} = \frac{B \cdot H^3}{12 \cdot \eta} \cdot \frac{\Delta p}{L} \tag{4.15}$$

As the shear rate is not constant and the viscosity is explicitly included in the flow equation (4.15), there is an influence of a variable viscosity in the pressure flow in comparison to the drag flow. This influence can be considerable. In several sections of this book examples with shear-thinning liquids are considered. Here, reference is made only to one example, Figure 4.18.

4.1.1.2.4 Combined Pressure and Shear Flow

In practice, it is rare that pure shear or pressure flows occur. At constant viscosity, both flows are combined linearly. This applies also to three-dimensional flows, not only to the here considered simple plane flow.

As per Equation (4.9), the throughput for the drag flow is:

$$\dot{V}_{drag} = \frac{1}{2} \cdot B \cdot H \cdot u_H \tag{4.16}$$

An addition of the drag portion equation (4.16) with the pressure portion equation (4.15):

$$\dot{V}_{total} = \dot{V}_{drag} - \dot{V}_{pressure} \tag{4.17}$$

provides, when a pressure build-up in the extruder contains a positive algebraic sign:

$$\dot{V}_{total} = \dot{V} = \frac{B \cdot H \cdot u_H}{2} - \frac{\Delta p \cdot H^3 \cdot B}{12 \cdot L \cdot \eta} \tag{4.18}$$

It should be noted that the medium product speed can no longer be calculated from the pure drag flow equation (4.7), but from the general equation (4.8). The overlapping of the two flow forms is described below several times, for example in Section 4.5.

The above equations for pressure and throughput describe the flow process "along the axis", which is why they are referred to as one-dimensional models. In general, the models do not describe dependencies over the cross-section. However, these are not models with plug flow, because gradients across the cross-section are used for their derivation. The same applies to the "real one-dimensional models", which will be introduced from Section 4.1.1.3 onwards.

Equation (4.18) is suitably converted for two reasons:

1. Generalized representation in a dimensionless form
2. Because of the graphical progress in the throughput-pressure coordinate system in a so-called axis intercept form.

$$\frac{\dot{V}^*}{A_1} + \frac{\Delta p^*}{A_2} = 1 \tag{4.19}$$

The following abbreviations were introduced as dimensionless key figures:

$$\dot{V}^* = \frac{\dot{V}}{B \cdot H \cdot u_H}; \qquad \Delta p^* = \frac{p' \cdot H^2}{u_H \cdot \eta}; \qquad A_1 = \frac{1}{2}; \qquad A_2 = 6 \tag{4.20}$$

Figure 4.8 shows the linear relation equation (4.19) with the key figures graphically.

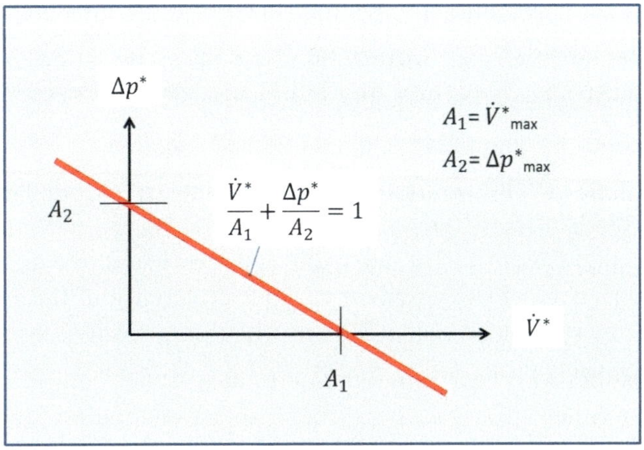

Figure 4.8 Relation between pressure and drag flow

The axis intercept point A_1 is the drag flow and can be interpreted as the maximum dimensionless "self-throughput". The axis intercept point A_2 describes the maximum dimensionless pressure build-up. Thus, dimensioned:

$$\dot{V}_{max} = \dot{V}_{drag} = A_1 \cdot B \cdot H \cdot u_H \tag{4.21}$$

$$\Delta p_{max} = \frac{A_2 \cdot u_H \cdot \eta \cdot L}{H^2} \tag{4.22}$$

This allows a further type of standardization which is summarized in Figure 4.9.

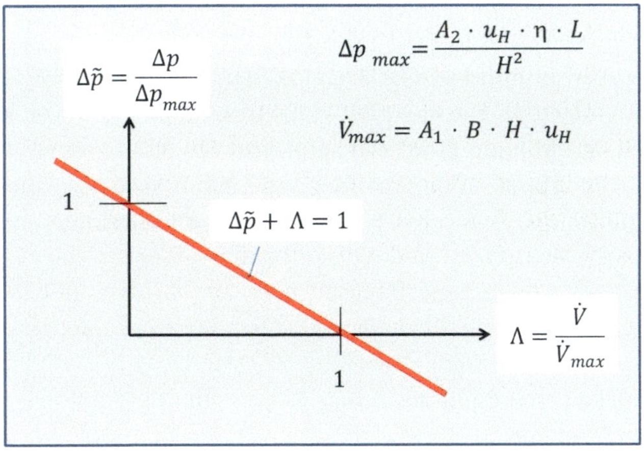

Figure 4.9 Summary of the throughput-pressure-relation of the plane flow with constant viscosity

The above depicted considerations of simple plane flow are intended for:

- the development of the basic understanding for the types of flow in extruders
- descriptive derivation of dimensionless key figures
- the derivation of the qualitative relation of the key figures

The relations of the simple plane flows and the depicted key figures can be formally transferred to real extruders. The velocity u_H can be interpreted as speed (rpm) and H as flight depth and in the tip area also as clearance δ. In the following, additional descriptive considerations with direct reference of the individual key figures on extruders shall be made. This results in a pressure basic equation for extruders in the same form.

4.1.1.3 Extruder Key Figures and Pressure Basic Equation for Extruders

4.1.1.3.1 The Most Important Key Figures of a Screw Element

The "pressure-less self-conveying capacity" of a screw section will be considered.

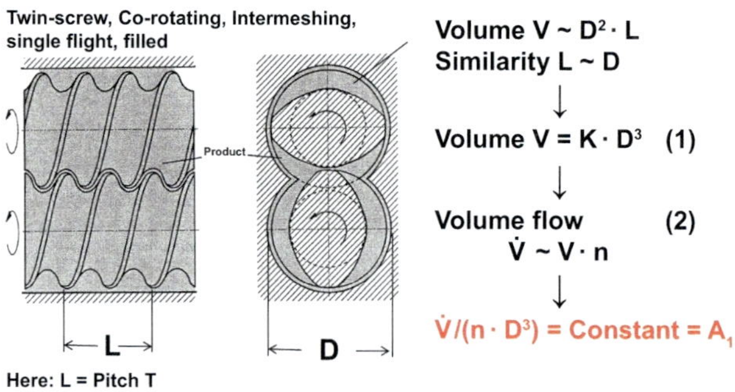

Figure 4.10 Considerations regarding the self-throughput of a screw element

The key figure derived in Figure 4.10 is identical to the key figure from Equation (4.11) for the annular gap. The meaning of the throughput, however, is different. In Equation (4.23), the volume flow corresponds to the axial volume flow and thus describes the "throughput" of the extruder and a screw section, respectively. In the annular gap, the volume flow corresponds to that in the circumferential direction.

$$\frac{\dot{V}}{n \cdot D^3} = A_1 \tag{4.23}$$

The dimensionless throughput according to Equation (4.23) is dependent on the specific geometry at the respective point along the extruder. Normally, diverse screw elements of different geometry are used. Of course, each screw element has a different conveying parameter A_1, which is also called profile parameter. The conveying capacity of a screw element is described by the conveying parameter. Therefore, this is the most important key figure of a screw element. The meaning of the conveying parameter will become clearer in the following sections. The most important key points are:

- A_1 is the most important key figure of a screw element
- A_1 is not dependent on process factors (throughput and speed) and not on material data, but only on the geometry of the screw (that is why it is also referred to as profile parameter).
- A_1 is with regard to geometrically similar screws (thereby, at a scaling the linear measures behave proportionally) independent of the screw size
- A_1 describes the maximum dimensionless self-throughput (throughput without axial pressure gradient; in the cross-section pressure gradients occur)

How can the self-throughput A_1 of an extruder conveyor element be measured or calculated?

In the past the parameter was mainly measured. At Bayer, a measuring screw with bearings [2] on both sides served for this, specially manufactured by M. Ullrich. A short description of the measuring screw is given in Section 1.2. For a better observation the housing was made of Plexiglass. Model liquids (silicone oils) are usually used to determine the conveying parameters, as these allow examinations at room temperature.

In recent times, the three-dimensional calculation models and the calculation times have enabled a calculation of the conveying parameter A_1 and further parameters still to be introduced.

Figure 4.11 shows the rough calculation of the self-throughput. The required free cross-section area is given in Sections 2.1 and 1.2. This is the octagon-shaped housing cross-section minus the two screw cross-sections.

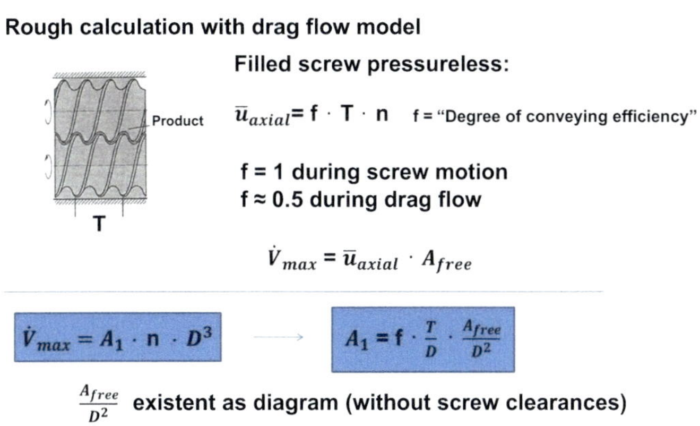

Figure 4.11 Rough calculation of the maximum self-throughput

4.1.1.3.2 Pressure Throughput Behavior

In the previous considerations in Section 4.1.1.2, the flow between two plates was used as a tool for the introduction of the drag and pressure flow and the "axis intercept form". To introduce a dimensionless pressure loss in screw sections, a simple pipe flow can initially be considered as a helpful idea. The dependence of pressure loss or pressure build-up with screw shafts will be formally like a flow through a pipe, amplified by the "fine geometry" of the shafts and the speed of the shafts.

$$\Delta p = f\left(\dot{V}, \eta, D, L, screw\ geometry, n\right) \tag{4.24}$$

Then, as is known, the pressure loss results in:

$$\frac{\Delta p}{\dot{V}} \cdot \frac{D^4}{L \cdot \eta} = \frac{128}{\pi} \tag{4.25}$$

The pressure loss in an extruder with stationary shafts will be, of course, higher. There is also the speed of the shafts as a further influencing factor. The dependency is linear at constant viscosity, as already mentioned in Section 4.1.1.2.4. In dimensionless form, the result is shown in Figure 4.12 [3].

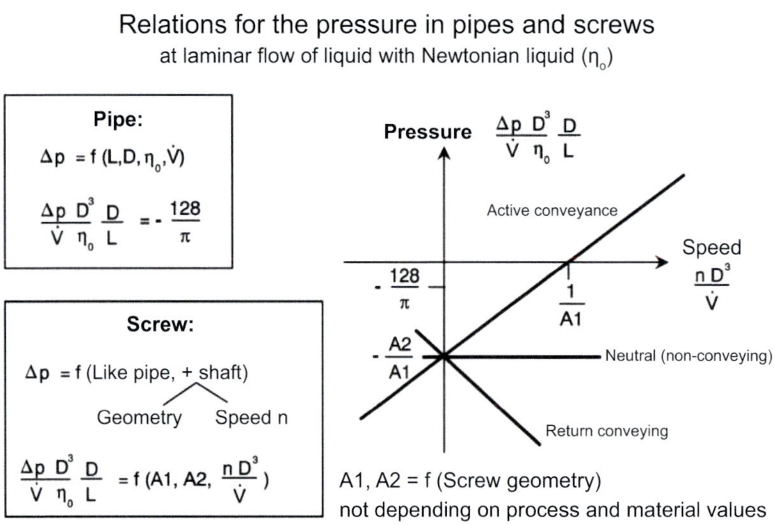

Figure 4.12 Comparison pressure loss in a pipe with the pressure characteristic of an extruder

In Figure 4.12 $(n \cdot D^3/\dot{V})$ is the dimensionless speed. Only for values > $1/A_1$ can a pressure be built up with a conveying active geometry. This fact is more clearly expressed below by the conveying key figure.

The representation as per Figure 4.12 makes sense for neutral (non-conveying) screw geometries (elements) because these are pure pressure consumers. We denote this pressure loss then with A_0 instead of A_2/A_1 because a pressure consumer does not supply self-conveying parameter A_1.

The usual display is, in dependency on the dimensionless throughput, the reciprocal speed variable, thus ($\dot{V}/(n \cdot D^3)$). Furthermore, the dimensionless pressure according to Equation (4.25) is depicted by multiplication with the dimensionless throughput. By combining two dimensionless key figures, a new key figure can always be created. This presentation then goes along with the previous versions. In axis intercept form this results then in the key figures according to Figure 4.13.

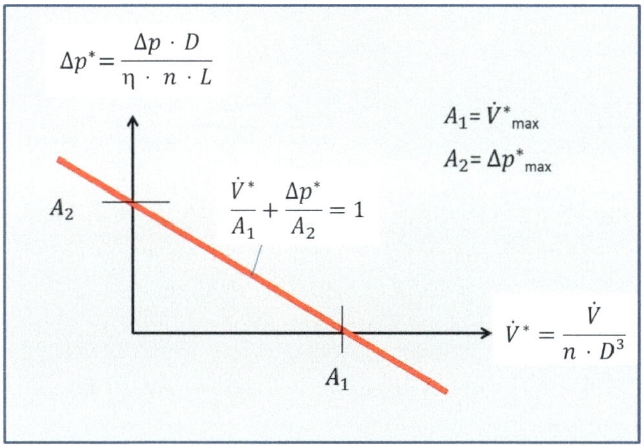

Figure 4.13 Usual "Bayer presentation" for the pressure throughput characteristic

The shown Bayer presentation was introduced by Pawlowski for single-screws and by Ullrich used for twin-screw extruders. Ullrich has done a lot of measurements for different screw elements. The linear relation has always been confirmed. Example measurements are listed in Section 1.2, among others. From the application of the measured values the axis intercept points will result. Böhme has shown theoretically the linear relation also for twin-screw extruders [4].

The representation with the throughput on the abscissa und the dimensionless pressure on the ordinate is not usual in the literature. Usually the pressure is displayed on the abscissa, e.g. [1, 5–8].

In mathematics, the independent variable is usually represented on the abscissa; this is the throughput of a twin-screw extruder, as this is usually externally dosed. On the ordinate, the dependent variables, such as pressure, capacity, mixing quality etc. are then applied. Regarding the pressure, the usual Bayer presentation as per Figure 4.13 results.

The axis intercept points indicate the "maximum values": A_1 maximum self-throughput, A_2 maximal possible pressure build-up, at throughput zero against a closed die. With dimensions this results for the maximum values in:

$$\dot{V}_{max} = A_1 \cdot n \cdot D^3 \tag{4.26}$$

$$\Delta p_{max} = A_2 \cdot \eta \cdot n \cdot L / D \tag{4.27}$$

A further standardization can be carried out. If the dimensionless parameters pressure and throughput are referenced to the axis intercept values, then this is resulting in the presentation according to Figure 4.14. This equation can be termed a basic pressure equation for extruders. The same presentation is obtained when the dimensioned values are related to the maximal values Equations (4.26) and (4.27):

The dimensionless pressure $\Delta\tilde{p}$ is described by:

$$\Delta\tilde{p} = \frac{\Delta p}{\Delta p_{max}} \tag{4.28}$$

$$\Lambda = \frac{\dot{V}}{\dot{V}_{max}} \tag{4.29}$$

The conveying key figure Λ is a very important one which is described in detail in Section 4.1.1.3.3. With these key figures, the basic equation for extruders has the dimensionless form:

$$\Delta\tilde{p} + \Lambda = 1 \quad \text{or} \quad \Delta p = (1 - \Lambda) \cdot \Delta p_{max} \tag{4.30}$$

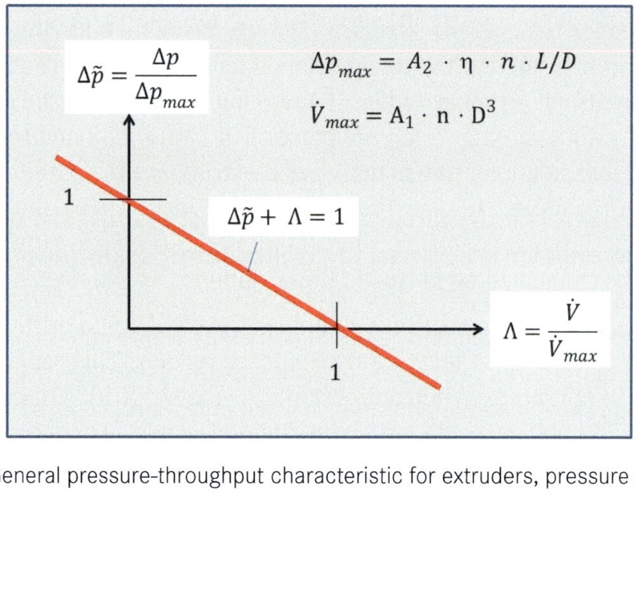

Figure 4.14 General pressure-throughput characteristic for extruders, pressure basic equation

The pressure basic equation for extruders has the simple form $x + y = 1$. The relation between drive and throughput provides in the following a drive basic equation for extruders of the same form.

For the calculation of Δp_{max} and $\dot{V}_{max}$, the numerical values for A_1 and A_2 are required. If these are not yet available then it makes sense to present and apply measurement or calculation values according to Figure 4.13. If the values are available, then the standardization to maximum values and the description according to Figure 4.14 are clearer. In feed active areas, all values are then between zero and one.

This applies, for example, to the key figures conveying key figure, filling level, and pump efficiency introduced in the following sections.

From the given relations, for example, the length of backing-up can be calculated. This is the length which is required for the build-up of a pressure difference Δp. Figure 4.15 clearly shows an example of the calculation of the backing-up length with the values shown in Figure 4.14. The maximum throughput and the maximum possible pressure for the operating conditions are indicated. By using the factors in the square brackets, the usual units of the example can be used (speed in 1/min, pressure in bar etc.), i.e. the conversion of the units can then be omitted.

Of course, the length of the backing-up can also be calculated explicitly without specifying the maximum values. It then results in Equation (4.12).

Given:

$\dot{M} = 170 \frac{kg}{h}$ $\quad D = 32\ mm$ $\quad \Delta p = 50\ bar$ $\quad n = 600 \frac{1}{min}$

$\rho = 1000\ kg/m^3$ $\quad \eta = 300\ Pa*s$ $\quad A_1 = 0.2$ $\quad A_2 = 3000$

$\dot{V} = \dot{M} / \rho = 0.17\ m^3/h$ $\qquad \dot{V}_{max} = A_1 \cdot n \cdot D^3\ [*60/10^9] = 0.236\ m^3/s$

$\Lambda = \dot{V} / \dot{V}_{max} = 0.72$ thus, 72% of the maximum throughput of the screw element

$\Delta\tilde{p} = 1 - \Lambda = 0.28$ 28% of the maximum throughput of the screw element

$\Delta p_{max} = \Delta p / \Delta\tilde{p} = 50\ bar/0.28 = 179\ bar$

With $\Delta p_{max} = A_2 \cdot \eta \cdot n \cdot L/D$ it follows that

Length of back-up

$L/D = \frac{\Delta p_{max}}{A_2 * \eta * n} * [60*10^5] \approx 2$

Figure 4.15 Numerical example for the calculation of the length of the back-up

The following two figures each show measurements [2] together with results of three-dimensional calculations [9] for A_1 and A_2. Further examples are listed in the following sections.

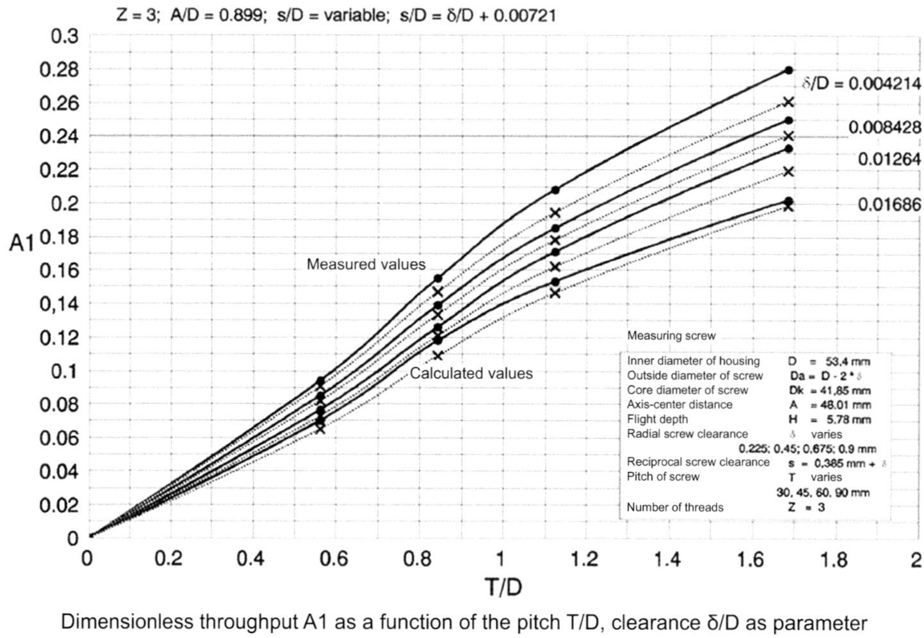

Figure 4.16 Maximum self-throughput A_1 of triple-flighted extruder conveying elements (scale-up parameters, conveying parameters)

Figure 4.16 gives more precise values than the rough calculation according to Figure 4.11. Among others it shows that there is no exact linear dependence on the pitch T.

As expected, the pressure build-up capacity decreases with increasing clearance of the screw (e.g., due to wear) and greater pitch. In the figures D is the diameter of the housing. In the above remarks, the housing diameter D serves as reference value, i.e. T/D, A/D, etc. In terms of dimensional theory, it is preferable to choose a reference value that changes the least. With a twin-screw extruder, this is the axis-center distance A. In the further development of triple-flighted extruders in the direction of double-flighted extruders with a larger product volume due to "boring", the axis-center distance remains constant, whereby the diameter increases. However, physically more useful and descriptive is a diameter as reference value, as many processes are volume proportional or dependent on the housing surface. The diameter which usually changes the least is the housing diameter. Changes are of course given by "bearing zones" and wear. Alternatively, the diameter of a screw element would be possible as a reference basis. However, screw elements are sometimes used with different clearances δ to the housing and are usually more subject to wear. The diameter as reference value compared to the axis-center distance has further the advantage that the results and key figures are easier to compare with single-screw extruders. Pawlowski has developed the key figures for single-screw extruders [10].

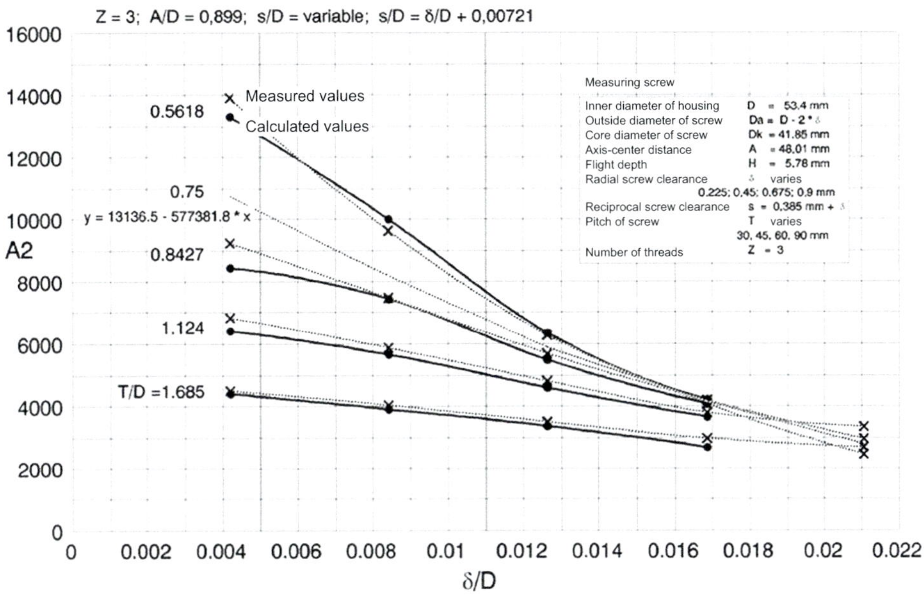

Dimensionless pressure A2 as a function of the clearance δ/D, pitch T/D as parameter

Figure 4.17 Maximum dimensionless pressure build-up of triple-flighted extruder conveying elements

In Section 1.5, the conveying parameters (profile parameters) A_1, A_2, B_1, and B_2 are given in diagrams for usual double-flighted conveying geometries.

At shear-thinning products, the viscosity decreases with the shear rate. The pressure gradient and the power input decrease in some cases considerably. Concerning the example of Section 4.7, Figure 4.75, the relation is shown in Figure 4.18.

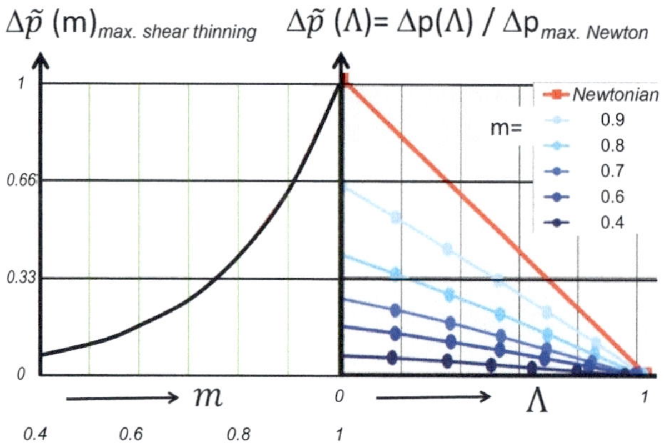

Figure 4.18
Example concerning the influence of the shear thinning on the pressure characteristic with the flow exponent *m*

In the three-dimensional calculations, the power law approach equation (3.5) in Section 3.1 was used with the flow exponent m. This approach is dimensional analytically "debatable", but is often used in practice. In the left half of Figure 4.18, it is shown on the ordinate how the maximum pressure gradient in point $\Lambda = 0$ behaves in dependence on the flow exponent m.

$$\Delta\tilde{p}_{max,\text{shear thinning}} = \frac{\Delta p_{max,\text{shear thinning}}}{\Delta p_{max,\text{Newtonian}}} \tag{4.31}$$

Thus, the pressure gradient falls below one-third for values $m < 0.75$. The Newtonian case is described as $m = 1$. The dimensionless pressure gradient according to Equation (4.31) can be described by the following formula:

$$\Delta\tilde{p}_{max,\text{shear thinning}} = 1 - a \cdot \left(e^{b} - e^{b*m}\right) \tag{4.32}$$

The constants stand for the example $\alpha = 0.01127$ and $b = 4.4816$. As the dependence of the pressure gradient on the conveying parameter Λ as per Figure 4.18 is nearly linear, the "curves" can be shifted to the straight line of the pressure basic equation. This is shown in Figure 4.19. The "shift factor" is $1 / \Delta\tilde{p}_{max.\ shear\ thinning}$. In other cases of shear thinning, the pressure characteristic is more non-linear. Then the influence of the conveying parameter must be considered in the shift factor.

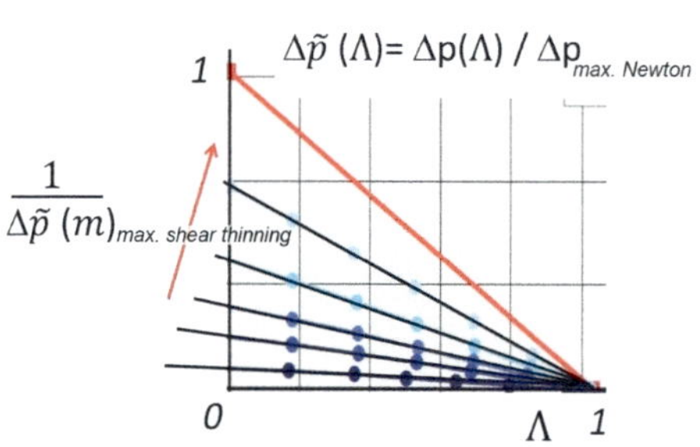

Figure 4.19
Leading back of the shear-thinning behavior to the pressure basic equation of the Newtonian case with the shift factor (left)

4.1.1.3.3 Conveying Key Figure

An important key figure is that already used on the abscissa in Figure 4.14, the conveying key figure Λ, which can be expressed in many cases by the previous definitions.

$$\Lambda = \frac{\dot{V}}{\dot{V}_{max}} = \frac{\dot{V}^*}{\dot{V}^*_{max}} = \frac{\dot{V}^*}{A_1} = \frac{1}{A_1} \cdot \frac{\dot{V}}{n \cdot D^3} \tag{4.33}$$

The conveying key figure adopts values between 0 and 1 for "normal" conveying elements. Particularly for an effective pressure build-up, there is an optimal value which is described in Section 4.3.2. For values greater than one, the screw element is "knocked over", it is a pressure consumer. For back-feeding elements, the conveying key figure is defined negatively. Figure 4.20 summarizes the operating areas of the conveying key figure.

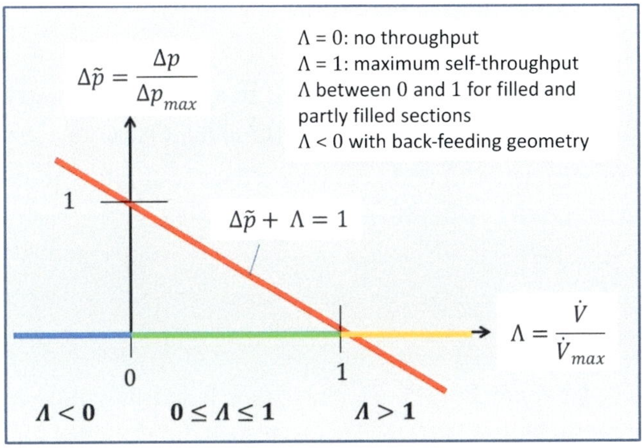

Figure 4.20 Operating areas of the conveying key figure

Figure 4.21 shows an example [2] of the operating areas of the conveying key figure in dependency on the pressure flow along the extruder. A more detailed description of the example is given in Section 4.5. In addition, in Section 4.3.3, Figure 4.40, the average temperature increase in the die area is considered.

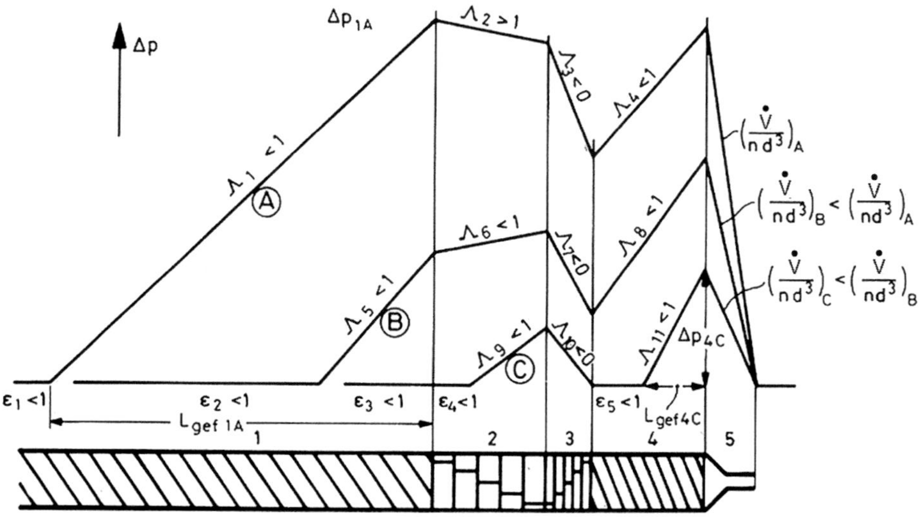

Figure 4.21 Example of the operation areas of the conveying key figure

Via A_1 the conveying key figure is connected to a specific geometry or a specific screw element, i.e. it can adopt different values along the screw.

In the literature [7] page 455 and [11] page 104, the throttle quotient φ is used instead of the conveying key figure.

$$\varphi = \frac{\dot{V}_{pressure}}{\dot{V}_{drag}} = 1 - \Lambda \tag{4.34}$$

In the literature, for example [11], a leakage flow is partially introduced additionally beside a drag und pressure flow. This is needed, for example, for channel models and indicates the backflow over the screw tips.

Then, this is:

Throughput = drag flow − pressure flow − leakage flow

This presentation has found its way into textbooks for the apprenticeship professions of the plastic industry [12], page 445. In the above equations with the conveying parameters A_1 and A_2 a drag flow is not needed. In Equation (4.30), all flows are integrally contained in the throughput. The conveying parameters A_1 and A_2 are not explicitly dependent on a flow form. The conveying parameters A_1 and A_2 depend only on the screw geometry, not on the throughput and material data. Therefore, they are also called profile parameters. The screw geometry (for example the clearance δ between tip and wall of the housing) influences the conveying behavior and this determines the values for A_1 and A_2. The conveying parameter of a certain screw geometry, as well as the value $\frac{128}{\pi}$ for dimensionless pressure loss

(Equation (4.25)), have universally fixed values for a flow through a pipe. Deviations from the logical numerical value $\frac{128}{\pi}$ are due to inaccurate measurements or calculations. Figure 4.12 illustrates the comparison between a pipe and an extruder.

4.1.1.3.4 The Most Important Extruder Key Figure

As described, the conveying key figure is connected to a screw geometry via the conveying parameter A_1 and can therefore change locally. More universal is therefore a dimensionless key figure without the parameter A_1, which is called the throughput key figure:

$$\dot{V}^* = Q = \frac{\dot{V}}{n \cdot D^3} \tag{4.35}$$

It describes the dimensionless throughput of an extruder regardless of its size. The following listing describes the advantages of the throughput key figure. The throughput key figure

- is the same at every point of the extruder (at constant volume flow)
- is independent of the extruder size
- should be the same at scale-up/down
- can indicate the economy of the application of the extruder (it describes the dimensionless throughput)

In a scale-up even more key figures are – process-dependent – to be considered, e.g. when degassing or when compounding.

4.1.1.3.5 Partial Filling, Filling Level, and Medium Product Speed

So far, only one operating condition without axial pressure gradient was considered, the case of the maximum self-throughput with the volume flow as per Equation (4.26). A further operating condition without pressure gradient exists in partly filled screw sections, for example in "degassing zones". If a screw element is charged with less throughput than the maximum possible self-throughput, then a partial filling may occur. The partial filling can be described by a filling degree. With regard to the throughput, the following throughput filling degree ε can be defined.

$$\varepsilon = \frac{\dot{V}}{\dot{V}_{\max}} \qquad \dot{V}_{max} = A_1 \cdot n \cdot D^3 \tag{4.36}$$

The throughput filling degree can also be referred to as conveyor technical filling degree. It is thus defined like the conveying key figure Λ equation (4.33). The throughput filling degree is applied to a partial filling and is between zero and one. This is the green area on the abscissa in Figure 4.20. The throughput filling degree can easily be calculated with the knowledge of the conveying parameter A_1. As different screw geometries, for example different screw elements, have different values of the conveying parameter A_1, the filling degree in the partly filled sections changes accordingly, even at the same throughput and same speed. The filling degree is a local size variable along the screw.

A further definition of the filling degree refers to the screw cross-section with product feed, Figure 4.22.

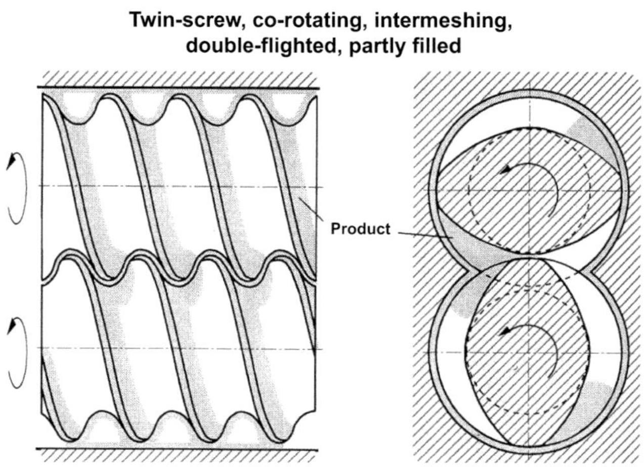

Figure 4.22 Cross-section area partly filled with product

If the maximum cross-section area to be filled with the product is indicated by A_{free} (eight-form cross-section surface of the housing minus cross-section areas of the screws) then the surface related filling degree can be defined.

$$\varepsilon_Q = \frac{A_{product}}{A_{free}} \tag{4.37}$$

This filling degree can be interpreted as cross-section filling degree ε_Q. At constant areas within a short length L it can also be identified as volumetric filling degree. The calculation of the free cross-section area is stated in Section 2.1 (basic geometry). The calculation of the cross-section area partly filled with product is more difficult because it depends not only on the geometry of the screw but also on throughput and material values. In a Newtonian liquid, "beads" form according to Figure 4.22.

To establish a relation between the throughput $\dot{V}$, a process factor, and the product cross-section $A_{product}$, the medium product speed $\overline{u}$ in a screw cross-section is considered according to Equation (4.8):

$$\overline{u} = \frac{\dot{V}}{A_{product}} \tag{4.38}$$

According to Equation (4.36) it follows:

$$\dot{V} = \varepsilon \cdot \dot{V}_{max} \tag{4.39}$$

and according to Equation (4.37) it follows:

$$A_{product} = \varepsilon_Q \cdot A_{free} \tag{4.40}$$

and thus the medium product speed is:

$$\overline{u} = \frac{\varepsilon}{\varepsilon_Q} \cdot \frac{\dot{V}_{max}}{A_{free}} \tag{4.41}$$

For $\varepsilon \approx \varepsilon_Q$, the medium product speed at partial filling depends only on the type of the screw element (A_1, A_{free}) and the speed (via the maximum throughput), not on the throughput. A change of the throughput (auxiliary dosing!) changes the throughput filling degree according to definition (4.36) in equal measure at constant speed and thus the medium speed remains the same.

Like the throughput key figure, the filling degree can adopt different values for different procedural tasks.

In the three-dimensional calculation of partly filled screws, the so-called free surfaces play an important role. In the near future it can be expected that the cross-sectional filling degree or volumetric filling degree can be calculated in dependence on the throughput filling degree. In particular, at non-self-cleaning geometries differences may occur. For example, in the case of a non-self-cleaning screw element, local backflow and, in particular, product "dead zones" can occur, resulting in different values for the defined filling degrees. This then has an effect on the dwell time behavior, Section 4.1.1.3.6.

When is a screw section partly filled, when filled? A partial filling can only occur when the conveying key figure Λ adopts values between zero and one. This condition is necessary but not sufficient. More information about "downstream" pressure is required. The pressure build-up Δp provokes a filled area with a back pressure length which can be calculated with Equation (4.30):

$$L_{filled} = \frac{\Delta p \cdot D}{A_2 \cdot \eta \cdot n} \cdot \frac{1}{1 - \Lambda} \tag{4.42}$$

Then, there is "upstream" a partial filling. An example is shown in Figure 4.15. In Section 4.6, further calculation examples regarding the back pressure length are given.

Theoretically, the question may be raised whether a pressure conveying without housing will be possible. Single-shaft extruders only convey with product contact to the wall. Counter-rotating twin-shaft extruders are conveying according to the displacement principle, similar to gear pumps. For both types of extruders it is not possible to convey the product without housing. In a co-rotating twin-shaft extruder, the product is transferred from one shaft to the other in the intermeshing zone, at least at a low filling level. A product conveying without housing is thus possible with low throughputs. We have called such an extruder "space screw" and tested it for degassing tasks with triple-thread geometry.

4.1.1.3.6 Medium Product Dwell Time in Filled and Partly Filled Screw Sections

From the view of the product, the dwell time of a product in a screw machine or in a screw section is more important than the filling level. Because of an axial velocity distribution in the screw cross-section, a dwell time distribution occurs. The medium product speed $\bar{u}$ can change along the screw, as the geometry and the filling degree are changing for example. If l defines a longitudinal coordinate, then the medium product speed at a position is:

$$\bar{u} = \frac{dl}{dt} \tag{4.43}$$

This results in the medium product dwell time for a particular screw section of length L:

$$\bar{t}(L) = \int_0^L \frac{1}{\bar{u}} dl \tag{4.44}$$

With:

$$\bar{u}(l) = \frac{\dot{V}}{A_{product}(l)} \tag{4.45}$$

Results in:

$$\bar{t}(L) = \frac{V_{product}}{\dot{V}} \tag{4.46}$$

$$V_{product}(L) = \int_0^L A_{product}(l)\, dl \tag{4.47}$$

Equation (4.46) is the well-known equation from the process technique of the "reactors".

In contrast to the local filling degree, the medium dwell time refers to a product volume and thus to a specific length L of a screw section or to the total extruder length.

In the following, a short screw section on the length L with a non-variable cross-section area is considered (e.g. a screw element length), then $V_{product} = A_{product} \cdot L$ and there are clear correlations. The average filling level may then be referred to as the volumetric filling level.

Filled Screw Sections

With the linguistic usage of filled screw sections it has come to mean screw sections totally filled with product. The product fills the entire free cross-section and is according to Equation (4.46):

$$\bar{t}_{filled} = \frac{V_{product}}{\dot{V}} = \frac{V_{free}}{\dot{V}} = f(throughput) \neq f(speed) \tag{4.48}$$

Thus, the medium product dwell time in filled screw sections depends only on the throughput, not on the screw speed.

Partly Filled Screw Sections

At constant areas of the length L is:

$$\varepsilon_Q = \frac{A_{product}}{A_{free}} \cdot \frac{L}{L} = \frac{V_{product}}{V_{free}} \tag{4.49}$$

And thus results from Equation (4.46):

$$\bar{t}_{partial\ filling} = \frac{\varepsilon_Q \cdot V_{free}}{\dot{V}} \tag{4.50}$$

If the throughput in this equation is expressed by means of the throughput filling degree equation (4.36), then the medium product dwell time is summarized as follows:

$$\bar{t}_{partial\ filling} = \frac{\Delta L \cdot A_{product}}{\dot{V}} = \frac{\varepsilon_Q \cdot V_{free}}{\dot{V}} = \frac{\varepsilon_Q}{\varepsilon} \cdot \frac{V_{free}}{A_1 \cdot n \cdot D^3} \tag{4.51}$$

For the case $\varepsilon \approx \varepsilon_Q$,

$$\bar{t}_{partial\ filling} \approx \frac{V_{free}}{A_1 \cdot D^3} \cdot \frac{1}{n} = f(speed) \neq f(throughput) \tag{4.52}$$

With increasing speed and at a larger conveying capacity of a screw element, the medium product dwell time decreases accordingly. The conveying capacity increases, for example, with the screw pitch *T*. With the approximation from Figure 4.11 results:

$$\overline{t}_{partial\ filling} \approx \frac{2}{n} \cdot \frac{\mathrm{L}}{\mathrm{T}} \tag{4.53}$$

In summary, some results of the information concerning the filling degree and the medium product dwell time in partly filled sections:

- At constant speed, the medium product dwell time is independent of the throughput. At a change in throughput, the filling degree changes to the same extent $\varepsilon_Q \sim \dot{V}$.
- The medium product dwell time changes proportionally with 1/speed. At constant throughput the filling degree changes as well with 1/speed. Then $\varepsilon \sim \overline{t}$ is valid.
- The medium product dwell time changes approximately proportionally with 1/screw pitch.
- The medium product dwell time refers to a certain screw section of the length *L*. The medium dwell time in an entire extruder is to add up according to the filled and partly filled sections.

4.1.1.3.7 Medium Dwell Time and Dwell Time Distribution

In practice, in addition to the medium dwell time, the dwell time distribution is important, for example when products are changed (keyword color change) and for considerations of the thermal damage of product. The topic product damage is dealt with in Section 3.4. In this book the dwell time distribution is discussed elsewhere, for example in Section 1.3 "Compounding.

Here, it should be mentioned again (see Section 4.1.1.2.4) that the above models and equations do not assume a plug flow. Essentially, one-dimensional models are concerned, but the effects above the screw cross-section are integrally contained. Thus, the equations for pressure and performance relating to the conveying parameters (A_1 etc.) take into account a dwell time distribution with backflow. Information regarding local distribution cannot be concluded from integral values. For example, the specification of a mixing quality refers to a finite volume. This does not imply the mix distribution in the volume. If there should be a plug flow in the extruder then there would be no longitudinal mixings. Longitudinal mixing in particular is an important feature of the co-rotating twin-screw.

Here, it should be briefly referred to the relation between a measured distribution density function *E*(*t*) and the medium dwell time.

$$\bar{t} = \int_0^\infty t \cdot E(t)\,dt \tag{4.54}$$

In practice, a "sum function" $F(t)$ is usually measured at the extruder outlet and $E(t)$ results from $E(t) = dF(t)/dt$. Of course, the measurement time is finite and the integral equation (4.54) is often approximately evaluated by a sum. Figure 4.23 shows some measurement results which have been executed in relation with tests of mixing and dwell time in twin-screws [13].

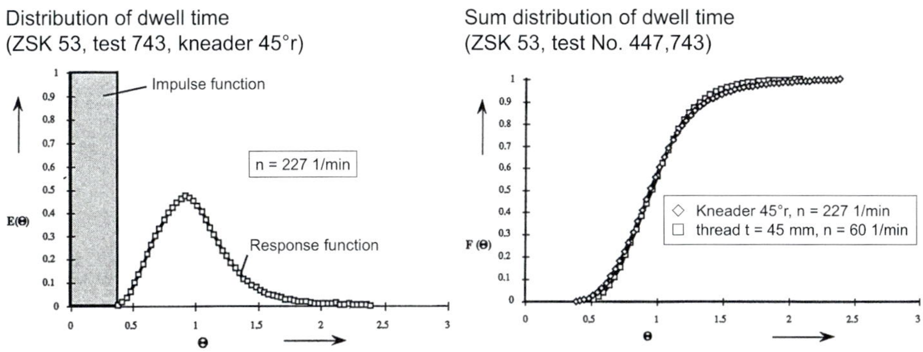

Figure 4.23 Measured distribution function and sum function for a trimming of kneading discs [13]

4.1.1.3.8 Economic Feasibility

The economic feasibility of an extruder depends largely on the medium dwell times and the product revenues – the high quality of a product. The product volume in an extruder in relation to the investment costs is relatively low compared with other process devices, for example reactors in the chemical industry. There are alternatives available for reactors. For long periods of dwell times, for example, special high-viscosity reactors from the companies Buss-SMS Ganzler GmbH or LIST Technology AG can be used. These are currently the only manufacturers on Konti-reactors for high-viscosity products with extensive self-cleaning and simultaneously with longer economic dwell times compared to extruders. Figure 4.24 shows a two-shaft machine in open state of the first mentioned company.

Figure 4.24 Large-volume reactor "Reasil" for a polymer reaction

Regarding short dwell times, there has not yet been any economic alternative to extruders for the melting of plastic granulates.

The product volume in an extruder can be calculated with the free cross-section areas and the local filling degrees with the corresponding lengths. If the medium dwell time is measured, for example by evaluating the dwell time distribution of the extruder, then the product volume in the extruder is calculated according to the equation:

$$V_{product\ in\ extruder} = \bar{t} \cdot \dot{V} \tag{4.55}$$

For a certain extruder with a certain process, the product volume is fixed, V_{fix}. The simple Equation (4.55) describes: If a process requires a certain dwell time, then the maximum throughput is:

$$\dot{V}_{max} = \frac{V_{fix}}{\bar{t}_{min}} \tag{4.56}$$

Example:

The medium dwell time in the extruder amounts to 15 seconds for a compounding process. If a minimum of 2 minutes are required for reactive extrusion to achieve the desired chemical conversion, then only 1/8 of the throughput can be achieved in the same extruder. A larger extruder would shift the cost function (cost = function of the throughput) to higher values, shown in Figure 4.25 on the right. The comparison presumes that the partly filled sections of both processes contribute approximately equal values to V_{fix}. Even if there is a deviation from this, the qualitative comparison will remain.

The medium dwell times in extruders are for many products relatively low with regard to an economical operation. For compounding processes, the dwell time "from granulate to granulate" is usually less than half a minute. In processes with reactions, the dwell time and dwell time distribution can become a challenge. For very high-quality products, e.g. from the pharmaceutical industry, longer dwell times can also be economical. To judge the economic feasibility, the revenues of the products manufactured by means of an extruder must be compared with the cost function. Figure 4.25 shows on the left side a typical course of a cost function, which results from initial investment costs and operation costs. Many applications, for example from the compounding area, are approximated depicted by the yellow revenue line E1. The green line E2 could describe a very economical pharmaceutical product. No economic feasibility can be achieved with the product of the red line E3. On the right side of the figure, a revenue line is drawn which can be applied to two products. For product 1 there is an economic feasibility from a certain throughput onwards. For example, the product 2 requires at the same throughput a significantly larger extruder, and therefore an economic feasibility will not be achieved. Here, as an example, it can be a reactive extrusion, which allows only very low throughputs in relation to the extruder size.

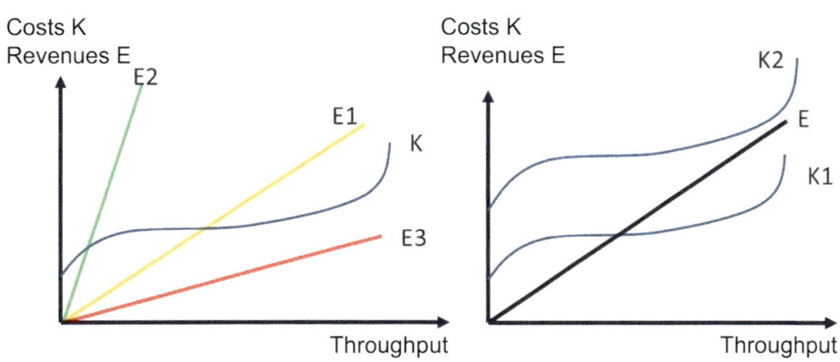

Figure 4.25 Schematic diagram concerning the economic feasibility with cost and revenue function

4.2 Introduction of the Performance Behavior of Highly Viscous Liquids in Extruders

Klemens Kohlgrüber

The basics of the conveying behavior of highly viscous liquids in extruders were described in detail in Section 4.1. For the derivation of the basics of the performance behavior, the same procedure as in Section 4.1 will be useful: first, the presentation of the solution of the simple plate-plate-model, then the definition of a dimensionless performance from the contemplations at the annular gap. In this section, therefore, the performance behavior is presented compactly.

4.2.1 Throughput Performance Behavior of the Plane Flow between Two Plates

In Section 4.1, the drag flow and the pressure flow have been considered at first separately. These two flows were then superimposed. This approach increases the understanding of the context. Of course, a derivation of the superimposed flow can be done without this explicit breakdown. The derivation with the dependency of the required performance for this flow is summarized in Figure 4.26.

$$\tau = \eta \cdot \dot{\gamma} \qquad \dot{\gamma} = \frac{du}{dy} \qquad \frac{d\tau}{dy} = \frac{dp}{dx} = P' = \frac{p_2 - p_1}{L}$$

$$u(\tau) = \frac{1}{2\eta p'}\left(\tau^2 - \tau_0^2\right) \qquad \tau(y) = p' \cdot y + \tau_0$$

$$u(\tau) = \frac{1}{2\eta p'}\left((p'y)^2 + 2p'\tau_0 y\right) \qquad \tau_0 = \frac{\eta \cdot u_H}{H} - \frac{P' \cdot H}{2}$$

$$\tau_H = \frac{\eta \cdot u_H}{H} + \frac{P' \cdot H}{2}$$

$$\dot{V} = \frac{B \cdot H}{2} \cdot \left(u_H - \frac{p' \cdot H^2}{6\eta}\right)$$

$$\frac{\Delta p^*}{A_2} + \frac{Q}{A_1} = 1, \quad \Delta p^* = \frac{p'H^2}{u_H \eta}, \quad Q = \frac{\dot{V}}{B \cdot H \cdot u_H} \qquad A_1 = \frac{1}{2}, \quad A_2 = 6$$

power: $P = F_H \cdot u_H = \tau_H \cdot B \cdot L \cdot u_H$

$$\frac{P^*}{B_2} + \frac{Q}{B_1} = 1, \quad P^* = \frac{P}{\eta u_H^2}\,\frac{H}{B \cdot L} \qquad B_1 = \frac{2}{3}, \quad B_2 = 4$$

Figure 4.26 Pressure and performance characteristic for a plane flow between two plates

The depicted linear dependence applies to the Newtonian liquids. The equation for the performance is given in form of axis intercept, which was already introduced at the pressure. The axis intercept points B_1 and B_2 describe the dimensionless throughput and a dimensionless maximum performance for the limit case with throughput zero. A detailed explanation is given in Figure 4.29.

$$\frac{\dot{V}^*}{B_1} + \frac{P^*}{B_2} = 1; \quad P^* = \frac{P}{\eta \cdot u_H^2} \cdot \frac{H}{B \cdot L}; \quad B_1 = \frac{2}{3}; \quad B_2 = 4 \tag{4.57}$$

4.2.2 Performance Key Figure for an Annular Gap

Considered is an annular gap of the gap width H, at which the outside cylinder is rotating. This is fluidic equivalent to the case where the inner cylinder is rotating. This presentation is chosen because D denotes the diameter of the housing. Then follows for the performance (drive power) according to Figure 4.27:

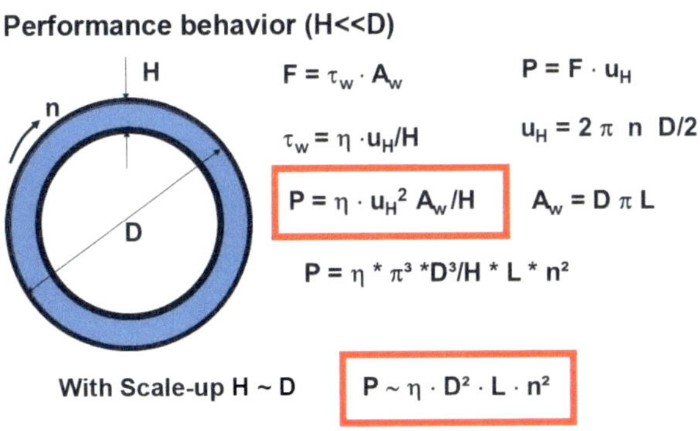

Figure 4.27
Drive power for an annular gap

Thus, the drive power is proportional to:

1. the viscosity
2. the speed squared
3. the volume $D^2 \cdot L$
4. $1/$*gap width* H, respectively

The last dependency states that small gaps cause a high energy input. Although the absolute amount is somewhat compensated by the low product volume, but a high energy input can result in thermal product damage. Therefore, the screw clearances should not be chosen too low. The following Figure 4.28 shows measured values [2] for the dimensionless performance for a "nearly empty screw" with throughput zero for the study of the power input. The figure shows that the performance increases significantly with decreasing gap width. According to the dependence on the gap width as per point 4, an asymptotic $1/\delta$ course is expected.

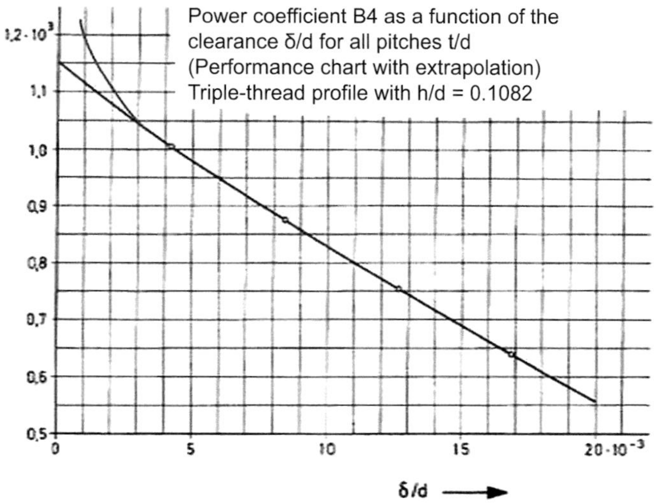

Figure 4.28 Measured power input of a triple-thread screw in dependence on the clearance to the wall of the housing

As the annular gap does not transport the liquid axially in the gap, the proportionality factor can be interpreted with the maximum value of the dimensionless performance in extruders. This is denoted in Figure 4.26 with B_2.

$$\frac{P}{\eta \cdot n^2 \cdot D^2 \cdot L} = B_2 \tag{4.58}$$

If the performance produced in a product by shear force is related to the product volume and the speed is replaced by the local shear rate, the following specific performance results:

$$\dot{q} = \eta \cdot \dot{\gamma}^2 \tag{4.59}$$

This performance can also be referred to as punctual or local performance if in a three-dimensional flow field the local shear rate is considered. If no further energy flows occur, then the drive power equation (4.59), supplied via the shaft(s) in partly filled sections, will be transferred into dissipation. In fully filled sections, the drive power is converted into dissipation and a pressure performance with pressure gradient when other energy flows are missing. In the case of a pressure build-up the latter is an actual power output. In Section 4.3.1 the dissipation performance is discussed again.

In Section 4.7, Equation (4.131), the local performance distribution equation (4.59) is shown in a screw cross-section (Figure 4.77).

4.2.3 Basic Equation of the Performance Characteristic of Extruders

The above relation can be generalized directly to any screw geometries. For filled screw sections then the representation according to Figure 4.29 applies.

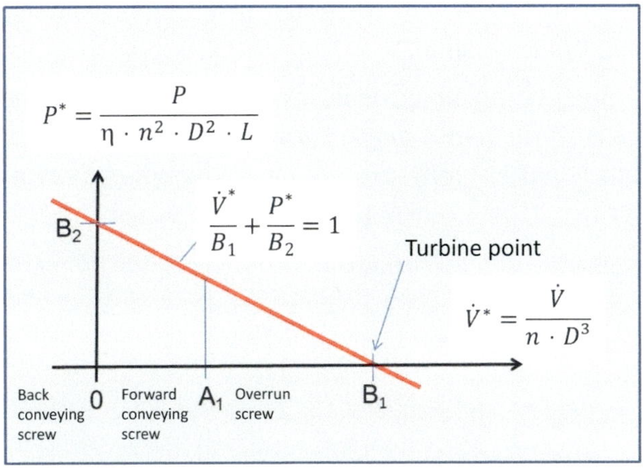

Figure 4.29 Dimensionless representation of the performance behavior

The representation according to Figure 4.29 is particularly suitable if the performance parameters B_1 and B_2 are not known. By measurements or three-dimensional calculations, points of the straight line can be determined and applied which then results in the performance parameters at the intersection points of axes. In the past, many measurements were carried out by Ullrich [2] and the following figure shows an example of the pressure and performance characteristic.

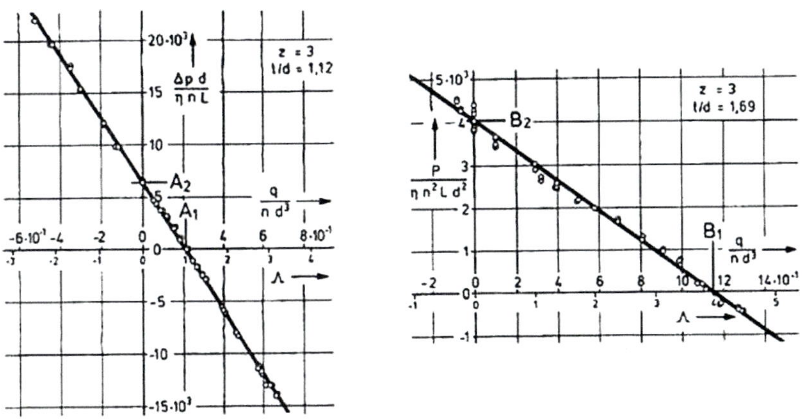

Figure 4.30 Measured pressure and performance characteristic

The figure shows that measurements were made for a large throughput or speed range. At dimensionless throughputs greater than B_1, the screw even delivers power. Therefore, the point B_1 is also called turbine point. This application is usually not found at co-rotating screws, but occasionally in the case of counter-rotating screws. However, for the measurement of the model screw already described, the range makes sense in order to more accurately determine the intersection points of axes of the straight line equation.

The numerical values of the dimensionless performance P^*, for example in Figure 4.30, are less clear. If the performance parameters are known, then a dimensionless presentation is purposefully carried out of the power P with respect to the maximum power P_{max}, which is denoted by $\tilde{P}$:

$$P_{max} = B_2 \cdot \eta \cdot n^2 \cdot D^2 \cdot L \tag{4.60}$$

$$\tilde{P} = \frac{P}{P_{max}} \tag{4.61}$$

The conveying key value Λ was introduced with Equation (4.33). The parameter Λ_B which is defined by:

$$\Lambda_B = \frac{B_1}{A_1} \tag{4.62}$$

results in a performance basic equation for extruders:

$$\tilde{P} + \frac{\Lambda}{\Lambda_B} = 1 \quad \text{resp.} \quad P = \left(1 - \frac{\Lambda}{\Lambda_B}\right) \cdot P_{max} \tag{4.63}$$

The following Figure 4.31 shows the relations graphically.

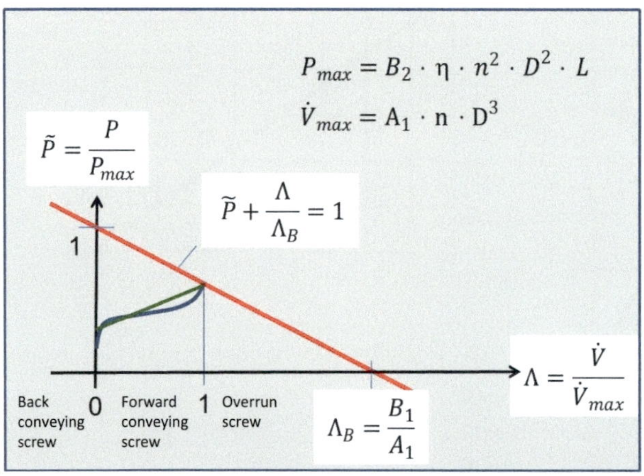

Figure 4.31
General performance characteristic for extruders; red line: performance basic equation

The performance relation between throughput and drive power for filled screw sections in Figure 4.31 has the form $X + Y = 1$, like the basic pressure equation for extruders $x + y = 1$.

In Figure 4.31, two processes for partly filled screw sections are incorporated, one straight line and one "S curve". A linear connection as with filled screws could not be shown theoretically. As newer calculations show (Section 4.7.3, Figure 4.93) there is a non-linear connection. The previous partial measurements [2] are more located on the shown straight line. When simplified straight lines are taken as a basis at a partial filling, then the relations will result according to the following Figure 4.32.

Numerical examples in Section 4.6 are simplified in terms of linear progression. Figure 4.31 and Figure 4.32 show the relations for the ideal Newtonian liquids. For shear-thinning liquids arise intersection points with the ordinate being considerably below the value one in Figure 4.31. Examples are given in the following sections:

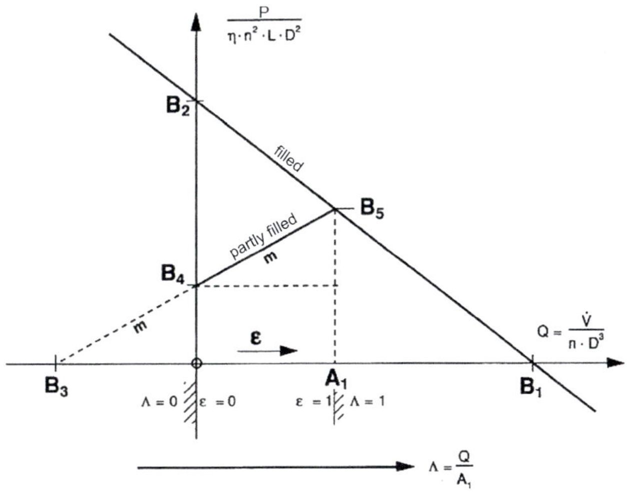

Definitions:
(with clear significance)

$$A_1 = \frac{\dot{V}_{\Delta p=0}}{n \cdot D^3} \qquad B_1 = \frac{\dot{V}_{P=0}}{n \cdot D^3}$$

$$B_2 = \frac{P_{\Lambda=0}}{\eta \cdot n^2 \cdot L \cdot D^2}$$

$$B_4 = \frac{P_{\varepsilon=0}}{\eta \cdot n^2 \cdot L \cdot D^2} \qquad B_5 = \frac{P_{\varepsilon=1}}{\eta \cdot n^2 \cdot L \cdot D^2} = \frac{P_{\Lambda=1}}{\eta \cdot n^2 \cdot L \cdot D^2}$$

$$m = \frac{B_5 - B_4}{A_1} = \frac{B_4}{-B_3}$$

Definition for B_3:
(without clear significance)

$$B_3 = -\frac{B_4 \, A_1}{B_5 - B_4} \quad \text{(always negative)}$$

Figure 4.32 Definition at simplified representation of the performance at partial filling as straight line [2]

4.3 Dissipation, Pump Efficiency Degree, Temperature Increase, and Heat Transfer

Klemens Kohlgrüber

4.3.1 Dissipation

In thermodynamics the increase of internal energy by friction is called dissipation. Related to the extruder: If there are no further process factors (for example heat flows over the housing) and enthalpy flows (for example due to reaction energy) then the drive power is only converted into "dissipation" (temperature increase) and pressure performance. This results from the first main theorem of thermodynamics for open systems.

For screw sections with pressure gradient the drive power is P and the pressure performance $\Delta p \cdot \dot{V}$, and usually the "dissipation performance" is defined as P_{diss} [10]:

$$P_{diss} = P - \Delta p \cdot \dot{V} \tag{4.64}$$

In partly filled areas, the drive power is only converted into dissipation, i.e. into the increase of the internal energy and thus into the increase of the product temperature. The power P is calculated according to Equation (4.63) and the pressure difference according to Equation (4.30). The corresponding local performance has already been described by Equation (4.59). With further findings not shown here [3, 14] further relations between the equations can be made and simplifications can be carried out for the dissipation performance.

4.3.2 Pump Efficiency Degree

Co-rotating screws have, from thread number two, a largely open intermeshing zone. This creates "open channels" and the pressure build-up occurs due to drag flow. The effectiveness of the pressure build-up will therefore be worse than with systems with "chambers" and displacement flow. The effectiveness can be expressed by the pump efficiency degree.

$$\eta_p = \frac{pressure\ performance}{drive\ power} = \frac{\Delta p \cdot \dot{V}}{P} \tag{4.65}$$

Regarding this equation it should be noted that the power P refers to the screw length, which corresponds to the pressure build-up Δp. This length is calculated from Equation (4.42). This results for the pump efficiency degree with the previously introduced relations in:

$$\eta_p = A_1 \cdot \frac{A_2}{B_2} \cdot g(\Lambda) = \dot{V}_{max} \cdot \frac{\Delta p_{max}}{P_{max}} \cdot g(\Lambda) \tag{4.66}$$

Thereby $g(\Lambda)$ is the function:

$$g(\Lambda) = \Lambda \cdot \frac{1-\Lambda}{1-\dfrac{\Lambda}{\Lambda_B}} \tag{4.67}$$

For $\Lambda = 0$ and $\Lambda = 1$ the pump efficiency degree is zero. The course of the pump efficiency degree results from the function $g(\Lambda)$ and the maximum between $\Lambda = 0$ and $\Lambda = 1$ depends only on the value Λ_B equation (4.62).

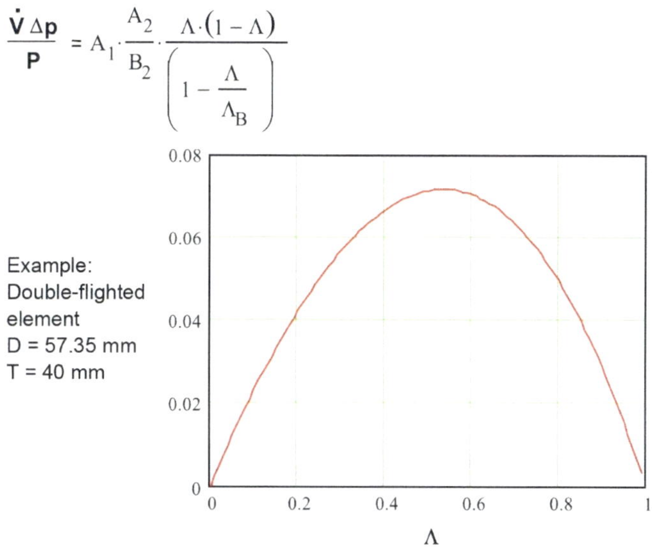

Figure 4.33 Example for the course of the pump efficiency degree

Figure 4.33 shows an example of the course of a pump efficiency degree. The maximum is slightly above $\Lambda = 1/2$. The result is important for the optimum operating point of an extruder, so that the temperature increase of the product remains minimal. The maximum value is relatively low at around 8%, i.e. around 90% of the drive power within the pressure build-up length is dissipated.

With shear-thinning products higher pump efficiency degrees arise. The dimensionless maximum throughput A_1 in Equation (4.66) remains approximately constant, the quotient A_2/B_2 becomes larger as measurements and three-dimensional calculations show, example in Section 4.7. This means that the maximum performance at shear-thinning viscosity decreases more than the maximum pressure. In summary for shear-thinning materials in comparison with Newtonian:

- The maximum throughput $\dot{V}_{max}$ changes rarely.
- The maximum pressure build-up ability Δp_{max} decreases.
- The maximum drive power P_{max} decreases more in relation to the pressure.
- The pump efficiency degree η_p increases.

It is interesting to observe which values the pump efficiency degree adopts for the simple plane flow between two plates. Figure 4.34 shows the results for the plate-plate model. The optimal pump efficiency degree amounts here to 33%.

$$\eta_{p,\,opt} = \frac{1}{3} \; at \; \Lambda_{opt} = \frac{2}{3} \tag{4.68}$$

It should be noted that the plate-plate model as processing of an annular flow formally describes the pressure in the circumferential direction; the screw, however, conveys axially and can have axially only lower values at a predominant drag flow (at point A_1 at classical double and multi-flighted geometries).

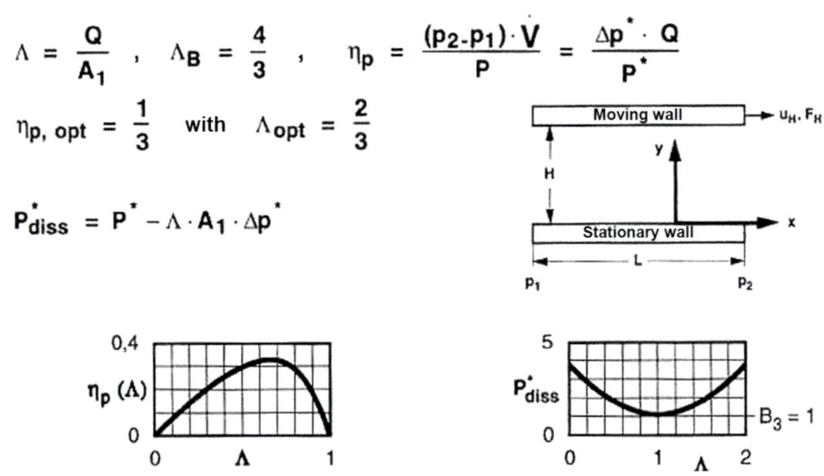

Figure 4.34 Pump efficiency degree and dissipation performance at a plane flow

For different extruders, maximum pump efficiencies and the operating point of the maximum are shown in Figure 4.35 [15]. Classic single-flighted screw elements have indeed a better pump efficiency degree and thus a comparatively lower medium product temperature increase, but due to the wider tips, the maximum temperature can be significantly higher. Simple calculation equations for increasing the temperature in the tip area are given in Section 4.7.

In the chapter on the history of co-rotating twin-screw extruders, a further Figure 1.40 for the maximum pump efficiency degree is given, there applied as reciprocal value. In numerous measurements, among others the screw pitch was varied and the maximal possible pump efficiency degree compared with counter-rotating screws.

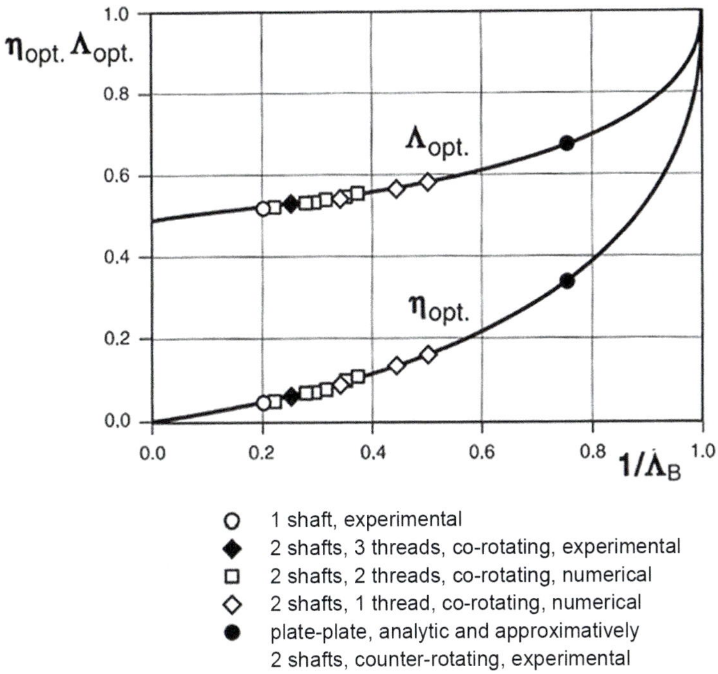

Figure 4.35 Example for the optimal pump efficiency degree with corresponding operation point at screw machines [15]

4.3.3 Temperature Increase

The temperature, like pressure, is a scalar, unsteady field, i.e. depending on place and time. In the case of pressure distribution, the splitting off of a pressure gradient in the cross-section pressure distribution and axial pressure flow succeeds in a stationary case through the exploitation of symmetries:

$$p(\textit{cross-section}, 1) = p_Q(\textit{cross-section}) + \int p'(l)\,dl \tag{4.69}$$

Here, p' is the axial pressure gradient which was described in detail in Section 4.2. As the equations apply for a constant viscosity, p' can be expressed by $\Delta p/L$. Due to this splitting off, the considerations of the pressure are simplified. Regarding the product temperature, a simplification is achieved by considering the energetic medium temperature of a screw cross-section. This is considered in this chapter. Because of the axial flow, the energetic medium temperature must be formed with the velocity distribution:

$$\int\limits_{\substack{\text{product}\\\text{cross-section}}} T \cdot u \, dA = \overline{T} \cdot \int\limits_{\substack{\text{product}\\\text{cross-section}}} u \, dA = \overline{T} \cdot \dot{V} \tag{4.70}$$

With:

$$\int\limits_{\substack{\text{product}\\\text{cross-section}}} u \, dA = \overline{u} \cdot A_{\text{product}} = \dot{V} \tag{4.71}$$

The average energy temperature according to Equation (4.67) corresponds to the temperature that results if the product would be collected in a container and homogeneously mixed.

An energy balance on a screw section is provided by the main theorem of thermodynamics for open systems:

$$P + \dot{H}_{in} = \Delta\dot{U} + \dot{V} \cdot \Delta p + \dot{H}_{out} + \dot{Q}_{housing\ wall} \tag{4.72}$$

With the change of the internal energy $\Delta\dot{U}$ and the specific heat capacity c at a constant volume:

$$\Delta\dot{U} = \dot{V} \cdot \rho \cdot c \cdot \left(\overline{T}_{out} - \overline{T}_{in}\right) \tag{4.73}$$

The two temperatures are the medium inlet and medium outlet temperatures of the considered screw section with the length L. The heat flow over the housing is

$$\dot{Q}_{housing\ wall} = U_G \cdot \int_0^L \alpha(l) \cdot \left[\overline{T}(l) - T_{housing\ wall}(l)\right] dl \tag{4.74}$$

The internal energy is a thermodynamic state variable; a change is dependent only of the initial and final value. In contrast, the heat flow over the housing (4.74) is a process factor. In the following Section 4.3.4, the heat transfer coefficient α will be discussed. $\dot{H}_{in}$ and $\dot{H}_{out}$ describe enthalpy flows for example by products that are supplied or discharged (mixtures, degassing etc.). Without these additional enthalpy flows follows:

$$\rho \cdot c \cdot \dot{V} \cdot \Delta\overline{T} = P - \dot{V} \cdot \Delta p - U_G \cdot \Delta L \cdot \alpha \cdot \left(\overline{T}_m - T_{housing\ wall,m}\right) \tag{4.75}$$

The conversion according to the temperature delivers:

$$\Delta\overline{T}_{product} = \Delta\overline{T}_{drive\ power} - \Delta\overline{T}_{pressure} - \Delta\overline{T}_{housing} \tag{4.76}$$

The individual shares on the change of the energetic medium product temperature are:

$$\Delta\overline{T}_{drive\ power} = \frac{e}{c}, \quad e = \frac{P}{c \cdot \rho} \tag{4.77}$$

$$\Delta\overline{T}_{pressure} = \frac{\Delta p}{\rho \cdot c} \tag{4.78}$$

$$\Delta\overline{T}_{housing} = \frac{A_G \cdot \alpha}{\rho \cdot c \cdot \dot{V}} \cdot \left(\overline{T}_m - T_{housing\ wall,m}\right) \tag{4.79}$$

The equations can be applied to the entire screw or to a screw section. If a screw section is considered in which the product is melted, then the performance P can be calculated from the performance basic equation (4.63) and the pressure difference Δp from the pressure basic equation (4.30). Thereby, the performance and the pressure difference relate to the energy balance limits. A positive pressure difference corresponds to a pressure build-up in Equation (4.77). In the equations, $\overline{T}$ denotes the energetic medium product temperature above the screw cross-section and in Equation (4.79); $\overline{T}_m$ additionally denotes the average value within the length ΔL. The sign in Equation (4.79) is defined in such a way that a product cooling takes place at $T_{housing\ wall} < \overline{T}_m$ and otherwise a heating.

As described, the equations can be applied to the entire screw or to a screw section. When applied to short screw lengths, an axial progress of product temperature, performance, pressure, filling degree, dwell time, etc. can be calculated. Variable material data and the medium temperature $\overline{T}_m$ can thus be better taken into account. This “difference method” for extruders will be perfected when the formulation of the equations will be carried out as differential equation system, thus $\Delta L \rightarrow 0$. Then follows $\overline{T}_m \rightarrow \overline{T}$ and the problem of average value determination is omitted. In particular, with the description by a differential equation system, substantially changing material values along the extruder can be taken into account. The side on the right of the differential equation system changes depending on changes along the screw, as the screw geometry changes and it must be differentiated between filled and partly filled sections. Backing-up lengths are variable and will also be calculated. For a partly filled screw section, Figure 4.36 shows an example for the formulation as differential equation. Further advice for the solution of the differential equations is given in Section 4.4.

$$\underbrace{\dot{M}\, c_{p,mix} \frac{d\overline{T}}{dl}}_{\text{temperature change}} = \underbrace{\left(1 - \Lambda \frac{A1}{B3}\right) B4\, \eta_{mixP} n^2 D_G^2 U_F}_{\text{mechanical energy input}} - \underbrace{h_{solvent} \left|\frac{d\dot{M}_{solvent}}{dl}\right|}_{\text{evaporation solvent}}$$

$$- \underbrace{\frac{\dot{M}_{mix}}{\rho_{mix}} \frac{dp}{dl}}_{\text{pressure build up}} - \underbrace{A_K\, \alpha \left(T - T_{wall}\right)}_{\text{wall heat transfer}}$$

$$\frac{dp}{dl} = (1 - \Lambda)\, A2\, \eta_{mixD} \frac{n}{D_G} \quad = 0 \text{ in the partly filled section}$$

$$\frac{d\dot{M}_{solvent}}{dl} = \ldots$$

$$\eta_{mixP} = f\left(product\,, c, T, \dot{\gamma}\right)$$

Figure 4.36 Example for formulation as differential equation system to realize continuous variable factors along the extruder

In the following example, the energy balance is applied to the whole screw. Then it can be spoken of as a zero-dimensional model. Figure 4.37 shows a total energy balance for a stationary-operated screw. For a further simplified but very clear consideration, the total heat flow that is supplied to and discharged via the housing should be negligible compared with other energy terms. Furthermore, no energy should be applied for the evaporation of a material. For amorphous products, the crystalline melting enthalpy is omitted. Then the energy supplied via the shafts into the product is converted into temperature increase and pressure build-up.

$$\boxed{P_{shafts} = \Delta\dot{U} + \dot{V}\cdot\Delta p + \underbrace{\dot{Q}_{housing}}_{\approx 0} + \underbrace{\dot{H}_{evaporation}}_{\approx 0} + \underbrace{\dot{H}_{cryst.\ melt.}}_{\approx 0}} \quad \text{Example:}$$

$$P_{shafts} = \underbrace{\Delta\dot{U}}_{=\dot{M}\cdot c_V\cdot(T_{out} - T_{in})} + \dot{V}\cdot\Delta p = \dot{M}\cdot c_V\cdot(T_{out} - T_{in}) + \dot{V}\cdot\Delta p$$

$$\Rightarrow T_{Aus} - T_{in} = \frac{P_{shafts} - \dot{V}\cdot\Delta p}{\dot{M}\cdot c_V} = \underbrace{\frac{P_{shafts}}{\dot{M}}}_{=e\left[\frac{kWh}{kg}\right]} \cdot \frac{1}{c_V} - \frac{\dot{V}\cdot\Delta p}{\underbrace{\dot{M}}_{=\rho\cdot\dot{V}}\cdot c_V}$$

$$\boxed{T_{out} - T_{in} = \frac{e}{c_V} - \frac{\Delta p}{\rho\cdot c_V}}$$

Figure 4.37 Example relating to the energy balance at the whole screw (T is the energetic medium temperature, for sake of simplicity, here without horizontal bar)

The pressure difference in the equation is the pressure difference between the pressure at the tip of the screw (e.g., at the end of the screw shaft) and the pressure at the product inlet. How the pressure runs along the screw, e.g. with more pressure peaks, does not matter. It depends only on "the state variables" at the balance limits. It is advisable to include the die in the balance limit as the differential pressure is generally zero. This results in a quite simple equation for the calculation of the medium temperature increase of the product (e/c_V). The results of this example are given in Figure 4.38.

$$e = \frac{P_{shafts}}{\dot{M}} = \frac{P_A \cdot \eta_A}{\dot{M}} = \frac{\overbrace{M_D \cdot 2\pi \cdot n}^{=P_A} \cdot \eta_A}{\dot{M}} = \frac{457\ \overbrace{\text{Nm}}^{=\text{J}=\text{Ws}} \cdot 2\pi \cdot 600 \frac{1}{\text{min}} \cdot 0.76}{170 \frac{\text{kg}}{\text{h}}} = 0.128 \frac{\text{kWh}}{\text{kg}}$$

$$\frac{e}{c_V} = \frac{0.128\,\text{kWh/kg}}{2\,\text{kJ/(kg K)}} = 230\,\text{K}$$

$$\left.\frac{\Delta p}{\rho \cdot c_V}\right|_{\text{Screw without die}} = \frac{50\,\text{bar}}{1000\,\text{kg/m}^3 \cdot 2\,\text{kJ/(kg K)}} = 2.5\,\text{K} \qquad \left.\frac{\Delta p}{\rho \cdot c_V}\right|_{\text{Screw with die}} = \frac{0\,\text{bar}}{\ldots} = 0\,\text{K}$$

$$T_{out} - T_{in} = \frac{e}{c_V} - \frac{\Delta p}{\rho \cdot c_V} \begin{cases} \approx \dfrac{e}{c_V} & \text{Screw without die} \\ = \dfrac{e}{c_V} & \text{Screw with die} \end{cases}$$

Figure 4.38 Temperature increase in a whole screw by dissipation

By measuring the drive power, the specific energy input e = drive power/throughput can be formed, which is also called specific drive power. This is a characteristic number of many processes and is often used for comparisons of extruders.

For a specific process with a specific product, the specific energy input at different screw trimmings is usually within a narrow range. As an example of a compounding process, the numerical value 0.128 kWh/kg is given. With the equation for increasing the temperature of the product shown in Figure 4.37 and Figure 4.38, it is clear that this is directly proportional to the specific energy input. Based on the example, a temperature increase of over 200 K occurs. With the preconditions given in Figure 4.37 (adiabatic housing, temperature after the die), the following simple relation can be formulated according to Equation (4.76).

$$\Delta \bar{T}_{product} = \Delta \bar{T}_{dissipation} = e_{drive\ power} / c_m \tag{4.80}$$

Instead of the medium specific heat capacity c_m, the specific enthalpy of the product can also be considered. Then applies:

$$\Delta h_{product} = e_{drive\ power} \tag{4.81}$$

By measuring the specific driving power, the product temperature increase can be read in an enthalpy diagram. Figure 4.39 shows the specific enthalpies for different products [16]. The unit kJ/kg can be converted into the unit kWh/kg: 1000 kJ/kg = 0.278 kWh/kg. In the example given, the driving power of the screw is thus converted exclusively into "dissipative product heating".

In considering the specific drive power for the melt area, the temperature increase results in:

$$\Delta \bar{T}_{melt} = \frac{\eta_m}{c_m \cdot \rho_m} \cdot \bar{t}_{melt} \cdot \dot{\gamma}_m^{\,2} \tag{4.82}$$

The equation shows the influencing variables. Higher viscosities and dwell times result in higher temperatures, as expected. The shear rate has a quadratic influence, as well as the screw speed as an "integral value". Equation (4.82) can also be applied to local small product volumes in order to consider specific influences.

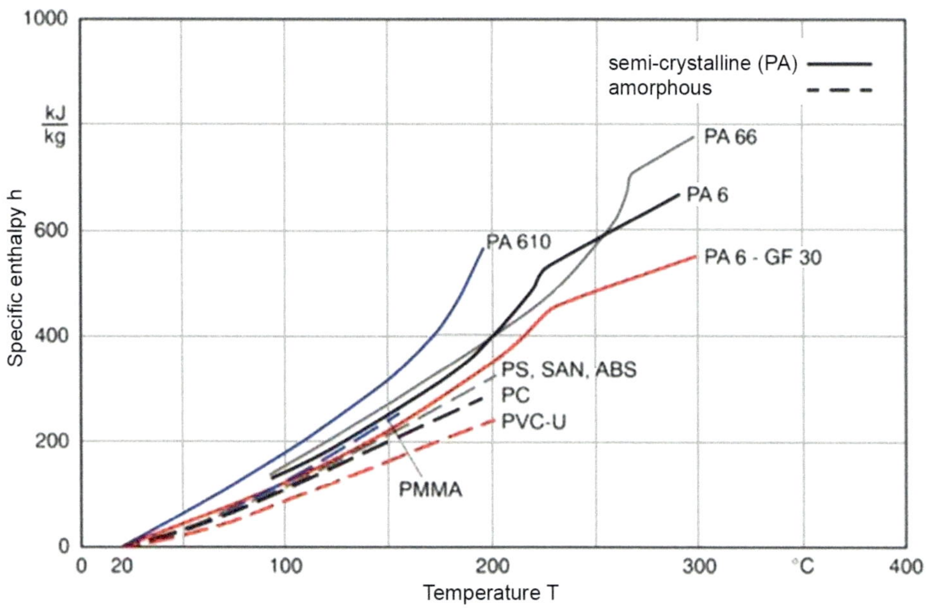

Figure 4.39 Enthalpy diagram for several plastics [16]

The equations given in Figure 4.37 and Figure 4.38 can still be applied to a screw section. The power input of the shafts for a screw section can be calculated for a melt according to the performance basic equation. For technical measurement reasons, the specific energy input for a section is generally unknown. However, the energy input in a pressure build-up zone can be expressed by means of the pump efficiency degree. This exists – at least it should – for a screw geometry from measurements or calculation as characteristic line. The characteristic line can be calculated by means of the pressure and power basic equations.

Figure 4.41 shows how the equation for the temperature increase in the pressure build-up zone ΔT_S with the pump efficiency degree can be formulated. Together with the temperature increase in the die ΔT_D, the total temperature increase is obtained. The individual shares are important:

$$\Delta \overline{T}_D = \frac{\Delta p_D}{c \cdot \rho} \tag{4.83}$$

$$\Delta \overline{T}_S = \frac{\Delta p_S}{c \cdot \rho} \cdot \left(\frac{1}{\eta_p} - 1 \right) \tag{4.84}$$

Example:

A pump efficiency degree of 10% causes 90% of the temperature increase in the pressure build-up zone. For the example with 50 bar from Figure 4.38 a total temperature increase of 25 K is obtained for the energetic medium temperature, see Figure 4.41. As a rule of thumb, the reader can learn from the example: a pressure build-up of 100 bar in a die corresponds approximately to a medium temperature increase of 5 K.

Due to the necessary pressure build-up and when using the usual conveyor elements, an approximate temperature increase of 50 K occurs when the extruder is operated at the correct operating point Λ_{opt}. Otherwise, the temperature increase is even higher for building up the pressure. It should be noted that at a higher temperature the viscosity decreases and thus the required pressure falls off. The relations can be calculated by means of the equations listed in Section 4.1.

From Equation (4.84) it can be seen that an increase of the pump efficiency degree from 10% to 20% lowers the temperature increase by 56%. This refers to the energetic medium temperature. When using screw geometries with a higher pump efficiency degree, for example classic single-flighted screw elements, it should be noted that the temperature peaks can be higher in comparison. In addition to the temperature, further procedural criteria (throughput pulsations at the outlet, ...) must be observed.

Assumption: adiabatic walls

Temperature increase in screw section

$$\Delta T_S = \frac{e_S}{c_V} - \frac{\Delta p_S}{c_V \cdot \rho}$$

With pump efficiency degree in the pressure build-up zone

$$\Delta T_S = \frac{\Delta p_S}{c_V \cdot \rho}\left(\frac{e_S \cdot \rho}{\Delta p_S} - 1\right) = \frac{\Delta p_S}{c_V \cdot \rho}\left(\frac{1}{\eta} - 1\right)$$

Temperature increase in die

$$\Delta T_D = \overbrace{\frac{e_D}{c_V}}^{=0} - \frac{\Delta p_D}{c_V \cdot \rho} = -\frac{\Delta p_D}{c_V \cdot \rho}$$

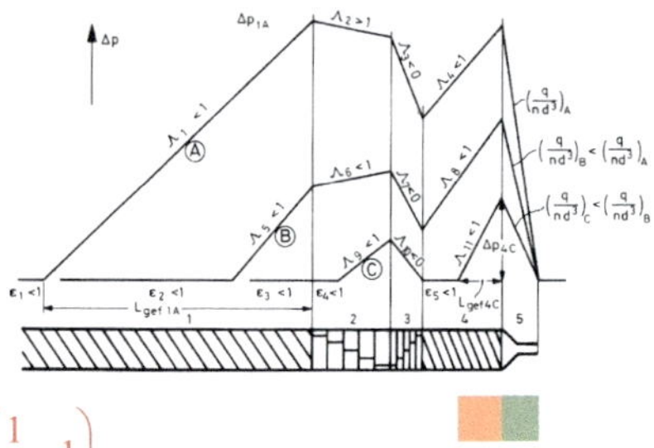

Figure 4.40 Temperature increase by pressure build-up and loss of pressure in a die

Total temperature increase

with $\Delta p_S = -\Delta p_D = \Delta p$

$$\boxed{\Delta T_{ges} = \Delta T_S + \Delta T_D = \frac{\Delta p}{c_V \cdot \rho} \cdot \frac{1}{\eta}}$$

Example: total temperature increase discharge

$$\Delta T_{ges} = \Delta T_S + \Delta T_D = \frac{\Delta p}{c_V \cdot \rho} \cdot \frac{1}{\eta} = \frac{50\,\text{bar}}{2\,\text{kJ/(kg K)} \cdot 1000\,\text{kg/m}^3} \cdot \frac{1}{0.1} = 25\,\text{K}$$

Figure 4.41 Total temperature increase at the screw discharge

Seen over the screw cross-section or the cross-section of the product flow outlet, very large temperature gradients can occur. The 25 K calculated in the example may still be admissible, but not temperature peaks which can be considerably higher. These differences in the cross-section are calculable with three-dimensional models. In Section 4.7, three-dimensional calculations are performed regarding the temperature and simple equations for the estimation of the maximum temperatures in the screw tip are stated.

In the above equations, for example in Equation (4.74), the so-called heat transfer coefficient α is needed. Details in the following Section 4.3.4.

4.3.4 Heat Transfer

The energy balance, for example described by Equation (4.72), refers to a particular balance limit. If the balance limit in an extruder is the product, then the balance limits in a cross-section are the screw shafts and the housing. At these balance limits, the heat transfer must be specified among others by so-called boundary conditions. As a rule, three boundary conditions are distinguished:

- *Boundary condition Type 1 (Dirichlet boundary condition):* The temperature at the boundary (housing, shafts) is specified.
- *Boundary condition Type 2 (Neumann boundary condition):* The temperature gradient is given. This is coupled via the heat conductivity with the heat flow density: $\frac{\partial T}{\partial n} = -\frac{\dot{q}}{\lambda}$ (Fourier heat conduction). A special case is the so-called adiabatic boundary condition, in which the heat flow is zero.
- *Boundary condition Type 3 (mixed boundary condition):* Combination of the two boundary conditions which is expressed by a heat transfer coefficient as a convective heat transfer: $\dot{q} = \alpha \cdot \left(T - T_{housing\ wall}\right)$.

In Section 4.7, the first two boundary conditions are used in the examples. For the limit case of the adiabatic wall, a simple formula for the approximate calculation of the maximum temperature in the screw tips can be stated.

In practice, the third boundary condition can generally be assumed. Then the heat transfer coefficient α must be known. It should be noted that the heat transfer coefficient depends on the temperature T used and the specific process. This is often ignored. Finally, the heat transfer coefficient is – when applied to other cases – only an approximation. If the order of magnitude of the heat transfer coefficient hardly changes, the application of this approximation is justified. Regarding larger extruders it should be noted that the proportion of energy transferred via the housing walls is low in relation to the other energies (energy input by shear/wall cooling proportional $1/D$). The shafts are normally solid. For special applications there are special designs with temperature-controlled shafts. With these hollow shafts no high torques can be transmitted.

We have calculated the heat transfer coefficient for different cases [17]:

- *Case 1:* thermally designed; the heat source is the dissipation of the liquid,
- *Case 2:* thermally designed, constant heat source in the liquid (without dissipation),
- *Case 3:* thermal supply process (stationary), without dissipation,
- *Case 4:* thermal supply process (stationary), with dissipation.

The main goal was to calculate the order of magnitude of the heat transfer coefficient. Therefore, we assumed on basis of the boundary conditions: no heat flow in the shaft and constant housing temperature. Furthermore, we had then modeled only a single-shaft extruder with two-shaft geometry. Figure 4.42 shows some calculated results.

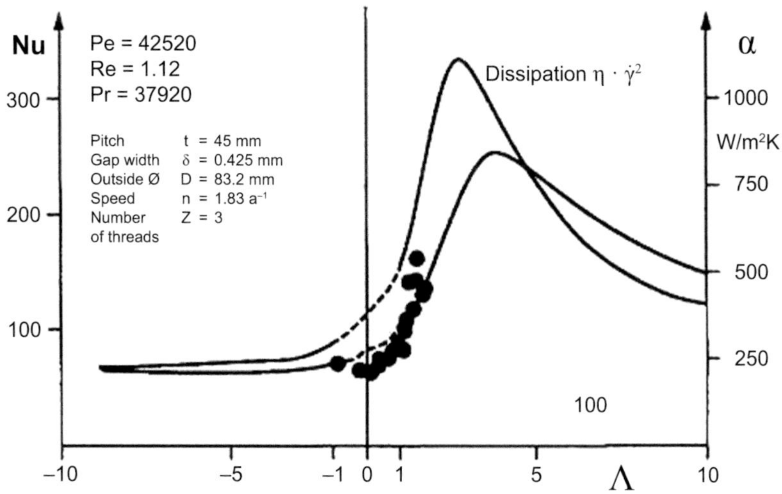

Figure 4.42 Nusselt number Nu = $\alpha \cdot D/\Lambda$ (left) and the heat transfer coefficient α (right, for a silicone oil) as function of the conveying key figure Λ

In the usual forward conveying area with a conveying key figure Λ between 0 and 1, the heat transfer coefficient is of the order of 250 W/(m^{2}K). It is interesting that the heat transfer coefficient in this area decreases with increasing speed – i.e. smaller Λ. This is confirmed by experimental test results with single-screw extruders [10] and co-rotating twin-screw extruders [18], although in both studies a different reference temperature was used for the product.

As already described in the introduction, it is important which reference temperature is used for the calculation of the Nusselt number or the heat transfer coefficient. We have used the practice-relevant energetic medium temperature in Equation (4.70). This temperature is important for the energy balances. The user wants to know the temperature at which the product exits the extruder. Therefore, the weighting with the speed as indicated makes sense.

4.4 Prospect to the Sections 4.1, 4.2, and 4.3

Klemens Kohlgrüber

The above introduction refers to a large extent to products with constant viscosity. The most important key figures and the essential basics were presented. Here a short outlook on the following extensions:

- In the following Section 4.5, the dimensionless key figures for pressure and power are additionally explained regarding dimension technique. In Section 4.6, simple numerical application examples are listed. For conventional conveyor geometries, the conveying parameters for pressure A_1 and A_2 and for performance B_1 and B_2 are given in Section 1.5.
- An extended consideration of the conveying parameters allows further simplifications for Newtonian liquids, which can be applied, for example, to the dissipation energy [3, 14].
- The extension to shear-thinning materials leads to further key figures; the equations become non-linear and more complex. Liesenfelder shows examples in the following Section 4.5. Instead of non-linear equations, the equations for constant viscosity can be applied approximately if representative viscosities are used. It should be noted that the representative viscosities for pressure and performance differ and are not easily determinable. In further developments for description of the relations in extruders in the presence of shear thinning, it should be managed in good approximation to lead back these to the basic equations for extruders with the formula $x + y = 1$.
- In practice, the parameters (e.g., temperature) will change and material data (e.g., temperature-dependent viscosity) will be mostly continuously along the extruder. At strong axial gradients, the presented equations are then advantageously formulated as a differential equation system; an example is given is Figure 4.36. There are very good numerical solvers for such systems which calculate exactly the solution to a given number of digits. The author has very good experience using the program DIFSY1 of Bulirsch/Stoer [19], see also literature [20]. The differential equation system for temperature, pressure, performance, filling degree, dwell time, etc. is formulated as a boundary value problem with break points. The right sides of the equations, for example, change due to the different screw elements, housing, and backing-up lengths. Unfortunately, an extruder computer program with the basics of this chapter and the described computational accuracy was and is not available on the market. Therefore, the author has developed at Bayer a corresponding program for the degassing of screws as from the end of the eighties.
- The above equations are "one-dimensional", i.e. they provide mostly average values concerning the screw cross-section and dependencies along the extruder.

Sometimes more information is necessary. If, for example, the flow behavior of products for different screw elements shall be studied and compared in detail then the three-dimensional flow simulation is a good albeit complex tool. For details in this regard see Section 4.7.1. Today the conveying parameters A_1, A_2, B_1 etc. can be calculated thereby. The calculation of partly filled sections is not yet standard. Section 4.7.3 shows the state of research and the latest findings.

- In the future, three-dimensional calculations will make it possible to determine the cross-sectional filling level as a function of the throughput filling level for non-self-cleaning screw geometries. Furthermore, extended calculations for the heat transfer coefficient are possible. For partly filled sections with temperature calculation is this a great challenge.

References for Sections 4.1, 4.2, and 4.3

[1] White, J. K., Kim, E. K., *Twin Screw Extrusion,* 2nd ed., Hanser (2010)

[2] Ulrich, M., diverse Bayer reports on measurement of the pressure and performance characteristics with dimensionless pressure and performance temperatures (1970–1985)

[3] Kohlgrüber, K., *Weiterentwicklung von Verfahrensschnecken,* internal report, Bayer (1992)

[4] Böhme, G., *Theoretische Betrachtungen über schleichende Strömungen,* Festschrift (1995), pp. 27–40

[5] Schenkel, G., *Kunststoff Extrusionstechnik,* 2nd ed., Hanser (1963)

[6] Rauwendaal, C., *Polymer Extrusion,* 5th ed., Hanser (2014)

[7] Röthemeyer, F., Sommer, F., *Kautschuk Technologie,* 2nd ed., Hanser (2006)

[8] Köster, L., Perz, H., Tsiwikis, G., *Praxis der Kautschukextrusion,* Hanser (2007)

[9] Kohlgrüber, K., Beck, D., *Neue Erkenntnisse über Transportvorgänge in Verfahrensschnecken,* internal report, Bayer (1995)

[10] Pawlowski, J., *Die Ähnlichkeitstheorie in der physikalisch-technischen Forschung,* Springer (1971)

[11] Schuler, W., *Der Doppelschneckenextruder,* 4th ed., VDI-Verlag (1998)

[12] *Kunststofftechnik,* Europa Lehrmittel (2009)

[13] Elsner, T., *Misch- und Verweilzeitverhalten in Doppelschnecken,* 2. Teil Knetelemente, Bayer report (1991)

[14] Ullrich, M., personal communication

[15] Kohlgrüber, K., Bayer report on "product dissipation" in extruders (1994)

[16] Baur, E., Brinkmann, S., Osswald, T. A., Rudolph, N., Schmachtenberg, E. *Saechtling Kunststoff-Taschenbuch,* 31st ed., Hanser (2013)

[17] Beck, Kohlgrüber, CIT 66, No. 8 (1994), pp. 1073–1076

[18] Reichle, L., *Wärmetechnische Auslegungsunterlagen für Doppelschnecken mit Gewindeelementen,* final report (1985)

[19] Kohlgrüber, K., Energetische Optimierung von Industrieöfen durch optimale Steuerung, Dissertation RWTH Aachen, Germany (1984)

[20] Werner, H., Arndt, H., *Gewöhnliche Differentialgleichungen,* Springer (1995)

4.5 Pressure Generation and Energy Input in the Melt

Ulrich Liesenfelder

4.5.1 Operating Conditions of Conveying Screw Elements

The previous sections have shown how screw machines are comprised of successive zones that perform specific process tasks. The most common process task is a combination of transporting materials and generating pressure. At the head of the machine, higher pressure needs to be applied, in particular where typically an extrusion die or a nozzle plate is attached; however, most mixing and kneading zones are also pressure consumers that require pressurization of the melt.

A conveying screw element (Figure 4.43) provides axial transport of a viscous fluid by its rotation. If the viscosity is so high that there is no flow due to gravity, feeding will be independent of the viscosity. The resulting fill level in the machine is proportional to the material throughput and is inversely proportional to the revolution speed of the screws. The quotient of material throughput and revolution speed ("throughput per revolution") determines the fill level of the machine and is therefore an important indicator.

Figure 4.43 Triple-flighted pumping screw (left) and kneading discs (right)

Once the throughput per revolution is so high that the affected screw element is filled completely, the "inherent throughput" of the screw element is reached. If the screw element is run with a throughput that is greater than its inherent throughput, a pressure drop establishes across the screw element. This operating condition is described as "overrun".

Kneading discs are generally used in arrangements with several discs in succession. These are arranged at an offset angle to one another. This generally creates a conveying element. To illustrate this, imagine a line connecting all screw tip segments that define an alternative pitch. This does not apply to arrangements in which every second kneading disc has the same coverage, i.e., its offset angle is $\alpha = \frac{360^\circ}{2 \cdot Z}$. In this case, there is no feeding effect.

Kneading blocks with this configuration are described as "neutral" (non-conveying). They are always filled completely and create a pressure drop.

Screw elements and kneading blocks can also be designed so that their feeding effect acts against the material flow direction, however. These are known as "back-feeding elements". They are also always filled completely and create a pressure drop.

Conveying screw elements can actively generate pressure, provided that they are filled completely. If there is a continuous gas phase present, there can naturally be no axial pressure gradient. This means that back pressure always forms upstream of an obstruction to flow in a conveying screw element. The conveying screw element fills up until the pressure needed to overrun the obstacle has built up.

The different operating conditions are explained below with reference to an example (Figure 4.44). Here, we have a pumping screw with a large pitch (1), followed by a conveying kneading block (2), and a back-feeding kneading block (3), in addition to a pumping screw with a small pitch (4) that serves to generate pressure upstream of the nozzle (5). This is a standard screw machine configuration for a mixing task followed by extrusion.

The nozzle (5) and the back-feeding kneading block (3) are obstacles to flow that generate a drop in pressure. The screw sections upstream of this must build up the pressure needed to overrun the obstacle. The back-feeding kneading block (3) ensures that at least a part of the kneading zone is wholly filled, which promotes the mixing effect.

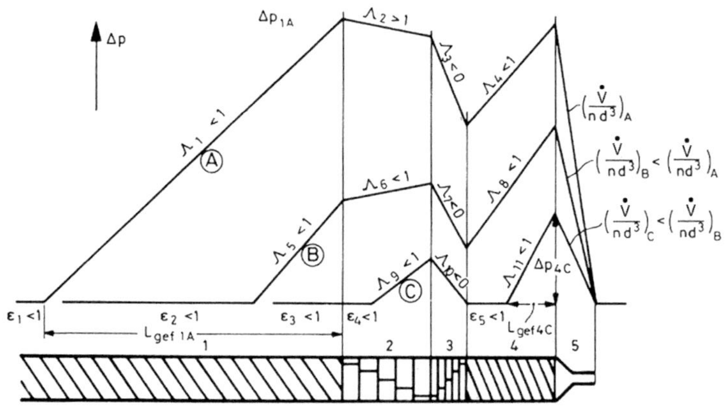

Figure 4.44 Operating states of a screw machine during pressure generation

In Figure 4.44, pressure curves for three different throughputs at a constant rotational speed are shown above the illustration of the screw set. For a small throughput (case C), zones (2) and (4) build up the pressure needed to overrun the flow obstacles (3) and (5). Zones (2) and (4) each consist of a partially filled and a filled section, because the back pressure upstream of the flow obstacle is shorter than the processing zone concerned. Case B can be generated if the machine is run at the same rotational speed as in case C, but with a considerably higher throughput. The length of the zone (4) is no longer sufficient to apply the nozzle pressure.

The back pressure at the nozzle then acts on the upstream zones. The same applies to the conveying kneading block (2). A part of the nozzle pressure must now be applied by pumping screw (1). Zones (2) and (4) are both still conveying, however, i. e., they help to generate the pressure. If the throughput is considerably increased once again (case A), zone (2) is overrun. The throughput is now larger than the inherent throughput of the kneading block. Zone (2) must now feed more per revolution than the maximum possible throughput during pressure-free conveying. The zone is overrun from the previous zone (1) that has a higher pumping capacity. Zone (2), which was conveying in the cases considered above, now creates additional pressure drop.

4.5.2 Illustration of Dimensionless Groups

The similarity theory for screw machines was introduced in the preceding chapters of this book. Here, the benefits of dimensionless thinking are highlighted and a clear interpretation of the dimensionless variables is provided. The following symbols are used:

V	Volume throughput
n	Rotational speed
Δp	Pressure difference
P	Power
D	Barrel internal diameter
L	Screw length
η	Viscosity

These seven quantities must always be known in order to describe the pressure generation and energy behavior of screw machines. Dimensional analysis reduces the seven influencing variables to three dimensionless groups:

$$\Delta p^* := \frac{\Delta p}{\eta \cdot n} \cdot \frac{D}{L} \qquad \text{Pressure number}$$

$$P^* := \frac{P}{\eta \cdot n^2 \cdot D^2 \cdot L} \qquad \text{Power number}$$

$$\dot{V}^{*} := \frac{\dot{V}}{n \cdot D^{3}} \qquad \text{Throughput number}$$

Of course, a screw profile cannot be described simply by specifying the barrel diameter and screw length. Other geometric parameters are needed, such as the centerline distance A, the clearance δ between the screw tip and barrel and the clearance s between the screws. These affect the pressure generation and energy input, and are thus influencing variables. Further dimensionless groups are obtained by dividing by the geometric reference variable, in this case the barrel diameter.

A	Centerline distance
δ	Clearance between screw tip and barrel wall
s	Clearance between screws

In practice, the throughput of a machine is generally predefined, while the energy input and pressure generation are target variables for the calculation. It is therefore sensible to formulate the dimensionless power and the dimensionless pressure generation as functions of their dimensionless influencing variables.

$$\frac{\Delta p}{\eta n} \cdot \frac{D}{L} = f\left(\frac{\dot{V}}{nD^{3}}, \frac{A}{D}, \frac{\delta}{D}, \frac{s}{D}\right)$$

$$\frac{P}{\eta \cdot n^{2} \cdot D^{2} \cdot L} = f\left(\frac{\dot{V}}{nD^{3}}, \frac{A}{D}, \frac{\delta}{D}, \frac{s}{D}\right) \tag{4.85}$$

Two screw profiles that are described by the same geometric parameters are known as geometrically similar. They can be converted into one another by scaling. The pressure and power numbers are dependent only on the throughput number if the screws are geometrically similar. The relationships are linear in the case of fluids of constant viscosity ("Newtonian fluids"). The complete pressure and energy behavior of a class of geometrically similar screws can thus be traced back to a simple linear relationship.

Since the meaning of the dimensionless groups is not immediately clear, we shall attempt to provide an interpretation here:

The *throughput number* is simply the machine's output per revolution in a dimensionless form. This is related to the volume of the machine, as represented by the diameter to the power of three. The displaced volume of a screw machine is scaled with D^3, so this gives a parameter that describes the flow status independently of the machine size.

According to practical experience a screw section can build up more pressure the higher the viscosity of the medium, the higher the rotational speed, and the longer the filled screw section that is available. It is therefore natural to relate the pressure generation to these variables. The parameter becomes dimensionless by multiplying it with the diameter. Between the throughput number and the pressure number there is a linear relationship. According to Pawlowski [1] it is represented by the axis intercept form of the linear equation with the intercepts A_1 and A_2.

$$\frac{\Delta p}{\eta n} \cdot \frac{D}{L} = \frac{A_2}{A_1}\left(A_1 - \frac{\dot{V}}{n \cdot D^3} \right) \tag{4.86}$$

The linear Equation (4.86) applies to all feasible operating states, and to all feasible machine sizes with geometrically similar profiles. For the scale-up, it can be derived that the same pressure generation per screw element can be obtained for the same rotational speed if the throughput per revolution increases in proportion to the diameter to the power of three.

Intercepts A_1 and A_2, as dimensionless profile parameters, apply to all geometrically similar screw profiles. Intercept A_1 identifies the inherent throughput of the screw element, i. e., the throughput per revolution that the machine can achieve if it is filled completely and is operated without back-pressure. Intercept A_2 identifies the pressure generation capacity for throughput zero, i. e., when the flow is totally restricted. The conveying capacity of the screw ranges between these two points. If the throughput per revolution is greater than the inherent throughput, then it is a screw element that is overrun by an element with a larger inherent throughput. In this case, the pressure difference is negative, so the screw element is a pressure consumer. The negative throughput indicator range means that the throughput and the feeding effect of the screw element are in opposite directions, as occurs with the back-feeding elements or backward-pumping screws. Figure 4.45 shows that the pressure drop at a back-feeding element is much higher than the pressure generation capacity of the corresponding conveying element.

According to Pawlowski [1], the kinematic flow parameter Λ is introduced:

$$\Lambda := \frac{\dot{V}}{A_1 \cdot n \cdot D^3} \tag{4.87}$$

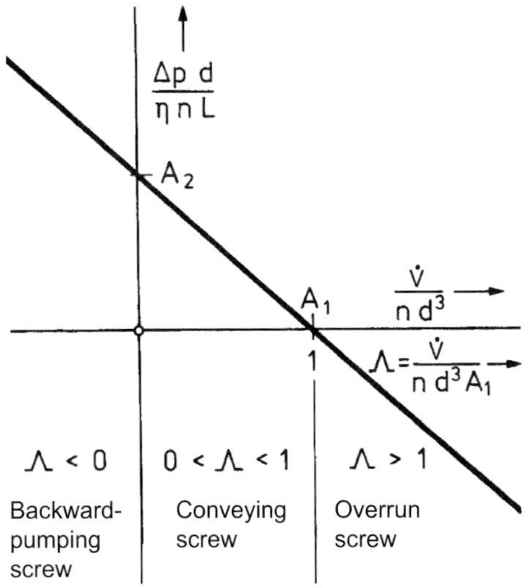

Figure 4.45
Pressure characteristic of a screw machine

The dimensionless power is ideally formed using the same reference variables as for the pressure number. It can be derived that the power input is proportional to the viscosity of the medium and to the screw section under consideration, as well as to the square of the rotational speed and to the square of the diameter. To this end, the power is derived from the torque M which results from the wall shear stresses τ_w that draw on the screw shafts.

$$\tau_w \sim \eta \cdot n \tag{4.88}$$

$$M \sim \tau_w \cdot A_w \cdot D \tag{4.89}$$

$$P \sim M \cdot n \tag{4.90}$$

$$P \sim \eta \cdot n^2 \cdot D^2 \cdot L \tag{4.91}$$

Here again, this gives a linear relationship between the dimensionless groups:

$$\frac{P}{\eta \cdot n^2 \cdot D^2 \cdot L} = \frac{B_2}{B_1} \cdot \left(B_1 - \frac{\dot{V}}{n \cdot D^3} \right) \tag{4.92}$$

Intercepts B_1 and B_2 form the characteristic energy parameters for the screw set. Point B_1 is the so-called turbine point, which is always way above the inherent throughput, i.e., in the overrun range. If the throughput per revolution is greater than at the turbine point, then energy is given off at the shaft. Although this range is technically possible and has been proven experimentally, it is not meaningful for practical operation. Intercept B_2 denotes the energy absorbed during total flow restriction. Figure 4.46 shows that the greatest energy absorption takes place in the domain of the back-feeding elements, and that there is still positive energy absorption even during pressure-free conveying. If the throughput per revolution is smaller than at point A_1 and there is no back-pressure, partial filling establishes in the machine. The fill level of the machine then corresponds to the kinematic flow parameter Λ. As the fill level tends to zero, the energy indicator assumes the value B_4. In this operating status, the shafts and barrel are wetted with melt, but no material is conveyed. The energy absorption increases linearly with the fill level until the machine is completely full, as indicated by point A_1.

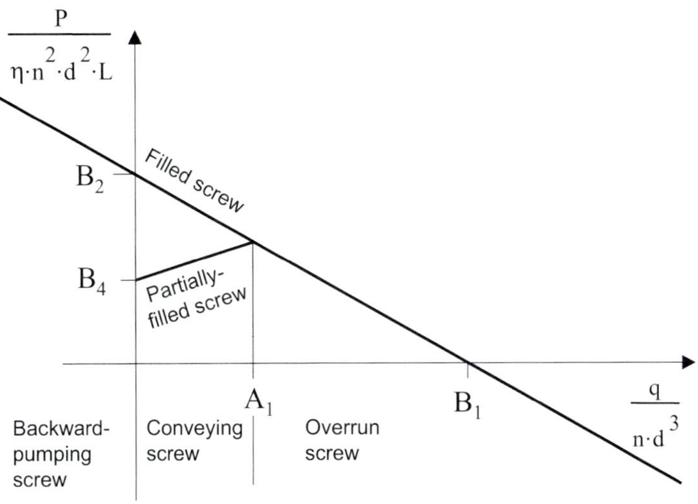

Figure 4.46 Energy characteristic of a screw machine

For the scale-up is it possible to derive that the energy absorption also rises in proportion to the volume, provided that the throughput is increased in proportion to the volume of the machine, i.e., the throughput number remains constant. Because the energy input increases to the same extent as the throughput, the same temperature profile along the machine and the same specific energy input are obtained. This applies if we ignore heat transfer processes and any non-isothermal cross-section.

The feed parameters A_1, A_2, B_1, B_2, and B_4 can be determined by measurement using a simple test set-up (Figure 4.47) [4].

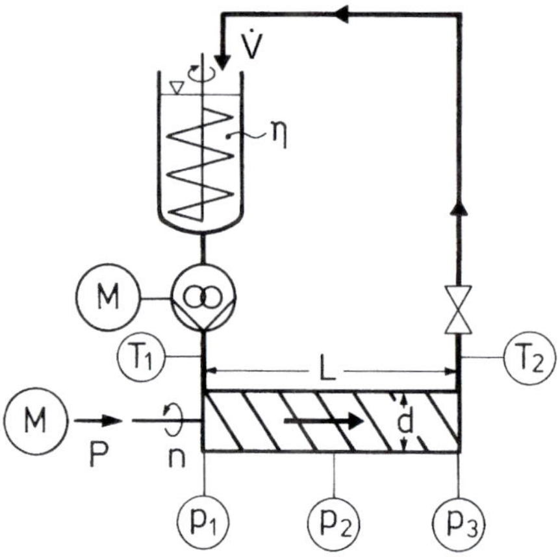

Figure 4.47
Test set-up for determining the screw parameters

In Figure 4.48 and Figure 4.49, the measured data and regression lines are shown for the pressure and energy characteristics of the example screw with 40 mm pitch presented in 4.5.5 [4].

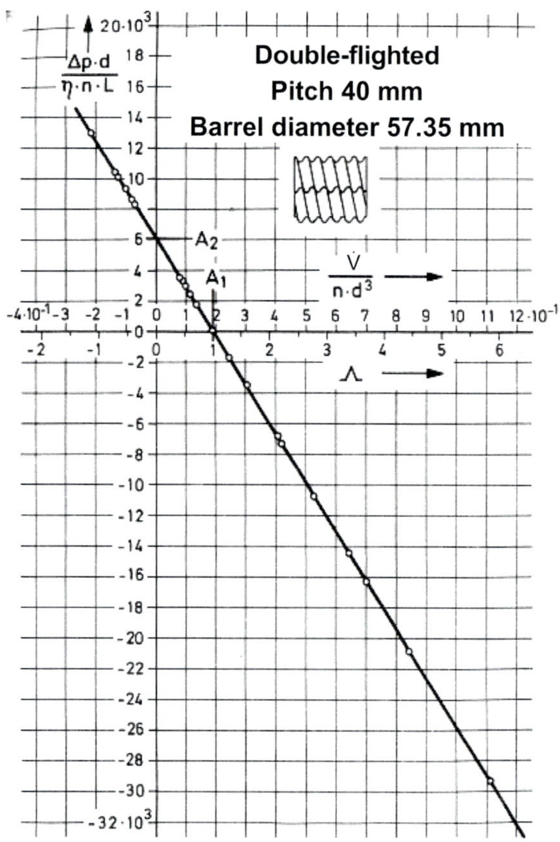

Figure 4.48
Measured pressure characteristic for a double-flighted screw [4]

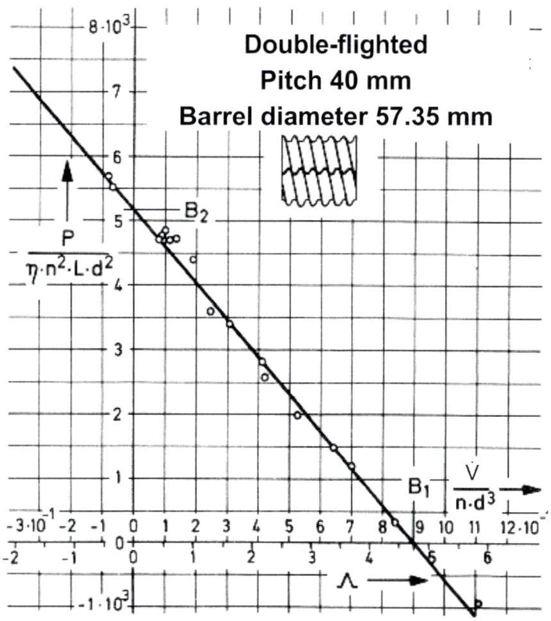

Figure 4.49
Measured energy characteristic for a double-flighted screw [4]

4.5.3 Calculation of the Back-Pressure Length

One question of practical significance concerns the back-pressure length upstream of a pressure consuming element. A totally filled area is created in the back-pressure from the pressure consuming element. The length of this area is established according to the pressure generation capacity of the screw element so that the required pressure is generated. The back-pressure length is an important design criterion for pressure build-up zones. If the pressure build-up zone is not long enough, flooding may occur in the upstream zone. This is often a venting zone with vent openings.

The back-pressure length can be derived from the pressure Equation (4.86) with a simple transformation:

$$L_r = \frac{\Delta p \cdot D}{\eta \cdot n} \cdot \frac{1}{A_2 \cdot (1 - \Lambda)} \tag{4.93}$$

A combination of a conveying screw element and a back-feeding element is often used to block off zones of different pressure with a plug of melt during devolatilization. If the right-hand and left-hand elements are of the same design, then feed parameter Λ applies to the conveying element and parameter $-\Lambda$ to the back-feeding element.

Figure 4.45 shows that the pressure dissipated in the back-feeding element is much greater than the pressure generated in a conveying element of the same type and same length. The following generally applies to a right/left combination of elements of the same type:

$$\frac{L_r}{L_{left}} = \frac{1+\Lambda}{1-\Lambda} \tag{4.94}$$

The optimum efficiency during pressure generation is generally achieved at $\Lambda \approx 0.5$. In this case, three times the length in relation to the left-hand element is required to generate the pressure. The back-pressure rises fast as the throughput per revolution increases. At $\Lambda = 0.8$, the back-pressure is already 9 times as long as the left-hand element.

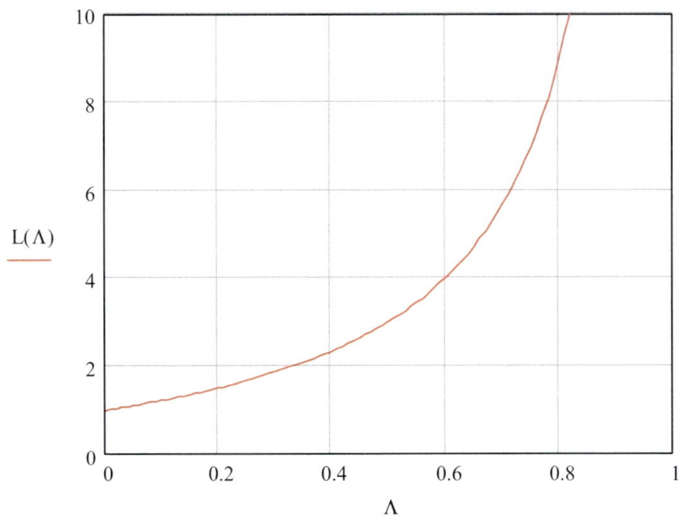

Figure 4.50 Back-pressure length in relation to the length of the back-feeding element for a right/left combination of elements of the same type

4.5.4 Efficiency during Pressure Generation

Efficiency during pressure generation is just as important as the back-pressure length for the design of a pressure build-up zone. By efficiency we mean the relationship between the pressure energy that is stored in the flow at the end of the pressure build-up zone and the energy that is introduced into the pressure build-up zone.

$$e = \frac{\dot{V} \cdot \Delta p}{P} \tag{4.95}$$

The efficiency during pressure generation can be calculated from Equation (4.86) and Equation (4.89) with a few transformations.

$$e = A_1 \cdot \frac{A_2}{B_2} \cdot \frac{\Lambda \cdot (1 - \Lambda)}{1 - \frac{A_1}{B_1} \cdot \Lambda} \tag{4.96}$$

It is sensible to define the efficiency using Equation (4.92) in the conveying range, i.e., between $\Lambda = 0$ (no throughput) and $\Lambda = 1$ (no pressure generation); in both cases the efficiency is assuming the value zero. Between these two, the efficiency passes through a maximum that is generally between 50 and 60% of the inherent throughput.

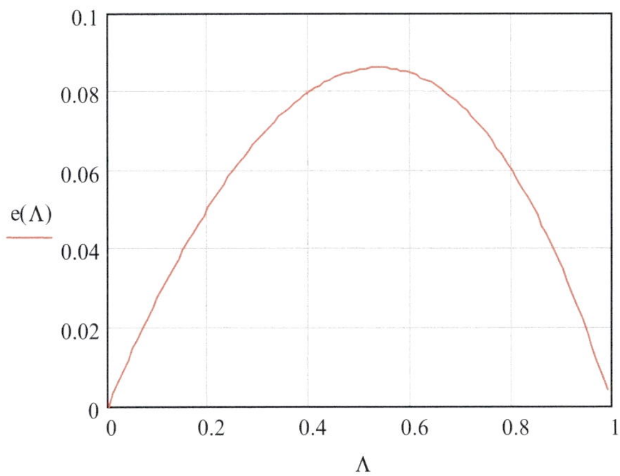

Figure 4.51 Efficiency of a triple-flighted screw during pressure generation in relation to the kinematic parameter of flow Λ

The achievable pumping efficiency of co-rotating twin screws differs considerably. Counter-rotating twin screws can achieve efficiency levels of 50%, while co-rotating shafts generally remain below 25%. The efficiency of the double-flighted screws normally used barely exceeds 10%, even when optimally designed. It is precisely for this reason that the efficiency should be considered when designing the pressure build-up zone. An efficiency of 10% means that ten times the pressure energy is dissipated in the polymer. It is not always possible to operate a pressure build-up zone under optimum conditions, although it should not be operated too far from the optimum in order to protect the product.

A numerical example is used in the following to illustrate this:

The temperature in the product is calculated by

$$\Delta t = \frac{\Delta p}{\rho \cdot c \cdot e} \tag{4.97}$$

and, for an efficiency of e = 10% and with typical physical properties such as density ρ = 1000 kg/m^3 and specific heat capacity c = 2000 J/kg·K, a temperature rise of 25 K occurs. If according to unsuitable design, build-up of 50 bar. The machine has to be operated at an efficiency of just 5%, a temperature rise of 50 K will occur.

The size of the machine and the throughput (and often the revolution speed as well) are determined by other factors, so the pressure build-up zone should be optimized by selecting the pitch and gap [2].

4.5.5 Example for the Design of a Pressure Build-up Zone

The following numerical example illustrates the effect of pitch selection on energy dissipated in the pressure build-up zone. A nozzle pressure of 50 bar has to be generated, the viscosity of the polymer at the machine's head is 200 Pa·s and a double-flighted machine with the following properties should be considered:

Table 4.1 Geometric Parameters for the Selected Screw Machine

Barrel internal diameter D	57.350 mm
Centerline distance A	48.000 mm
Radial screw clearance δ	0.170 mm
Reciprocal screw clearance s	0.525 mm

There are screw elements available with a pitch of 40 mm and 60 mm. The feed parameters of the two screws were determined experimentally [4].

Table 4.2 Dimensionless Feed Parameters

Pitch	40 mm	60 mm
A_1	0.19	0.292
A_2	6100.00	3630.000
B_1	0.9	1.340
B_2	5165.00	5040.000

In Example 1, a volume throughput of 300 l/h is pumped at a rotational speed of 300 rpm. Equation (4.86) to Equation (4.97) and the typical physical properties of density ρ = 1000 kg/m^3 and specific heat capacity c = 2000 J/kg·K are used to calculate the back-pressure length, pumping efficiency, energy input and temperature rise in the polymer for the two available screw elements.

Table 4.3 Quantities Calculated for Example 1

Pitch	60 mm	40 mm
Back-pressure length	110 mm	90 mm
Pumping efficiency	4.8%	6.25%
Energy input	8.7 kW	6.7 kW
Temperature increase	52 K	40 K

In all respects, the smaller pitch is the better choice. If a throughput of 500 l/h is conveyed with values that are otherwise the same, the picture is different (Example 2).

Table 4.4 Quantities Calculated for Example 2

Pitch	60 mm	40 mm
Back-pressure length	160 mm	220 mm
Pumping efficiency	5.9%	4.5%
Energy input	12.3 kW	16.8 kW
Temperature rise	44 K	60 K

For a pitch of 40 mm, the throughput per revolution is now much higher than at the optimum energy level, resulting in an unfavorable operating point. In this case, a pitch of 60 mm would be the better choice. This example clearly shows that no screw element is best in all situations. Rather, the choice of suitable screw element depends significantly on the dimensionless throughput.

4.5.6 Pressure and Energy Behavior with Shear Thinning

If the material to be processed is subject to shear thinning, the linear relationships for the pressure and energy behavior illustrated above no longer apply. With shear thinning, there is a non-linear relationship between the shear rate and shear stress that is reflected in the "flow curve" (see Chapter 3). As a rule, the zero viscosity and one or two rheological time constants are enough to describe the flow curve with sufficient accuracy. The Carreau equation is often used; it contains a dimensionless flow exponent in addition to the zero viscosity and a rheological time constant.

$$\tau = \frac{\eta_0 \cdot \dot{\gamma}}{\left(1 + \Theta \cdot \left|\dot{\gamma}\right|\right)^m} \tag{4.98}$$

The dimensionless flow exponent m and the rheological time constant Θ are additional influencing variables that turn the one-dimensional problem for Newtonian fluids into a three-dimensional problem. The rheological time constant Θ, when multiplied by the revolution speed n, forms an independent dimensionless group (Deborah number).

The flow exponent m, which generally assumes values between 0.5 and 1 in practice, is regarded as constant in the following, making the problem two-dimensional. If test results that were obtained with a model material are transferred to a real material system, the results will only apply to materials with the same flow exponents. Here we can see that model theory is limited in its usefulness. With complicated material behaviors, the amount of experimentation required increases vastly. The only solution is to use a material law that contains just one time constant as a parameter in addition to zero viscosity. Although suitable material laws do exist, they often provide an inaccurate description of the flow curve.

The rotational speed, which only appeared as a parameter in the linear Equation (4.85) and Equation (4.88), forms now an independent dimensionless parameter in the form of the Deborah number $n{\cdot}\Theta$. While the dimensionless pressure generation and dimensionless energy only depend on the kinematic parameter of flow for Newtonian liquids, the dimensionless revolution speed appears as an additional influencing variable for shear thinning. This is plausible if we consider that the rotational speed is a measure of the shear stresses on the material, and thus influences the effective viscosity of the material. It is also to be expected that the interaction will assume a non-linear form since the flow curve is already non-linear.

We derive approximation equations in order to clarify the basic structure of the shear thinning behavior during pressure generation and energy input. They are also suitable for the approximation of measured values. A trough model is used to derive the relationships as a simplification of the screw channel, and the flow is regarded as the overlaying of a drag flow and a pressure flow in the opposite direction.

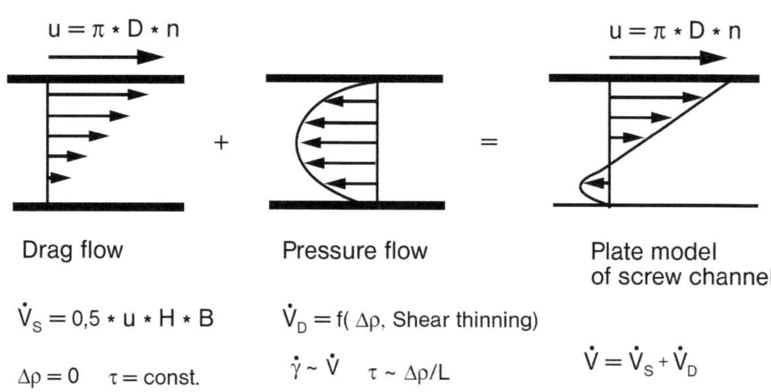

Figure 4.52 Overlaying of the drag/pressure flow

Since the flow is both stationary and in a creeping flow condition, the impulse conservation law for this arrangement is:

$$\frac{dp}{dz}=\frac{d\tau}{dy} \tag{4.99}$$

An equation in the form of the flow curve may be used to describe the pressure generation. Its derivation from the impulse balance is not shown here. The following boundary conditions must be fulfilled:

- For Newtonian fluids ($\Theta = 0$), should derive Equation (4.86)
- For purely drag flow ($\Lambda = 1$), should apply $\Delta p = 0$
- For purely pressure flow, i.e., revolution speed = 0, this gives a formula for the pressure flow with a volume shear rate proportional to throughput/d^3 the

Equation (4.100), resulting from these considerations may be used to approximate measured values (Figure 4.53).

$$\frac{\Delta p}{\eta_0 \cdot n}\cdot\frac{D}{L}=\frac{A_2\cdot(1-\Lambda)}{\left[1-A_3\cdot(n\cdot\Theta)\cdot(1-\Lambda)\right]^m} \tag{4.100}$$

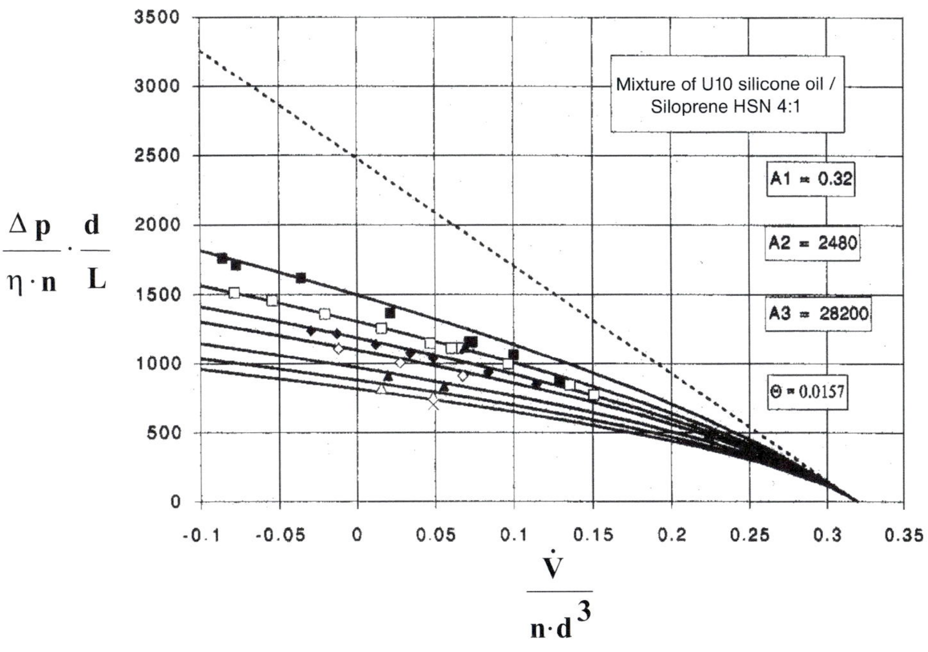

Figure 4.53 Approximation of the pressure characteristic on measured values for shear thinning

Information about the behavior of shear thinning melts in twin screws can be derived from Equation (4.100), in addition to the approximation of the measured values. The concept of representative viscosity based on Frederickson [3] is used for this purpose. The representative viscosity is the viscosity that provides the same pressure generation as a Newtonian fluid under the specified conditions. If we combine Equation (4.100) and Equation (4.86), we obtain for the representative viscosity:

$$\eta_r = \frac{\eta_0}{\left[1 - A_3 \cdot (n \cdot \Theta) \cdot (1 - \Lambda)\right]^m} \tag{4.101}$$

The representative viscosity represents a point on the flow curve Equation (4.95). It occurs when the representative shear rate is imposed on the shear thinning fluid. If we compare Equation (4.95) and Equation (4.101), we can see that the representative shear speed is proportional to the rotational speed and proportional to $(1 - \Lambda)$.

$$\gamma_r = A_3 \cdot n \cdot (1 - \Lambda) \tag{4.102}$$

One advantage of the concept of representative viscosity is that the measured flow curve is entirely sufficient to determine the pressure and energy behaviors. It is not necessary to determine the flow exponent m and the time constant Θ. As can be seen in Figure 4.54, the representative viscosity is determined with reference to the representative shear rate from the flow curve using Equation (4.102). This then replaces the constant Newtonian viscosity in Equation (4.86).

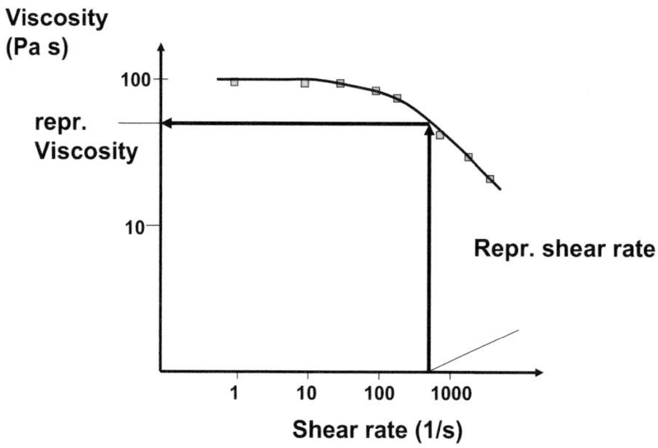

Figure 4.54
Determination of the representative viscosity from the flow curve

With pressure-free conveying, the influence of shear thinning disappears. This status is approximated by the shear flow between two plates, in which the shear speed and shear stress are constant over the channel height. In this case, the throughput is independent of the flow curve of the medium and is merely proportional to the plate speed (or rotational speed). As can be seen from the measured values in, the influence of shear thinning increases as the flow restriction increases and is greatest in the vicinity of the back-feeding elements. According to Equation (4.102), the representative shear rate increases linearly with the flow restriction (1 – Λ).

To discuss energy behavior with shear thinning, a single screw extruder is considered first. The required power results from the shear stresses occurring at the cylinder wall τ_w, which, in turn, are made up of pressure and drag flow fractions.

$$P = M \cdot \omega = \tau_W \cdot \pi \cdot D \cdot L \cdot \frac{D}{2} \cdot 2 \cdot \pi \cdot n \tag{4.103}$$

$$P = \left(\tau_{W.pressure} + \tau_{W.drag}\right) \cdot \pi^2 \cdot D^2 \cdot L \cdot n \tag{4.104}$$

$$\tau_{W.pressure} = \frac{\Delta p}{L} \cdot \frac{H}{2} \cdot \cos(\phi) \tag{4.105}$$

$$\tau_{W,drag} = \eta \cdot \frac{\pi \cdot D \cdot n}{H} \tag{4.106}$$

Here H is the channel depth and φ the pitch angle. This gives the following dimensionless form:

$$\frac{P}{\eta_0 \cdot n^2 \cdot D^2 \cdot L} = \frac{\Delta p}{\eta_0 \cdot n} \cdot K_1 + \frac{K_2}{\left(1 + K_3 \cdot n \cdot \Theta\right)^m} \tag{4.107}$$

For Newtonian fluids, Equation (4.86) and Equation (4.89) can be combined to give:

$$\frac{P}{\eta \cdot n^2 \cdot D^2 \cdot L} = \frac{A_1}{A_2} \cdot \frac{B_2}{B_1} \cdot \left(\frac{\Delta p}{\eta \cdot n} \cdot \frac{D}{L}\right) + B_2 \cdot \left(1 - \frac{A_1}{B_1}\right) \tag{4.108}$$

The constants K_1 and K_2 can be determined by comparing the coefficients with the Newtonian case. The third constant K_3 is used as a free fitting parameter, so it is designated as B_3.

$$\frac{P}{\eta_0 \cdot n^2 \cdot D^2 \cdot L} = \frac{B_2 \cdot \frac{A_1}{B_1} \cdot (1 - \Lambda)}{\left[1 - A_3 \cdot n \cdot \Theta \cdot (1 - \Lambda)\right]^m} + \frac{B_2 \cdot \left(1 - \frac{A_1}{B_1}\right)}{\left(1 + B_3 \cdot n \cdot \Theta\right)^m} \tag{4.109}$$

Equation (4.107) states that the dimensionless energy consists of a part that is proportional to the dimensionless pressure generation and a second part that is dependent only on the rotational speed. The energy characteristics do not intersect like the pressure characteristics in $\Lambda = 1$. Rather, the effect of shear thinning can be clearly seen, even for a pressure-free screw ($\Delta p = 0$). The effect of shear thinning continues to grow as the flow in the screw is increasingly restricted. The effect of shear thinning is again at its greatest in the vicinity of the backward-pumping screw or left-hand element.

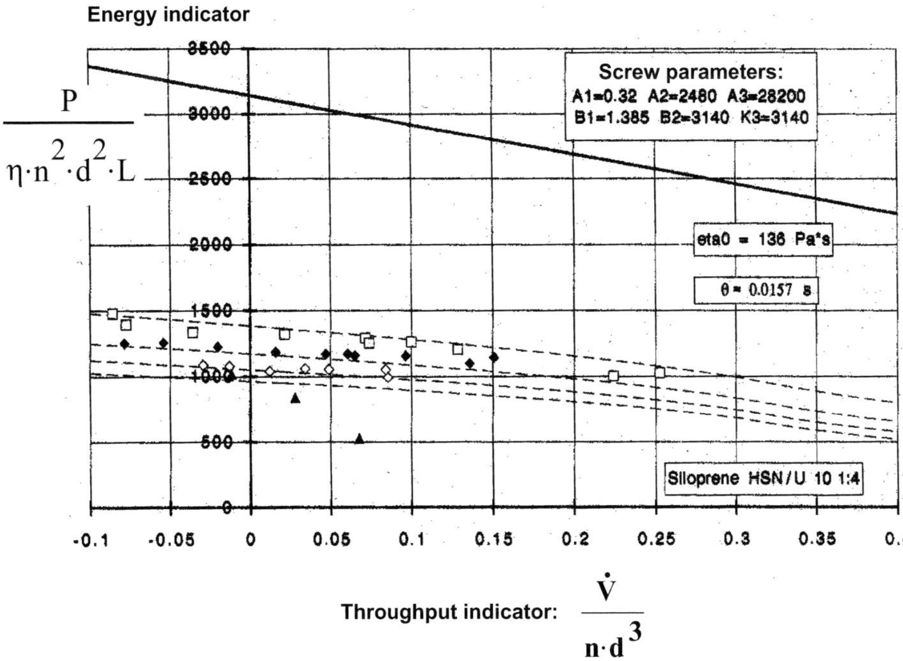

Figure 4.55 Approximation of the energy characteristic for shear thinning

As with the pressure behavior, a representative viscosity can also be specified for the energy behavior. The representative viscosity is the viscosity that gives the same energy absorption as a Newtonian fluid at the same operating point (Λ = idem). The formulae are derived as for the procedure for the pressure generation behavior. Only the result is shown here:

$$\frac{\eta_{r,energy}}{\eta_0} = \frac{A_1 \cdot (1-\Lambda)}{(B_1 - A_1 \cdot \Lambda)\left[1 - A_3 \cdot n \cdot \Theta \cdot (1-\Lambda)\right]^m} + \frac{B_1 - A_1}{B_1 - (\Lambda \cdot A_1)} \cdot \frac{1}{(1 + B_3 \cdot n \cdot \Theta)^m} \quad (4.110)$$

The representative viscosity for energy behavior is markedly lower than for pressure behavior, i.e., the shear thinning has a much greater effect here. In our example, the representative viscosity for pressure behavior is 3 to 6 times higher than that for energy behavior. The representative shear rate is consequently higher for energy behavior. The energy absorption is thus determined to much higher degree by the high shear between the screw tip and barrel wall or between the two screw shafts than the pressure generation.

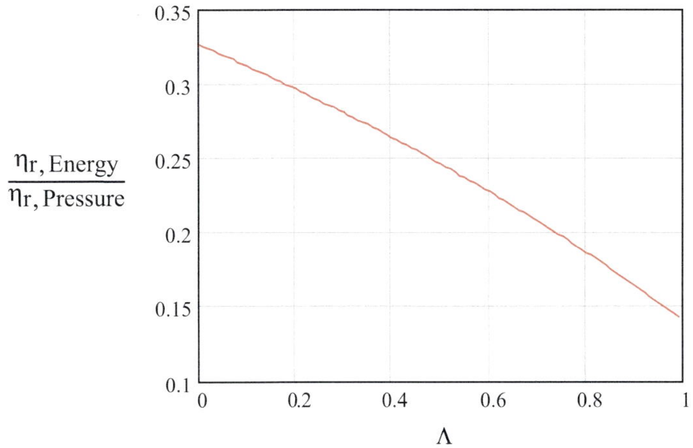

Figure 4.56
Ratio between the representative viscosities for energy and pressure

It follows directly from this that the pumping efficiency for shear thinning materials is higher than for Newtonian fluids. Figure 4.57 illustrates pumping efficiency values for two shear thinning fluids as calculated from measured data. The light symbols relate to a fluid with more strongly marked shear thinning than the black symbols. The continuous line shows the pumping efficiency for a Newtonian fluid for comparison. The maximum pumping efficiency increases the more marked the shear thinning is. The maximum also tends towards a greater throughput per revolution or larger Λ value.

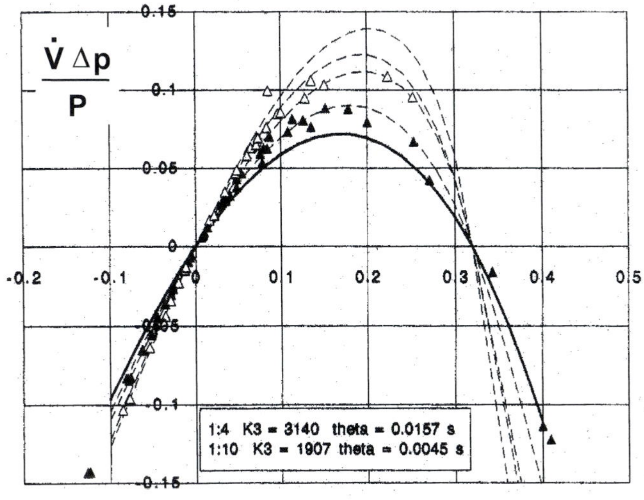

Figure 4.57
Efficiency for pressure generation with shear thinning

References for Section 4.5

[1] Pawlowski, J., *Die Ähnlichkeitstheorie in der physikalisch technischen Forschung,* Springer Verlag, (1971)

[2] Ullrich, M., Patents DE2464436, (1974), and DE2611908, (1976)

[3] Frederickson, A.G., Bird, R.B., *Ind. Eng. Chem.* (1958), 50, p. 347

[4] Ullrich, M., "Das Förderverhalten zweigängiger, gleichsinniger Doppelschnecken", internal report, Bayer AG, (1977)

4.6 Tasks Regarding the Power Input and the Back-Pressure Length

4.6.1 Task: Influence of the Flight Pitch

Carsten Conzen

The first task is to familiarize yourself with the formulas and the diagrams and to get a first impression of the influence of the pitch of the screw element. The diagram shows the power and pressure characteristics as a function of the dimensionless throughput. The power can be divided into a net power and a dissipation power. The formulas for this can be derived from the power and throughput equations and are displayed dimensionless next to the diagram. The dissipation power is very high and makes up the largest part of approx. 95–75%, depending on the design of the pressure build-up zone. The net power is very low and represents a downward open parabola with a maximum at $0.5 \cdot A_1$. In the first task, the net power and the pressure build-up capacity are to be calculated over a defined back-pressure length for two pitches.

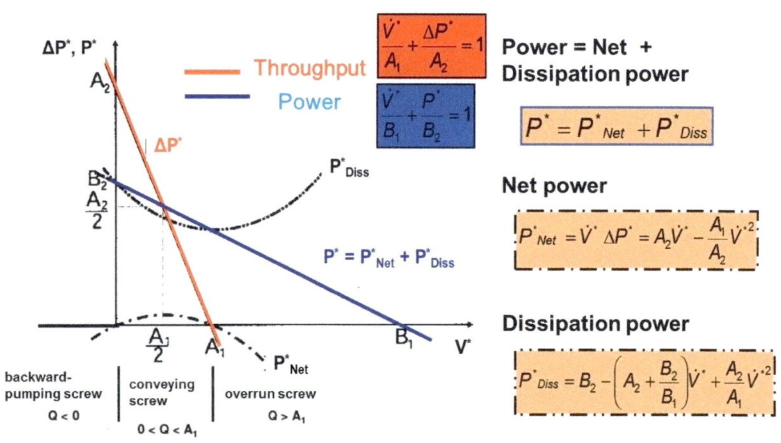

Figure 4.58 Power and pressure characteristics in dependence on dimensionless throughput

Calculate the following values for the two pitches of the machine:

Calculate the maximum effective power over a back-pressure length of $L = 4 \times D$. How much pressure can be built up at this operating point?

Operating and material data as well as A,B parameters are listed in the following tables

$\eta = 200$ Pas, $\rho = 1000$ kg/m^3, $c_v = 2000$ J/kg/K, $n = 300$ rpm

Table 4.5 Extruder Machine Data

Pitch	40 mm	60 mm
A_1	0.19	0.292
A_2	6100	3630
B_1	0.9	1.34
B_2	5165	5040
Barrel diameter	57.35 mm	
Axial distance	48 mm	
Screw-barrel clearance	0.17 mm	
Screw-screw clearance	0.525 mm	

The maximum net power point is at $0.5 \cdot A_1$. In the equation for the power input, $0.5 \cdot A_1$ must therefore be entered for the dimensionless throughput. The following results are obtained:

$$\begin{aligned} &60mm \rightarrow P^*_{Net,max} \Rightarrow P_L = \frac{A_1 A_2}{4} \eta_0 n^2 L d_G^2 = 1\,\text{kW} \\ &40mm \rightarrow P^*_{Net,max} \Rightarrow P_L = \frac{A_1 A_2}{4} \eta_0 n^2 L d_G^2 = 1.094\,\text{kW} \end{aligned} \tag{4.111}$$

The maximum net power is greater at the small pitch, so a larger pressure build-up is possible at the small pitch compared to the 60 mm pitch. In order to calculate the pressure build-up over the length of $4 \times D$, the equation for the pressure throughput characteristic is converted after the pressure and dimensionally calculated.

$$\begin{aligned} &60mm \rightarrow \Delta P^* = A_2 \left(1 - \frac{\dot{V}^*}{A_1}\right) \Rightarrow \Delta P = (1 - \Lambda) \frac{A_2 \eta n L}{D} = 72.6\ bar \\ &40mm \rightarrow \Delta P^* = A_2 \left(1 - \frac{\dot{V}^*}{A_1}\right) \Rightarrow \Delta P = (1 - \Lambda) \frac{A_2 \eta n L}{D} = 122\ bar \end{aligned} \tag{4.112}$$

This comparison shows that the small pitch can build up considerably more pressure with the same back-pressure length and the same dimensionless conveying parameter 0.5.

4.6.2 Task: Partial Filling

The next task refers to the partially filled area of the machine and compares the same elements as in Task 1 (Section 4.6.1). First, the power diagram in the partially filled area will be discussed. The purple straight line shows the course of the power characteristic in the partially filled area of the extruder. There, the extruder does not build up any pressure, but it does convey the melt. In order to become familiar with the equations, the power input over a length of 3 × D and the temperature increase should be calculated for the two elements.

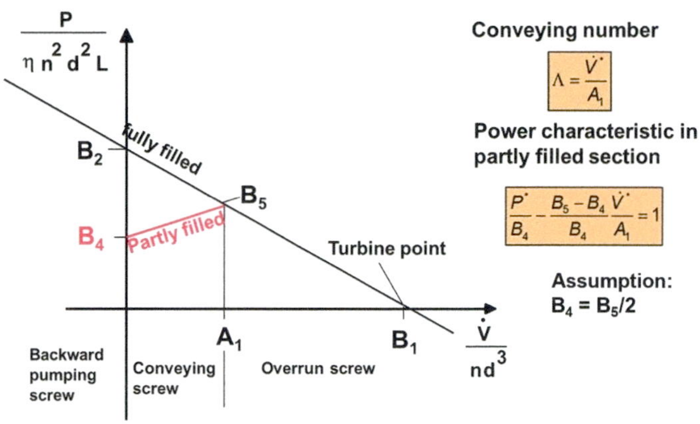

Figure 4.59 Power characteristics

Calculate B_5, the power input at A_1 and then the power input in the partially filled area over a length of 3 × D_G and the temperature increase.

The machine data, material data and operating parameters are the same as in Task 1. The throughput is 300 kg/h.

In order to be able to apply the equation for the power input in the partially filled area, point B_5 must be calculated, this is the power input at A_1. The value for B_4 is assumed to be $0.5 \cdot B_5$, as a good approximation.

$$\frac{\dot{V}^{\cdot}}{B_1} + \frac{P^{\cdot}}{B_2} = 1, \quad B_5 = P^{\cdot}(\dot{V}^{\cdot} = A_1) = \left(1 - \frac{\dot{V}^{\cdot}}{B_1}\right) B_2$$

$$60mm \rightarrow B_5 = 3942 = \frac{P_L}{\eta_0 n^2 L d_G^2} (\Rightarrow P_L = 11.15 \text{ kW})$$

$$40mm \rightarrow B_5 = 4075 = \frac{P_L}{\eta_0 n^2 L d_G^2} (\Rightarrow P_L = 11.53 \text{ kW}) \quad (4.113)$$

The smaller pitch has a higher B_5 value, so that the power line in the partially filled zone runs on a higher level. If you now calculate the power input for conveying the melt, you will get this result:

$$\Lambda = \frac{\dot{V}^*}{A_1} = \frac{\dot{V}}{A_1 n D^3} \quad \begin{matrix} 60mm \rightarrow \Lambda = 0.3025 \\ 40mm \rightarrow \Lambda = 0.465 \end{matrix}$$

Assumption:
$B_4 = B_5/2$

$$\frac{P^*}{B_4} - \frac{B_5 - B_4}{B_4}\frac{\dot{V}^*}{A_1} = 1 \Rightarrow P_t = \eta_0 n^2 L\, D^2 \left(1 + \frac{B_5 - B_4}{B_4}\frac{\dot{V}^*}{A_1}\right) B_4 \quad \begin{matrix} 60mm \Rightarrow P_t = 5.86\text{kW} \\ 40mm \Rightarrow P_t = 8.94\text{ kW} \end{matrix} \tag{4.114}$$

The 60 mm pitch delivers the same throughput with less power than the 40 mm pitch. The large pitch is therefore more efficient in the partially filled zone. In the temperature increase this makes almost 20 K between the elements. It should be noted that the viscosity is assumed to be constant. In reality, the viscosity would decrease with increasing temperature and the effect would be weaker.

$$\begin{matrix} 60mm \rightarrow \Delta T = \dfrac{P_L}{\dot{m}\, c_p} : 35.2\text{ K} \\ 40mm \rightarrow \Delta T = \dfrac{P_L}{\dot{m}\, c_p} : 53.6\text{ K} \end{matrix} \tag{4.115}$$

Conclusion: Basically it can be said that a small pitch builds up pressure better and a large pitch is more efficient in the partially filled area. The entire design of the pressure build-up zone and the partially filled areas must always be taken into account together with the operating point. The next tasks deepen this topic and get deeper into these questions.

4.6.3 Task: Design of a Pressure Build-up Zone with Uniform Pitch as Well as Fully and Partially Filled Areas

Now a complete zone with partially filled and fully filled areas at the end of the extruder in front of the die is to be calculated. Screw elements with the same pitch are to be used in the entire area. The elements used are from Task 1 (Section 4.6.1) with 40 mm and 60 mm pitch. In addition, two throughputs are compared with each other. The material data as well as A,B parameters are to be taken from Task 1. First, the following values for 300 kg/h are to be calculated:

Conveying parameter Λ, back-pressure length and the length of the partially filled area, pump efficiency e, power input in the partially and fully filled area, the temperature increase in the partially and fully filled area.

Calculation of the conveying parameter Λ:

$$\Lambda = \frac{\dot{V}^*}{A_1} = \frac{\dot{V}}{A_1 n D^3} \quad \begin{matrix} 60\text{ mm} \rightarrow \Lambda = 0.3025 \\ 40\text{ mm} \rightarrow \Lambda = 0.465 \end{matrix} \tag{4.116}$$

Calculation of the back-pressure length for both elements:

$$L=\frac{1}{1-\Lambda}\cdot\frac{1}{A_2}\cdot\frac{\Delta p\cdot D}{\eta\cdot n} \qquad \begin{array}{l}60\text{ mm} \rightarrow L = 190\text{mm}\\ 40\text{ mm} \rightarrow L = 150\text{mm}\end{array} \tag{4.117}$$

This partially filled length comes from the difference between the total length and the back-pressure length.

$$\begin{array}{l}60\text{ mm} \rightarrow L_t = 460\text{mm} - 190\text{mm} = 270\text{mm}\\ 40\text{ mm} \rightarrow L_t = 460\text{mm} - 150\text{mm} = 310\text{mm}\end{array} \tag{4.118}$$

The length of the partially filled area is needed to calculate the energy input in this area and then the temperature increase. Next, the pump efficiency is calculated. It is to be expected that the efficiency for the 40 mm pitch from this operating point is higher, as the value for lambda is closer to the maximum of 0.5:

$$e=\frac{\Delta P\dot{V}}{P_L}=A_1\frac{A_2}{B_2}\cdot\frac{\Lambda(1-\Lambda)}{\left(1-\frac{A_1}{B_1}\Lambda\right)} \qquad \begin{array}{l}60\text{ mm} \rightarrow e = 4.8\ \%\\ 40\text{ mm} \rightarrow e = 6.25\%\end{array} \tag{4.119}$$

Interim conclusion on the results: The small pitch has a higher efficiency and therefore a shorter back-pressure length. As a result, a lower temperature increase is to be expected during pressure build-up. The conveying parameter lambda is higher at the small pitch, which indicates that the energy input in the partially filled section should be higher. The next calculations show the exact results. Please note that we expect a high viscosity and that the material is Newtonian. This results in a very large temperature increase. In reality this would be much less because the viscosity decreases with the temperature. However, the trends and the comparison between the elements remain the same, but at a different temperature level.

Calculation of the power entry in the fully filled area:

$$\frac{1}{B_1}\cdot\frac{\dot{V}}{n\cdot D^3}+\frac{1}{B_2}\cdot\frac{P_L}{\eta\cdot n^2\cdot L\cdot D^2}=1$$

$$P_L=\left(1-\frac{\dot{V}}{nD^3B_1}\right)B_2\,\eta n^2LD^2 \qquad \begin{array}{l}60\text{ mm} \rightarrow P_L = 8.7\text{ kW}\\ 40\text{ mm} \rightarrow P_L = 6.7\text{ kW}\end{array} \tag{4.120}$$

Calculation of the power input in the partially filled area:

$$P_t=\eta_0 n^2L_tD^2\left(1+\frac{B_5-B_4}{B_4}\frac{\dot{V}^*}{A_1}\right)B_4 \qquad \begin{array}{l}60\text{ mm} \rightarrow P_t = 5.37\text{ kW}\\ 40\text{ mm} \rightarrow P_t = 9.2\text{ kW}\end{array} \tag{4.121}$$

Resulting temperature increase in the fully filled

$$\Delta T = \frac{\Delta P}{\rho c e} \qquad \begin{array}{l} 60\text{ mm} \rightarrow \Delta T = 52.6\text{ K} \\ 40\text{ mm} \rightarrow \Delta T = 40.4\text{ K} \end{array} \tag{4.122}$$

and partially filled area:

$$\Delta T = \frac{P_t}{\dot{m} c} \qquad \begin{array}{l} 60\text{ mm} \rightarrow \Delta T = 32.3\text{ K} \\ 40\text{ mm} \rightarrow \Delta T = 55.2\text{ K} \end{array} \tag{4.123}$$

The result shows that the 40 mm and 60 mm pitch in the fully or partially filled areas work differently efficiently at this operating point. The 40 mm pitch builds up the pressure more efficiently, while the 60 mm pitch provides less energy input in the partially filled areas. The total temperature increase in comparison is about 10 K higher at the 40 mm pitch:

$$\begin{array}{l} 60\text{ mm} \rightarrow \Delta T = 84.9\text{ K} \\ 40\text{ mm} \rightarrow \Delta T = 95.6\text{ K} \end{array} \tag{4.124}$$

Now the throughput should be increased to 500 kg/h and compared with the results of 300 kg/h. The results are given below. The pressure build-up conditions are now reversed and the 60 mm gradient builds up the pressure more efficiently than the 40 mm gradient. The back pressure length is therefore shorter at the 60 mm gradient. The amount of the back-pressure length is larger compared to 300 kg/h, because more material is conveyed at the same speed.

300l/h		500 l/h:
60 mm → Λ = 0.3025 40 mm → Λ = 0.465	**Conveying parameter**	60 mm → Λ = 0.504 40 mm → Λ = 0.775
60 mm → L = 190mm 40 mm → L = 150mm	**Backpressure length**	60 mm → L = 270 mm 40 mm → L = 350 mm
60 mm → e = 4.8 % 40 mm → e = 6.2 %	**Pumping efficiency**	60 mm → e = 5.9 % 40 mm → e = 4.7%

Figure 4.60 Comparison of operating points for T = 60 mm and 40 mm at 300 l/h and 500 l/h

The overall energy input is greater as more is conveyed. The 40 mm gradient now requires significantly more energy than the 60 mm gradient for the same amount at the same speed. This can also be seen in the temperature increases. With the pressure up the trend reverses itself between 40 mm and 60 mm in the temperature increase with throughput increase on 500 kg/h. The temperature increase in the partially filled area becomes lower overall, as the length in this area has become shorter due to the greater back-pressure length at 500 kg/h.

60 mm → P_L = 8.7 kW 40 mm → P_L = 6.7 kW	**Power filled**	60 mm → P_L = 11.76 kW 40 mm → P_L = 14.85 kW
60 mm → P_t = 8.17 kW 40 mm → P_t = 10.5 kW	**Power Partly filled**	60 mm → P_t = 6.43 kW 40 mm → P_t = 4.18 kW
60 mm → ΔT = 52.6 K 40 mm → ΔT = 40.4 K	**Temperature Increase full**	60 mm → ΔT = 42.3 K 40 mm → ΔT = 53.4 K
60 mm → ΔT = 49.0 K 40 mm → ΔT = 63.1 K	**Temperature Increase partly filled**	60 mm → ΔT = 23.2 K 40 mm → ΔT = 15.0 K

Figure 4.61 Calculation results for various gradients

When considering the total power and total temperature increase over both zones, partially and fully filled, it can be seen that the temperature development in the areas is very different depending on the operating point. A high throughput leads to a lower temperature increase, as the energy input is distributed over "more" product. In this case it is approx. 25–35 K less. The differences between the pitches are approx. 7–10 K. By looking at the individual zones it can be seen that it is necessary to equip the screw with different elements in the different zones in order to achieve an optimization. In addition, it is important to consider the operating point in order to work on a high pump efficiency. The next task deals with this issue.

300l/h		500 l/h:
60 mm → P = 16.9 kW 40 mm → P = 17.2 kW	**Power**	60 mm → P = 18.19 kW 40 mm → P = 19.3 kW
60 mm → ΔT = 101.6 K 40 mm → ΔT = 103.5 K	**Temperature increase over all**	60 mm → ΔT = 65.5 K 40 mm → ΔT = 68.4 K

Figure 4.62 Comparison of operating points for T = 60 mm and 40 mm at 300 l/h and 500 l/h

The diagram below shows how the operating point affects the used efficiency for both pitches. By shifting the operating point on the efficiency curve you can see how the optima of the calculated cases moves for 300 kg/h and 500 kg/h. The

40 mm gradient, for example, is at 300 kg/h close to the optimum and at 500 kg/h the operating point falls into a region with lower efficiency. Therefore, the elements must be selected in combination with the operating points.

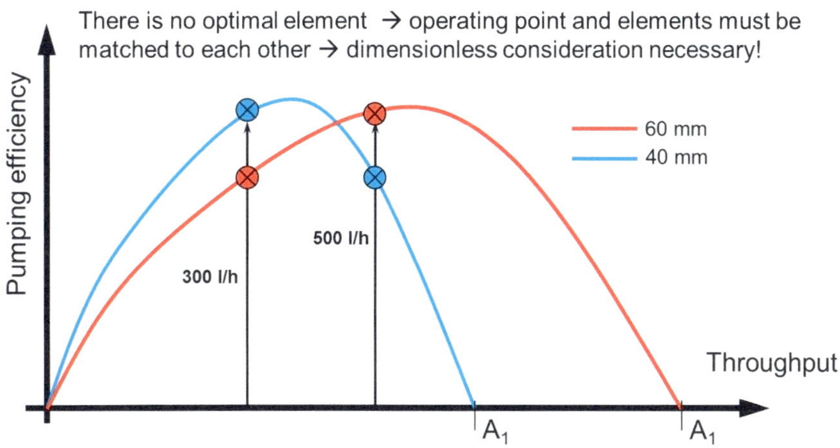

Figure 4.63 Pumping efficiency curves for two gradients

4.6.4 Task: Design of the Pressure Build-up Zone with Various Elements with 40 mm and 60 mm Pitch Combined

Combining the 40 mm pitch with the 60 mm pitch further optimizes the energy input. Assumption are four 40 mm elements for pressure build-up and the rest is equipped with 60 mm pitch, i. e. five elements. This corresponds to the total length of 460 mm. We consider the operating point at 300 kg/h and calculate the power input and the temperature increase. The back-pressure lengths can be taken from the first task and are approx. 160 mm for the 40 mm pitch, the remaining 300 mm are partially filled. The example was chosen as it fits well with the assembly of the elements. Otherwise, an exact calculation with case differentiation and pressure curve analysis has to be made, as already shown before.

The power inputs and temperature increases are summarized here. The 40 mm pitch is in the fully filled area and the 60 mm pitch is in the partially filled area.

40 mm → P_L = 6.7 kW
60 mm → P_t = 5.9 kW
40 + 60 mm → P = 12.6 kW

40 mm → ΔT = 40.4 K
60 mm → ΔT = 35.1 K
40 + 60 mm → ΔT = 75.5 K

Figure 4.64 Power and temperature increase

By the combination of the elements 10–20 K can be saved in the temperature increase for this example. Looking at the 40 mm as a comparison, the temperature reduction is 20 K and the energy saving is approx. 3 kW. Related to 15.8 kW this is a reduction in energy input of 20%. The values are here very high for the temperatures, since we consider the Newtonian case. A real product reduces the viscosity at a higher temperature, so that the energy input is lower as already mentioned.

4.6.5 Task: Impact of Shear Thinning Effects

Polymers often show a structurally viscous behavior. This means that the viscosity decreases with increasing shear rate. This must be taken into account when designing the extruder. In the following an approach is shown and an example is calculated. The models are one-dimensional. However, the shear rate in the extruder has a spatial distribution. This can be shown with CFD simulations. This distribution must be taken into account representatively by a single shear rate in the one-dimensional model. The shear rate in the extruder is influenced by the rotational speed. When speed increase means an increase of the shear rate. For a pipe the representative shear rate is shown below and analogously a proposal for a representative shear rate for an extruder using the sheet-sheet model. The representative shear rate corresponds to the tangential speed of the screw element at the outermost radius divided by half the flight depth.

Pipe flow

$$\dot{\gamma}_{rep} = \frac{\dot{V}}{r^3}$$

Plate model Of screw channel

$$\dot{\gamma}_{rep} = \frac{2\pi n d_G}{H} \qquad H = \frac{d_G - d_k}{2} \tag{4.125}$$

With the calculated representative shear rate the rep. viscosity can be read from a viscosity diagram or calculated by inserting into a viscosity model.

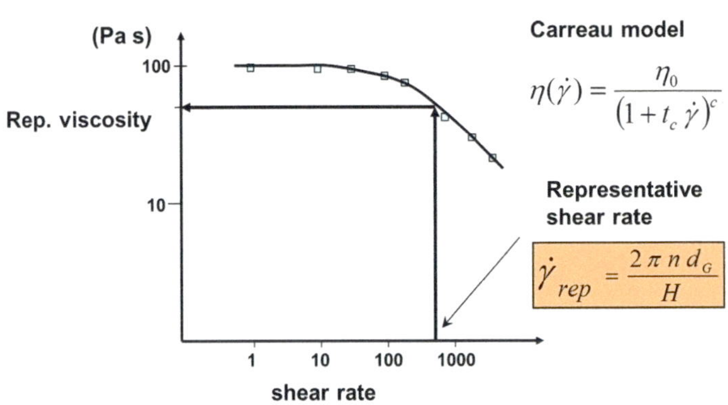

Figure 4.65 Simplified determination of the representative shear rate and the corresponding viscosity

Now calculate the back-pressure length for the Newtonian and viscous case for the following parameters and operating points of the extruder:

Operation and property data:

η_0 = 130 Pas, V = 300 l/h, pressure 5 bar, n = 120 rpm, H = 10 mm, c = 0.3, t_c = 0.0157 s

Table 4.6 Machine Data

A_1	0.32
A_2	2480
B_1	1.385
B_2	3140
Barrel diameter	57.35 mm
Axial distance	48 mm
Screw-barrel clearance	0.17 mm
Screw-screw clearance	0.525 mm

It can be seen that due to the reduced viscosity the back-pressure length increases compared to Newtonian case. This must always be considered and therefore good rheological measurements are a prerequisite for the design.

Newtonian

$$\frac{\dot{V}^*}{A_1} + \frac{\Delta P^*}{A_2} = 1 \Rightarrow \Delta P^* = (1-\Lambda)A_2 = \frac{\Delta P D}{\eta n L}$$

$$L = \frac{1}{A_2(1-\Lambda)} \cdot \frac{\Delta p \cdot D}{\eta \cdot n} = 0.143\,\text{m} \qquad \Lambda = \frac{\dot{V}^*}{A_1} = \frac{\dot{V}}{A_1 n d^3} = 0.69$$

Shear thinning

$$\dot{\gamma}_{rep} = \frac{2\pi n d_G}{H} = 72\ \text{s}^{-1} \qquad \eta_{rep}(\dot{\gamma}_{rep}) = \frac{\eta_0}{(1+t_c\,\dot{\gamma}_{rep})^c} = 104\,\text{Pas} \tag{4.126}$$

$$L = \frac{1}{A_2(1-\Lambda)} \cdot \frac{\Delta p \cdot D}{\eta_{rep} \cdot n} = 0.179\,\text{m}$$

4.7 Computational Fluid Dynamics

Michael Bierdel, Olaf Wünsch

4.7.1 Introduction to Computational Fluid Dynamics

Numerical methods are a working tool which we encounter almost every day in the fields of engineering and science. The complexity of numerical methods ranges from simple spreadsheets to the solution of non-linear differential equation systems, as they also occur in fluid dynamics. Especially in extruders with their complex geometries, these systems of equations can no longer be treated analytically. Here numerical simulations can provide a very good approximate solution. The aim of computational fluid dynamics, or CFD, is to obtain a deeper understanding of the flow processes taking place within an extruder in order to ensure, together with experiments, a safe and economical extruder design.

Correctly applied, computational fluid dynamics offers a range of benefits:

- **Problem understanding:** In many cases, experiments can provide only reliable integral values. In the case of twin screw extruders, for example, these are the shaft torque and the pressure and the temperature at the extrusion nozzle. Computational fluid dynamics, however, provide local information about pressure, velocity, and temperature within the overall computational domain. The calculation of gradients provides additional information about the shear rate or the heat transfer coefficients.
  In experiments, the measurement result is frequently obtained with an overlapping of different individual effects which cannot be entirely separated from one another. The complexity of a computational model, however, can be increased incrementally to enable the identification of process values decisive for product quality.
- **Specific optimization:** By varying the geometry and the operating conditions it is possible to optimize specific areas of the process. Since calculations are carried out using real system values, better scale-up and/or scale-down rules can be created which will improve the design reliability.
- **Smaller number of experiments:** Computational fluid dynamics limits the range of relevant process parameters and thereby reduces the number of experiments required. In particular, it is possible to identify critical system conditions and to eliminate them from the experiment in advance. This results in considerable time and cost savings.

In addition to the advantages described above, the disadvantages of computational fluid dynamics must also be mentioned and taken into account:

- **Limited computational domain:** In the case of complex three-dimensional, time-dependent fluid processes we cannot calculate the flow throughout the entire extruder because today's computers do not (yet) have the required computing power.
- **Boundary conditions:** Furthermore, in CFD frequently the fact is ignored that suitable boundary conditions are required at all flow boundaries. This is a major difficulty, for example, when calculating viscoelastic flow properties. We do not know the history of a fluid element before it entered the computational domain.
- **Model quality:** To date, there are no suitable computational models available for certain flow conditions such as melting. In many cases, however, it is possible to simplify the model to such a degree that computational fluid dynamics correctly reflects the trends, allowing movement in the right direction in the experiment.
- **Computational inaccuracies:** The calculations exhibit only limited accuracy, and model uncertainties exist. Numerical rounding and termination errors can lead to an incorrect appraisal of the flow processes. Such cases require an operator with a lot of experience in recognizing and avoiding such potential errors in good time.

A comparison of the advantages and disadvantages of computational fluid dynamics reveals that simulation cannot replace experimentation even in the long term (see also Section 1.4). However, effective application of "numerical" and "real" experiments will minimize the overall expenditure on process development.

Workflow of a Computational Fluid Dynamics Process

A computational fluid dynamics process can be divided into three steps: pre-processing, flow computation, and post-processing (Figure 4.66). The users are faced with a multitude of commercial programs for each step. Newcomers in particular find it difficult to choose a suitable program, in part because the development of codes is advancing at a rapid pace. For many years, *www.cfd-online.com* has provided an excellent overview of CFD (suppliers, literature, further links).

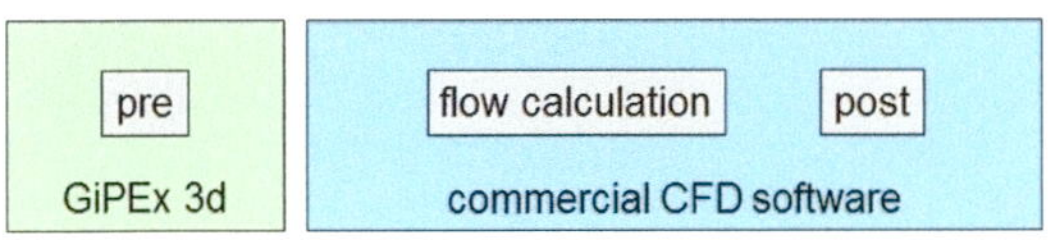

Figure 4.66 Subdivision of a flow simulation into the three main stages of pre-processing, flow calculation, and post-processing

Pre-Processing

The term pre-processing refers to geometry and grid generation. Currently, there are no suitable commercial codes available for twin screw extruders. While the geometry is easy to define in general, it is very difficult to generate a computational grid for a twin screw extruder in practice. It requires the resolution of the interaction both between the screws and between the screws and the barrel by means of a grid, which correctly reflects the steep flow gradients. Automatic grid generation is achieved only with a tetrahedral grid. This process, however, quickly requires several million nodes to illustrate the interactions involved, resulting in unacceptable computation times.

A hexahedral grid is the optimum solution. Commercial grid codes tend to fail in the complex intermeshing area between the two screws.

Therefore, the program GiPEx 3d is under constant development, which automatically generates a grid on the basis of geometrical parameter inputs. Figure 4.67 shows the mesh on the surface of a right-handed and a left-handed conveying element with two flights. Figure 4.68 illustrates the grid in the cross-section of single- to four-flighted screw elements. A mesh with a million nodes takes less than a minute to generate. It requires only a fraction of the time taken to calculate a flow in the twin screw extruder.

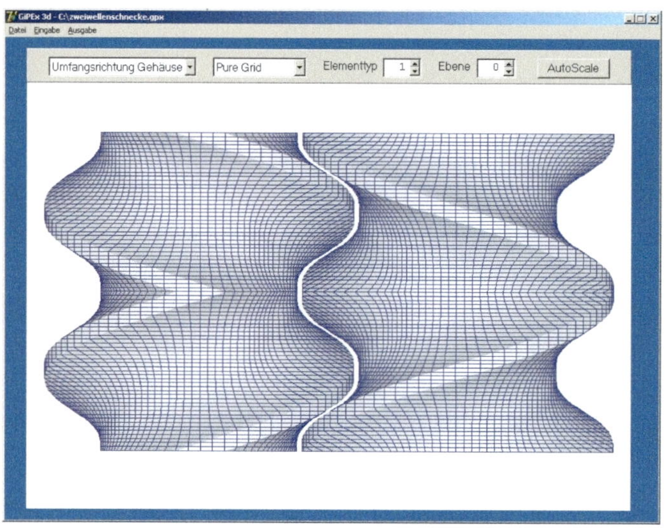

Figure 4.67 Grid on the surface of a right-handed and a left-handed conveying element with two threads

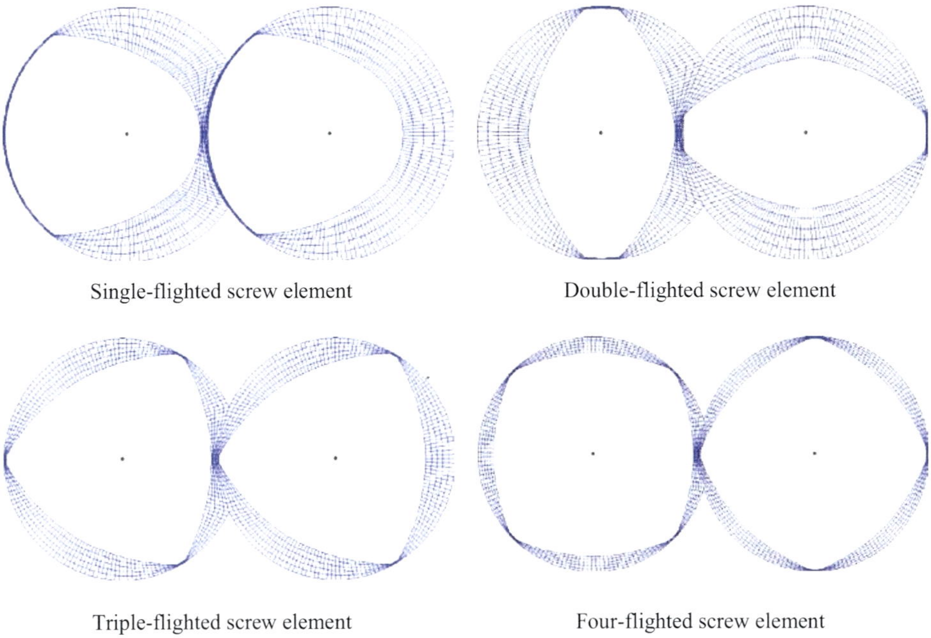

Figure 4.68 Grid in the cross-section of conveying elements with different numbers of threads

Flow Computation and Post-Processing

Once the grid has been generated, the required physical properties and operating conditions are fed into the solver for the flow computation. Depending on the selected flow model, it is also necessary to adjust the parameters of the solver to guarantee the fastest possible, yet still stable, convergence. The differential equation system is solved iteratively. The subsequent post-processing involves a comprehensive analysis of the calculated flow fields. The analysis is not trivial and well-founded previous knowledge about the behavior of twin screw extruders is of great value.

Several commercially available programs from different manufacturers are suitable in principle for the calculation of flows in twin screw extruders; they include but are not limited to: Cfx, Fluent, Polyflow (respectively Ansys), and Star-CD (CD-adapco). This list includes the programs tested and used by the author. The program manufacturers offer a number of introductory courses to instruct users in the effective use of the solver and post-processing. These are recommended for all CFD newcomers.

4.7.2 Fully Filled Screw Sections

The possibilities of CFD are demonstrated in this section on the basis of two examples. The focus is exclusively on post-processing; in other words, the analysis of the computation results.

4.7.2.1 Example 1

Geometry, Physical Properties, and Operating Conditions

The first example shows a simple conveying element (Figure 4.69). The geometry of the screw element is uniquely defined by the following six parameters:

▪ Number of flights	2
▪ Barrel diameter	61.0 mm
▪ Center distance	50.0 mm
▪ Screw-barrel clearance	0.5 mm
▪ Screw-screw clearance	1.0 mm
▪ Pitch	120.0 mm

Because the geometry of a double-flighted screw element repeats after half the pitch, it is sufficient to observe a length of 60 mm. Assuming a fully developed, isothermal flow, the upper and lower cross-section surfaces are linked together by periodic boundary conditions. As usual, the no slip condition applies to the flow at the barrel and the screws.

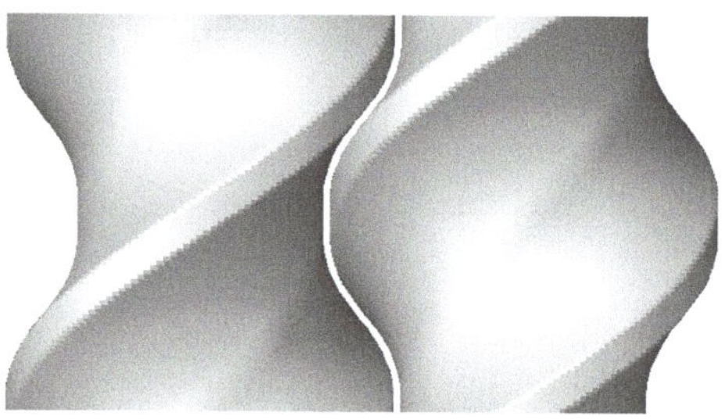

Figure 4.69 Geometry of the first example

The polymer melt used in this example has a density of 1000 kg/m^3. The following initially assumes a Newtonian flow behavior with a viscosity of 1000 Pa·s. In later computations, a more realistic shear thinning flow behavior is assumed, which can be described using the power law equation. The flow exponent n ranges between 0.4 and 0.9 and the consistency factor is 1000 Pa·s^n. The heat capacity is 2000 J/(kg K) and the heat conductivity is 0.2 W/(m K). The flow field was initialized with a uniform temperature of 300 °C at the start of the computational simulation. All walls are viewed as adiabatic, both those of the screws and of the barrel.

The influence of temperature on the physical properties is not taken into account. The speed of the two shafts is 95 rpm.

Conveying and Power Characteristic

Two cases are observed in more detail as specific operating points for a Newtonian flow behavior:

In the first case the axial pressure difference is 0 bar: This operating point corresponds to an extruder operation in which a process zone is situated between two degassing zones.

Figure 4.70 shows the pressure fields for a 90° rotation in 8 separate images. The color scale ranges from −35 bar (blue) to +35 bar (red)[1]. It shows that outside the intermeshing area, a maximum pressure of approx. 10 bar is generated upstream from each screw tip and a minimum pressure of approx. −10 bar downstream from each screw tip. Inside the intermeshing region, one screw tip of each shaft meet in a way that, together with the barrel, they enclose a volume (see 32.4° position and 57.6° position). Local pressure peaks of up to 25 bar occur if the enclosed volume is reduced by the rotation of the screws (32.4° position). If the enclosed volume is enlarged, however (57.6° position), the local pressure peaks can reach −25 bar. The remaining areas of the flow cross-section exhibit only very small pressure differences.

[1] Negative pressures are not uncommon in computational fluid dynamics. In the case of fluids with constant densities we are not concerned with the absolute pressure but with the pressure differences. The absolute pressure is determined with the addition of a static pressure. The static pressure must be at least 35 bar to avoid cavitation in the screw element under observation.

Figure 4.70 Pressure at a pressure difference of 0 Pa. The color scale ranges from −35 bar (blue) to +35 bar (red)

The distribution of the axial velocity is shown in Figure 4.71. The color scale ranges from −0.17 m/s (blue) to +0.32 m/s (red). An axial velocity of 0 m/s (turquoise) is present at the walls and also at the rotating screws due to wall adhesion. We observe a flow inside the channels of approx. 0.1 m/s in the positive z-direction, irrespective of the screw position. In the intermeshing region, also irrespective of the screw position, there is a clear velocity increase of 0.2 m/s compared to the channels. Over the screw tips, a backflow is observed. This is due to the pressure difference over the screw tips, which acts not only in the circumferential but also in the axial direction.

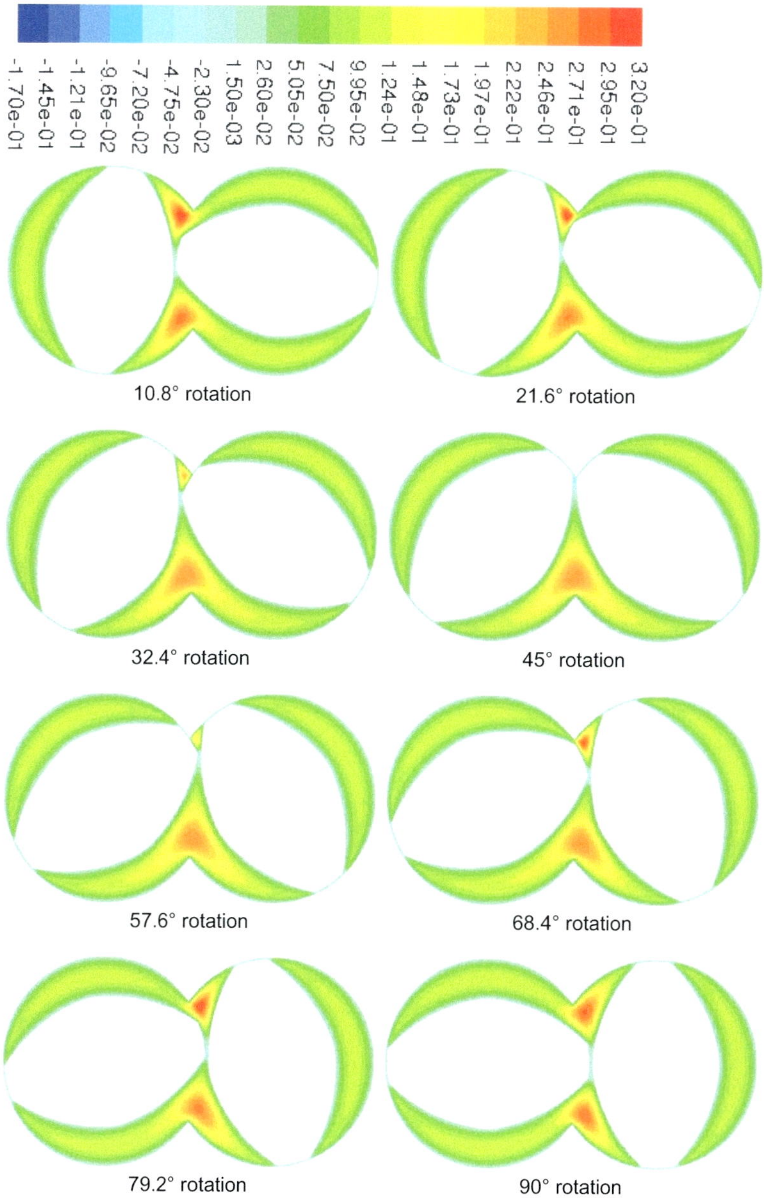

Figure 4.71 Axial velocity at a pressure difference of 0 Pa. The color scale ranges from −0.17 m/s (blue) to +0.32 m/s (red)

Since pressure and velocity fields are coupled, wherever there is a pressure gradient, a flow in the direction of the falling pressure results. Figure 4.72 illustrates this. In the upper image of Figure 4.72 the pressure field is shown in two selected sections through the screw element. The first section lies in the intermeshing region in axial direction and is represented by the central part of the image. The second section lies parallel to the first section in the axis of the right shaft and is represented by the left and right parts of the image. The position of the section planes and the line of sight are indicated by the pictogram. In the lower image of Figure 4.72 the axial velocity is displayed in the same way. It is shown that the pressure difference across the screw tips in the intermeshing region (50 bar) is considerably greater than in the barrel region (20 bar). However, this also means that the pressure gradient from tip to tip is considerably greater in the intermeshing region than in the barrel region. Due to the higher pressure gradient in the axial direction in the intermeshing region, a considerably higher axial velocity results here. This effect is increased by the fact that the free flow cross-section in the intermeshing region is significantly larger than in the barrel region.

In the second case the axial flow rate is 0 kg/s. This operating point corresponds to an extruder operation with closed discharge nozzle, an interesting case from a theoretical point of view, which is certainly not intended in practice.

Figure 4.73 shows the pressure fields for different screw positions. A back pressure of 47.8 bar is required to set a flow rate of 0 kg/s [2]. For comparison purposes, the color scaling is selected as shown in Figure 4.70. The pressure before and after the screw tips has increased to +15 bar and decreased to −15 bar respectively. A clear pressure gradient is now clearly visible in the channels of the two screws.

[2] The desired mass flow is generated numerically by applying a constant positive or negative pressure gradient along the screw axes. This pressure gradient is considered as a separate term in the momentum balances. In this and all other figures relating to pressure, the fraction based on the pressure gradient is not shown.

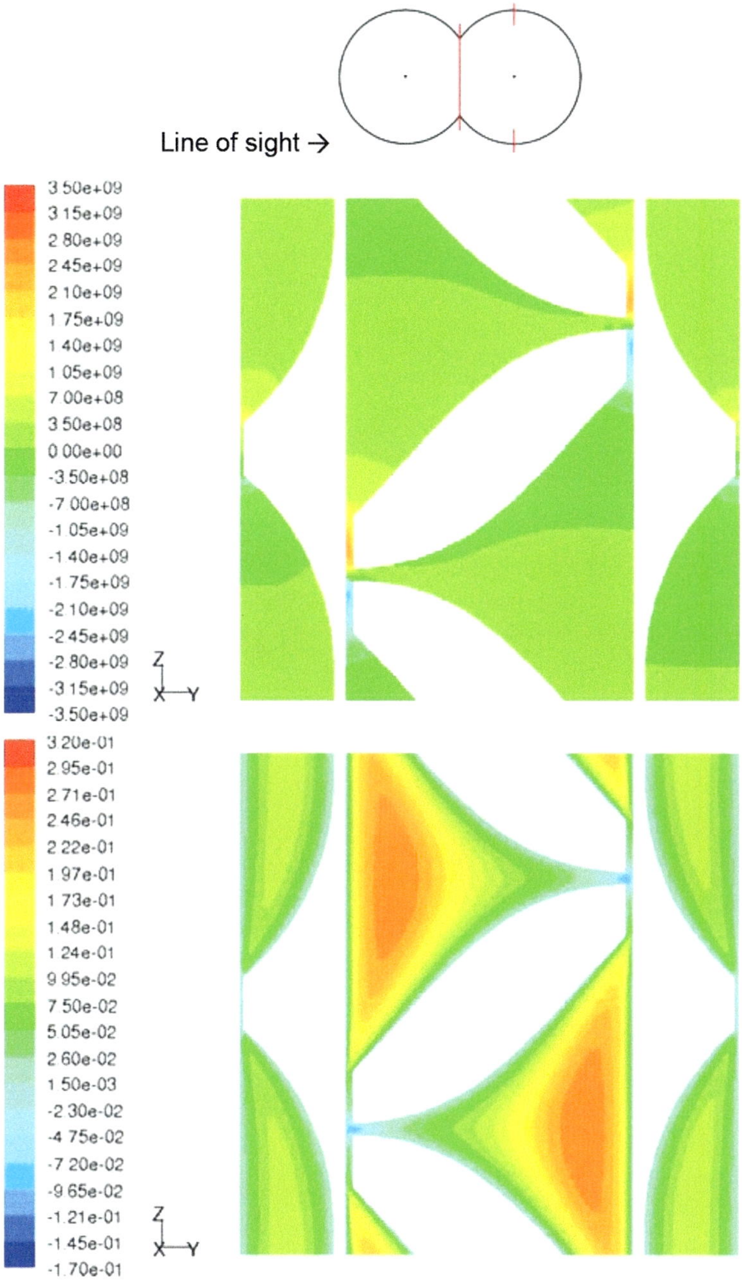

Figure 4.72 Pressure (top) and axial velocity (bottom) at a pressure difference of 0 Pa. Each figure shows two sections (pictogram comparisons). 1st section: section through the intermeshing area (central part of the figures). 2nd section: section through the axis of the right-hand shaft parallel to the first section (left and right parts of the figures). The pressure color scale ranges from −35 bar to +35 bar. The axial velocity color scale ranges from −0.17 m/s to +0.32 m/s

Figure 4.73 Pressure at a mass flow of 0 kg/s. The color scale ranges from −35 bar (blue) to +35 bar (red)

These changes in the pressure field lead to a completely different flow pattern. The axial velocities in Figure 4.74 (same color scaling as in Figure 4.71) illustrate this.

In the intermeshing region, the fluid is still pumped in the positive z-direction against the pressure gradient. In the channels, however, it flows back in axial direction, resulting in an overall zero mass flow over the cross-sectional area.

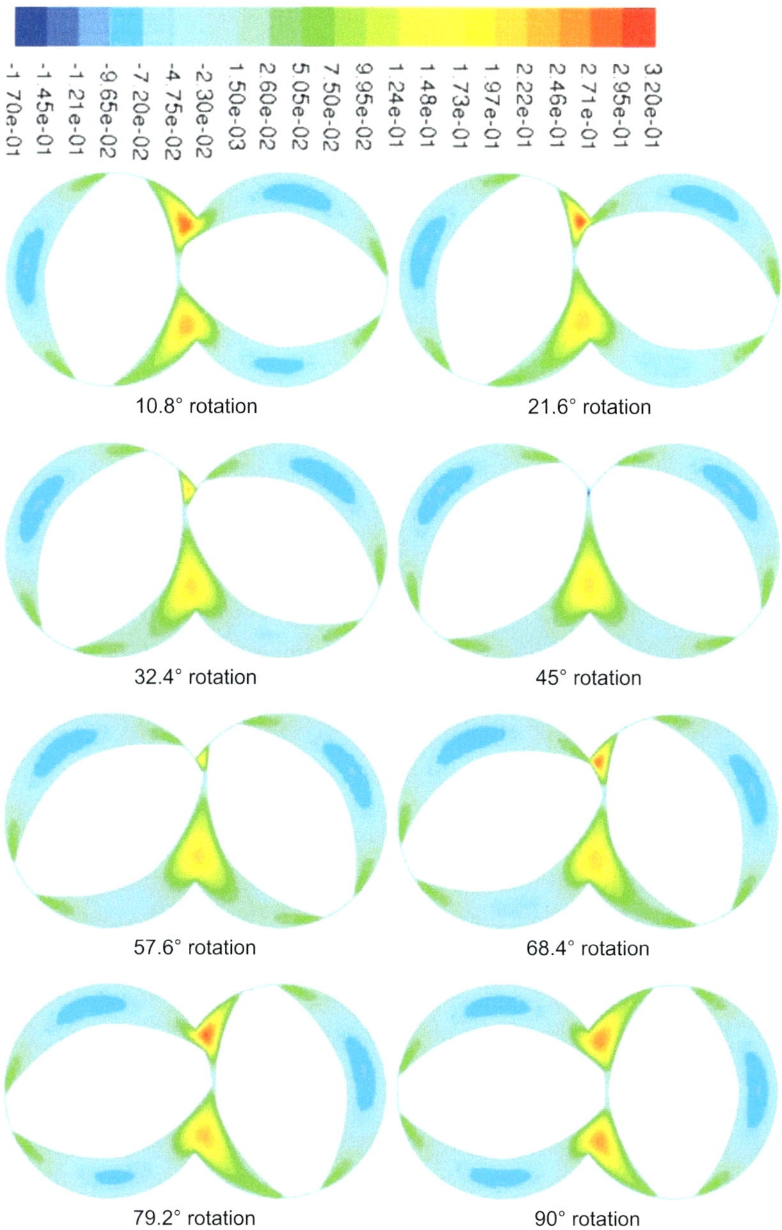

Figure 4.74 Axial velocity at a mass flow of 0 kg/s. The color scale ranges from −0.17 m/s (blue) to +0.32 m/s (red)

For a generally valid representation the calculation results are made dimensionless (see Sections 4.1 to 4.5). The dimensionless volume flow rate is:

$$\dot{V}^* = \frac{\dot{V}}{n\,D^3} \tag{4.127}$$

The dimensionless pressure is calculated as follows:

$$\Delta p^* = \frac{\Delta p\,D}{n\,\eta\,L} \tag{4.128}$$

And for the dimensionless power the following applies:

$$P^* = \frac{P}{n^2\,\eta\,L\,D^2} \tag{4.129}$$

with the barrel diameter D, the screw element length L, the speed n, the pressure difference Δp, the power P, the volume flow rate $\dot{V}$, and the viscosity η.

Since the Reynolds number is much smaller than 1 and a Newtonian flow behavior is considered initially, the Navier-Stokes equations convert into the Stokes equations and a system of linear differential equations for the flow calculation is obtained. This implies that there must be a linear correlation between the flow rate and the pressure as well as between the flow rate and the power. This becomes clear in Figure 4.75 and Figure 4.76, in which the dimensionless conveying and the power characteristic are displayed respectively. The red lines show the Newtonian cases and, as expected, linear correlations appear.

The power law equation is used for the viscosity for the other curves of the two diagrams. The flow exponent n is varied between 0.4 and 0.9. The choice of the power law equation results in a non-linear correlation between the flow rate and the pressure and the flow rate and the power (cf. Section 4.5). It is evident that the flow exponent has a very large influence both on the conveying characteristic and on the power characteristic.

As the flow exponent gets smaller, the polymer melt becomes increasingly shear-thinning and the pressure required to achieve a specific flow rate decreases (Figure 4.75). It is also noticeable that the intersection of the curves with the x-axis is in good approximation independent of the flow exponent. At the intersection there is no axial pressure gradient. A simple two-dimensional analogy model is the planar channel. If only the walls are moved and no pressure gradient is superimposed, according to the analytical solution the flow rate is independent of the flow properties of the fluid (cf. Section 3.1).

Also the power characteristic is strongly dependent on the flow exponent (Figure 4.76). The more shear thinning the polymer, the less power is required to transport the polymer melt. In contrast to the conveying characteristic, the intersection points of the individual curves with the *x*-axis lie far apart and are all located outside the diagram.

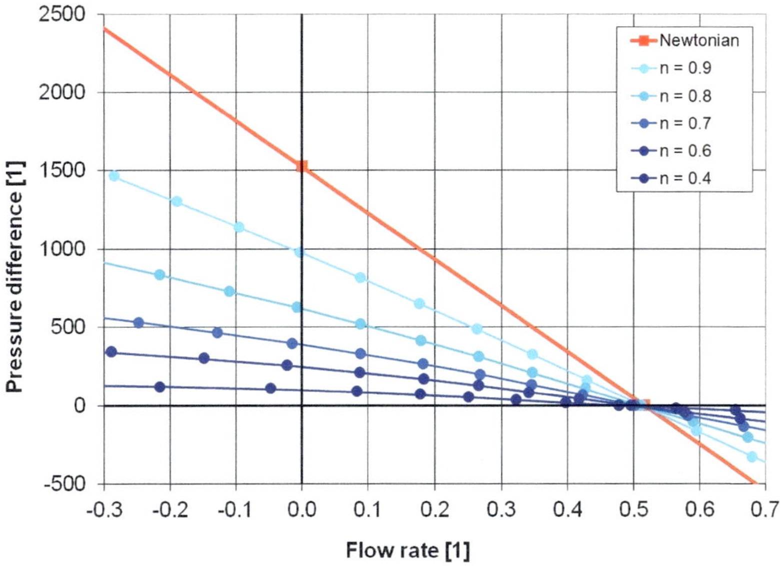

Figure 4.75 Conveying characteristic of a conveying element with Newtonian and shear-thinning viscosity

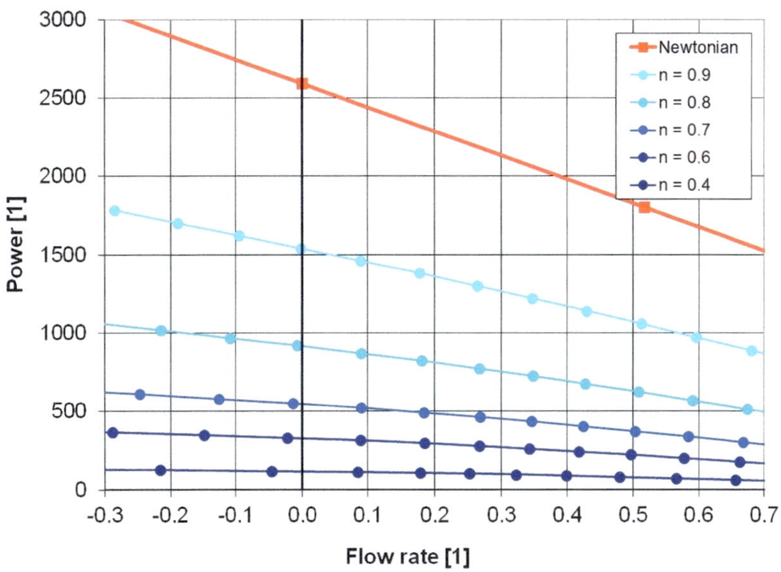

Figure 4.76 Power characteristic of a conveying element with Newtonian and shear-thinning viscosity

Mechanical Stress

In addition to the conveying and power characteristic, the mechanical and thermal stress of the polymer is of particular interest. The mechanical stress is characterized by the shear stress distribution in the polymer. The shear stress is calculated as follows

$$\tau = \eta(\dot{\gamma}) \cdot \dot{\gamma} \tag{4.130}$$

with the shear stress τ, the shear rate $\dot{\gamma}$, and the viscosity η, which generally depends on the shear rate. For a fluid with constant viscosity (Newtonian fluid), the shear stress is directly proportional to the shear rate.

Figure 4.77 Distribution of the shear rate within a conveying element, bottom: enlarged illustration around a screw tip. The color scale ranges from +21.5 s^{-1} (blue) to +857 s^{-1} (red)

For an axial pressure difference of 0 bar, the distribution of the shear rate and thus the distribution of the mechanical stress for a selected position of the screw element is shown in the upper part of the picture in Figure 4.77. The lower part of Figure 4.77 shows an enlargement of a screw tip. It can be seen that there is very little mechanical stress in the channels. In the intermeshing region it is already slightly increased. As expected, the greatest mechanical stresses occur in the clearance region between the two screws and between screw and barrel. Whether the mechanical stress between the two screws or between screw and barrel is greatest depends only on the clearances and the circumferential speeds of the screw tip and screw channel. In this case, the clearance between the two screws is twice as large as the clearance between the screw and the barrel. In contrast, the circumferential speed of the screw channel is only about 0.63 times the circumferential speed of the screw tip. Despite the opposite moving direction of the two screws in the intermeshing region, in this case the mechanical stress between the screw tip and the barrel is greatest. The lower picture with the enlarged screw tip shows that the shear rate in the gap between tip and barrel is not constant, as it would correspond to a simple shear flow. Instead, a pressure flow is superimposed on the shear flow. As a reminder, the pressure difference across the screw tip is approximately 20 bar. The superimposition of pressure and shear flow leads to a radial change of the shear rate. At the barrel wall, the shear rate is only about 75% of the shear rate at the screw tip.

Thermal Stress

The thermal stress is characterized by the distribution of dissipation in the polymer melt. For a Newtonian fluid, dissipation is calculated as follows

$$\dot{q}_{diss} = \eta \cdot \dot{\gamma}^2 \tag{4.131}$$

with the dissipation $\dot{q}_{diss}$, the viscosity η, and the shear rate $\dot{\gamma}$. The dissipation is proportional to the square of the shear rate. The distribution of dissipation in a cross-section of the screw element is shown in the upper part of the picture in Figure 4.78. The lower part of Figure 4.78 again shows an enlargement of the flow region around one of the screw tips. By squaring the shear rate, the regions of high dissipation are concentrated almost completely on the flow regions in the tip clearance compared to the shear stress. Analogous to the shear rate, a radial change of the dissipation across the tip clearance is also observed. Thus the dissipation at the barrel is only about 50% of the dissipation at the screw tip.

Figure 4.78 Distribution of dissipation in a conveying element, bottom: enlarged illustration around a screw tip. The color scale ranges from 4.64e5 W/m^3 (blue) to 7.34e8 W/m^3 (red)

The distribution of dissipation clearly shows where heat is generated in the flow domain. The question now arises where the heat is transported further and which local temperatures result from this. Figure 4.79 and Figure 4.80 show the temperature field development over time within the first two revolutions of the screws. Figure 4.79 illustrates this in the cross-section of the conveying element after half a revolution and after two revolutions, each with an enlargement of one of the screw tips. Figure 4.80 shows the temperatures directly on the screw surfaces. The color scale in the two illustrations ranges from 300 °C (blue) to 350 °C (red). In addition, Figure 4.81 shows a diagram which displays the maximum temperature on the screw tip over the rotation of the conveying element.

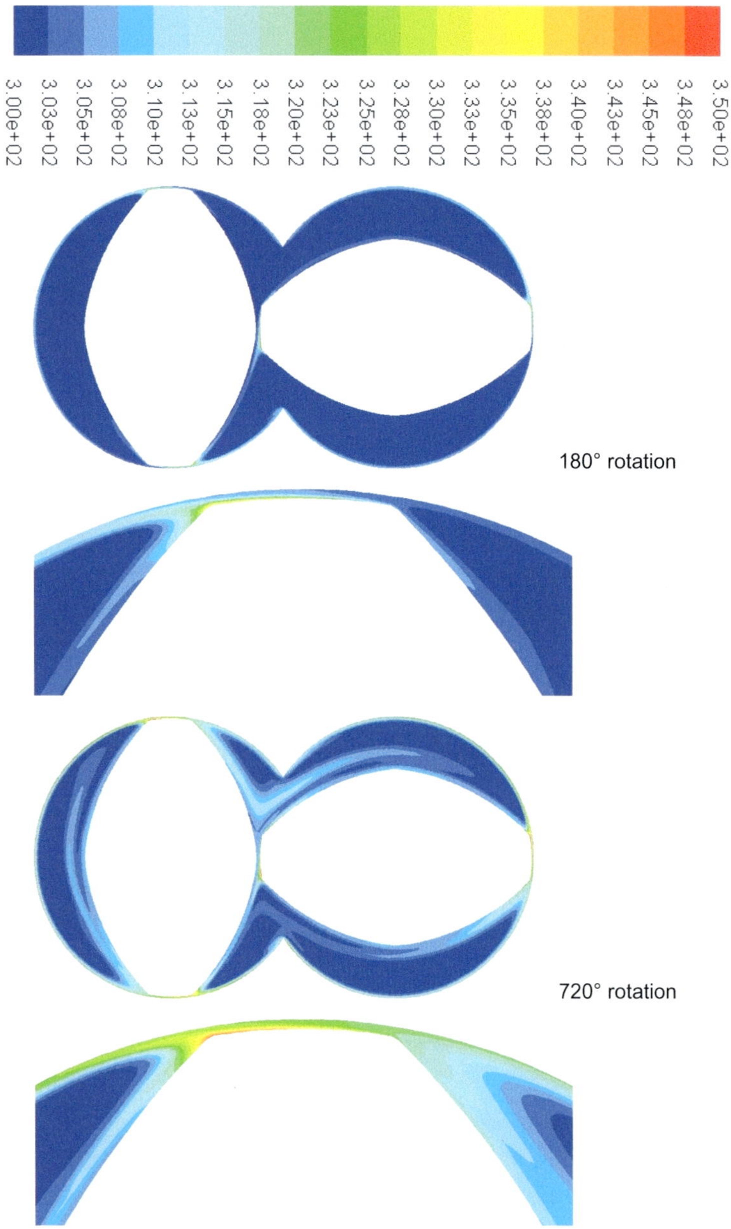

Figure 4.79 Temperature of the polymer melt in the cross-section of a conveying element with adiabatic walls after a half turn and after two turns, each with a detailed enlargement around one of the screw tips

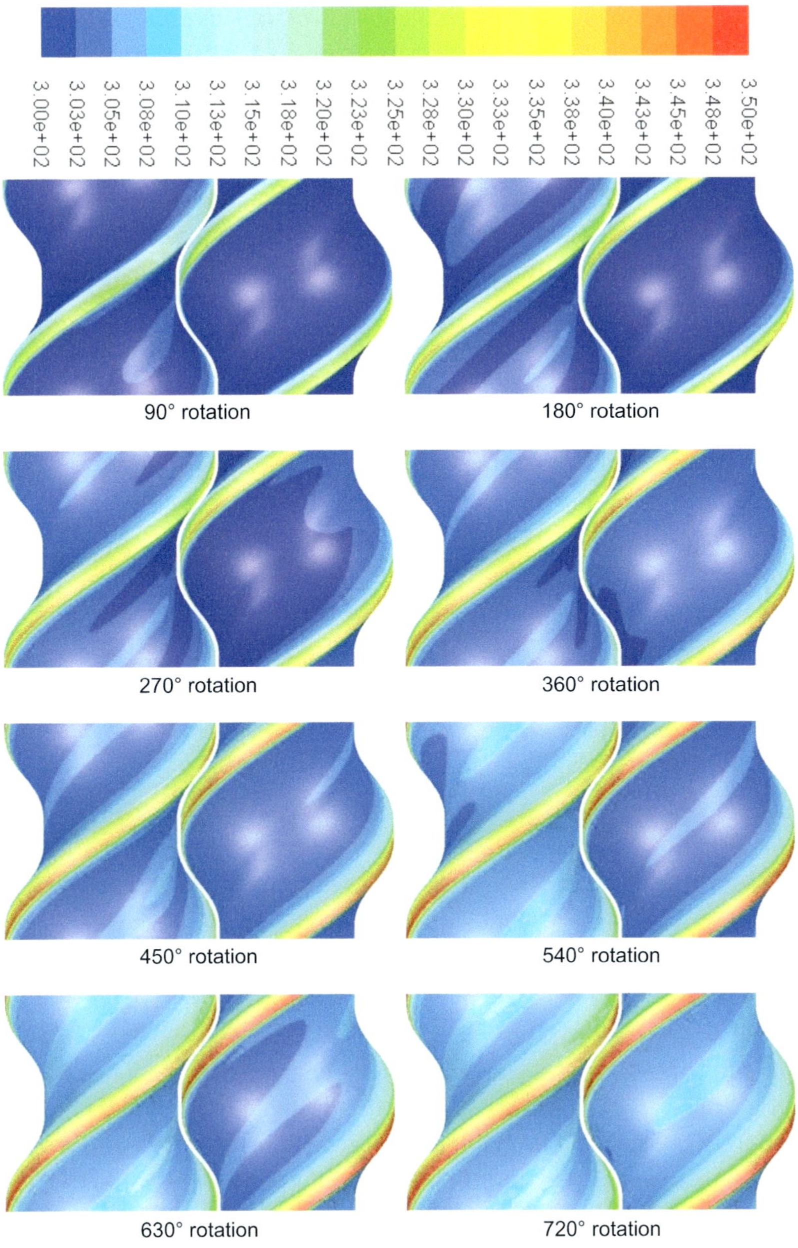

Figure 4.80 Temperature of the polymer melt on the screw surface of a conveying element within two revolutions

Since most of the heat is generated directly in the vicinity of the screw tips, the temperature initially rises to 339 °C within half a revolution. Afterwards, a significantly slower temperature rise can be observed on the surface of the screw tip. This is due to the fact that the heated fluid continuously leaves the tip area and contributes to a heating of the remaining flow field. It is noticeable that the heated polymer melt does not flow along the screw surface, but rather detaches from the screw tip and transfers to the barrel. In the intermeshing region there is lower dissipation and the screw tips cool down by convective heat transport and to a smaller extent by heat conduction. Figure 4.80 also shows that the screw surfaces do not heat up uniformly (symmetrically) in the channel area of the left and the right screw. This is due to the fact that the left screw first transports cold polymer melt into the intermeshing region. There the melt is heated by friction and transported out of the intermeshing region with the right screw.

In first approximation, the flow in the curved tip gap can be regarded as a flat shear field. In this case, [1] derives an analytical solution for the calculation of the temperature field. The maximum adiabatic temperature increase along the screw tip results in

$$T_{ad,max} = 1.43 \cdot \frac{\eta \dot{\gamma}^2 \delta^2}{\lambda} \tag{4.132}$$

with the temperature boundary layer thickness

$$\delta_K = \sqrt[3]{\frac{\lambda x}{\rho c_p \dot{\gamma}}} \tag{4.133}$$

and with the tip length x. In this example, the geometric data of the conveying element, the physical properties and the operating conditions result in a tip length of about 10 mm and a (mean) shear rate of about 600 1/s. This leads to a temperature boundary layer thickness of about 0.12 mm. The temperature boundary layer thickness is therefore only a quarter of the clearance between the screw tip and the barrel. The grid used must be chosen with sufficient resolution in the tip area in order to accurately capture the development of the temperature field along the screw tip.

Equation (4.132) calculates a maximum adiabatic temperature increase of 35 °C at the end of the screw tip. At first glance, this is somewhat lower than the 39 °C temperature increase from the flow simulation. However, if one considers that in the flow simulation after half a revolution the inlet temperature into the tip gap is no longer 300 °C but already 304 °C, the temperature increase over the tip length of 35 °C results here too. Flow simulation and analytical solution therefore match very well.

Since the viscosity in reality depends on both the shear rate and the temperature itself, the temperature increase will be significantly lower in practice after two revolutions. On the other hand, the transport of the polymer melt requires far more than two revolutions, resulting in further temperature increases. Overall, one should be aware that the temperature in the tip area can quickly be several 10 °C warmer than the temperature in the main flow. Depending on the thermal sensitivity of the product, this may already cause significant local polymer damage.

From Equations (4.132) and (4.133) the influence of wear on the local temperature development can be estimated directly. Inserting Equation (4.133) in Equation (4.132) returns

$$T_{ad,max} = 1.43 \cdot \sqrt[3]{\frac{\eta^3\, x^2\, \dot{\gamma}^4}{\lambda\, \rho^2\, c_p^2}} \tag{4.134}$$

The maximum adiabatic temperature increase is proportional to the shear rate high 4/3. If the screw clearance has increased by 50% due to wear, the shear rate in the tip area is reduced by 33% and the maximum adiabatic temperature increase is reduced by about 42%. Even if wear is generally seen negatively, it has at least the positive effect of lower local temperature peaks when operating the extruder.

In order to cool the polymer melt, in a further calculation a constant barrel temperature of 300 °C is assumed. During the first half turn, the temperature development in the conveying element is practically indistinguishable from that with the adiabatic barrel wall (see Figure 4.81). Due to the low thermal conductivity of the polymer and the very high shear rates in the tip gap, a thin temperature boundary layer forms, as already mentioned, which is significantly smaller than the barrel clearance. The polymer on the screw tip simply does not yet *feel* the cooling effect of the barrel. This only becomes noticeable with further rotation and the maximum temperature on the screw tip rises considerably slower.

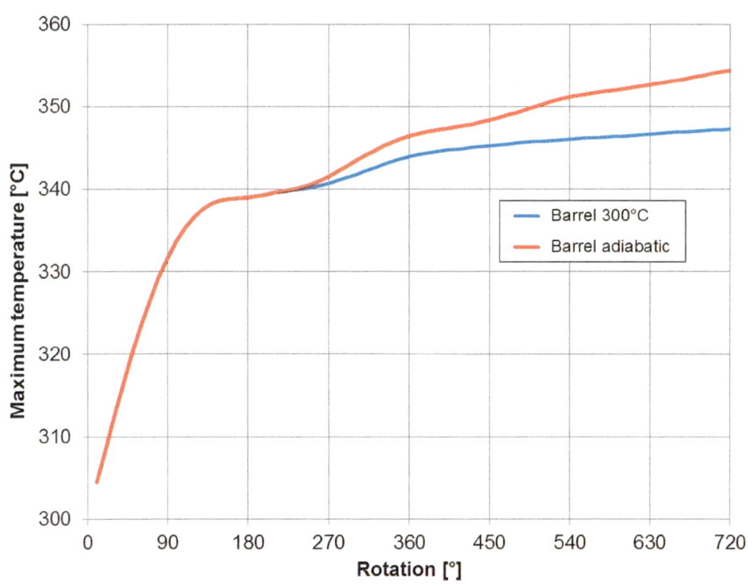

Figure 4.81 Development of the maximum temperature on the screw tip during two revolutions

Figure 4.82 shows the temperature of the polymer melt around one of the screw tips after two revolutions. The cold temperature boundary layer on the barrel wall is clearly visible. Due to the poor thermal conductivity of the polymer, it is very thin. Since there is practically no mass exchange on the barrel wall despite the rotation of the screws, the boundary layer thickness can be estimated with half the screw clearance between the screw tip and the barrel:

$$\delta_G = \frac{s_G}{2} \tag{4.135}$$

In this example, a temperature boundary layer thickness of about 0.25 mm is obtained and the grid must be selected along the entire barrel wall with adequate resolution in order to capture the temperature processes reliably.

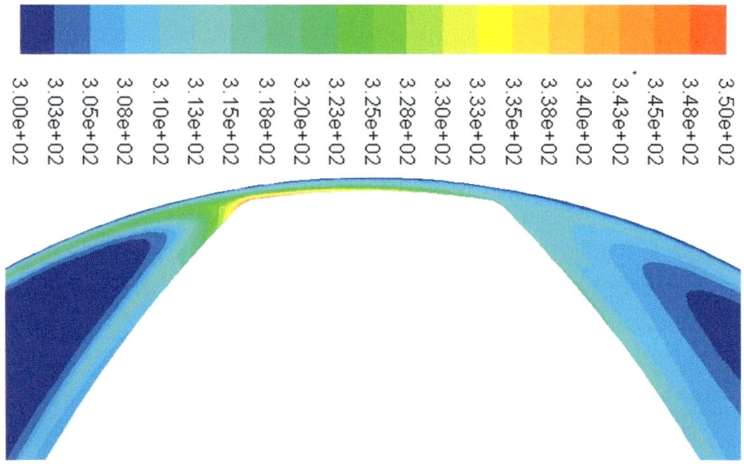

Figure 4.82 Temperature of the polymer melt in the cross-section of a conveying element with a wall temperature of 300 °C after two revolutions (detailed enlargement around one of the screw tips)

Based on the boundary layer thickness, a heat transfer coefficient can be estimated on the polymer melt side:

$$\alpha_{G,poly} = \frac{\lambda}{\delta_G} = \frac{2\lambda}{s_G} \tag{4.136}$$

Here the heat transfer coefficient is calculated to 800 W/(m^2 K). For the total heat transfer coefficient, of course, the heat conduction resistance in the barrel material itself and the heat transfer coefficient on the side of a coolant must also be taken into account.

4.7.2.2 Example 2

Geometry, Physical Properties, and Operating Conditions

The second example shows a pressure build-up zone followed by a left-rotating element (Figure 4.83). The same basic geometry is used for the screw elements as in the first example. Only the pitch, the length and the direction of rotation vary from element to element:

Element	Pitch [mm]	Length [mm]	Direction of rotation
1	60	60	right
2	45	22.5	right
3	30	15	right
4	30	15	left
5	60	60	right

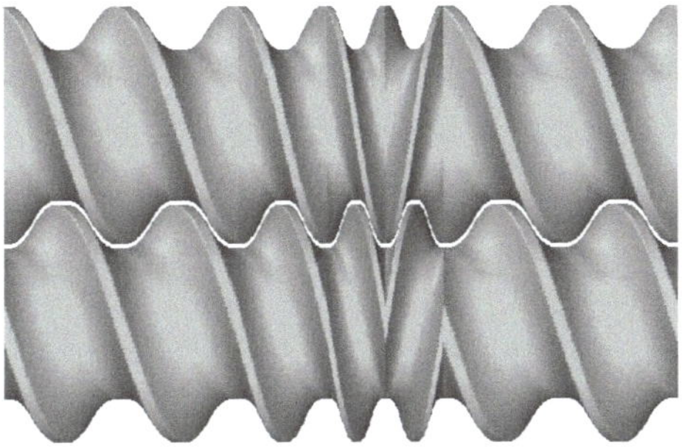

Figure 4.83 Geometry of the second example: series of conveying elements with different pitches and different rotational directions

The physical properties and operating conditions are also the same as in the first example. In the following, the flow is considered isothermal and the pressure difference is set to 0 bar. Beginning and end of the geometry are linked again by periodic boundary conditions in the flow simulation.

Pressure Field

The discussion of the pressure build-up zone is limited to the pressure field. Figure 4.84 illustrates the axial pressure profile averaged over the cross-sectional surface. The maximum pressure is approx. 38 bar, the minimum pressure approx. −20 bar. It can be seen that elements 5 and 1 cause a significant pressure build-up under the selected operating conditions. Element 2 is neutral with regard to the pressure build-up, while element 3 already leads to a slight pressure drop even though it is also right-hand rotating. The flow rate is already so high that the element is overrun. The 4th element is a left-hand rotating element and, as expected, a sharp pressure drop occurs. In addition, local pressure peaks occur at the transition from right-hand to left-hand rotation and at the transition from left-hand to right-hand rotation whose causes are not apparent by the average pressure profile.

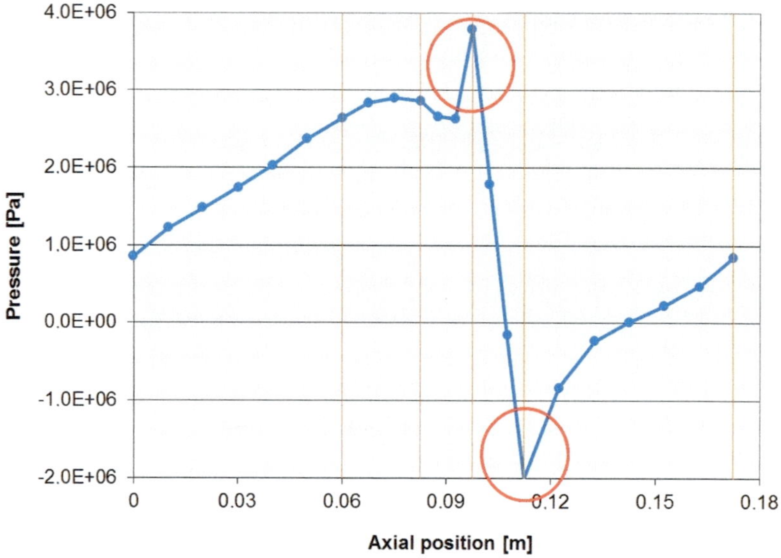

Figure 4.84 Average pressure along the axial position

Figure 4.85 and Figure 4.86 show the pressure on the screw surface viewed from the front (flow from left to right) and from the rear (flow from right to left). The color scale ranges from −95 bar (−9.5e6 Pa, blue) to 126 bar (1.26e7 Pa, red). The local maximum pressure peak is more than three times greater than the maximum pressure in the average pressure profile. The local minimum pressure peak is almost 5 times greater than in the average pressure profile. Both figures show that the two pressure peaks occur precisely at the change of rotation. In each case, the screw tips, the intermeshing of the two screws and the barrel in the intermeshing region create a largely enclosed volume. The polymer melt is squeezed out (maximum pressure) or sucked in (minimum pressure) through the small available clearances. It is easy to see that the polymer is both thermally and mechanically stressed to a specific degree in these local regions.

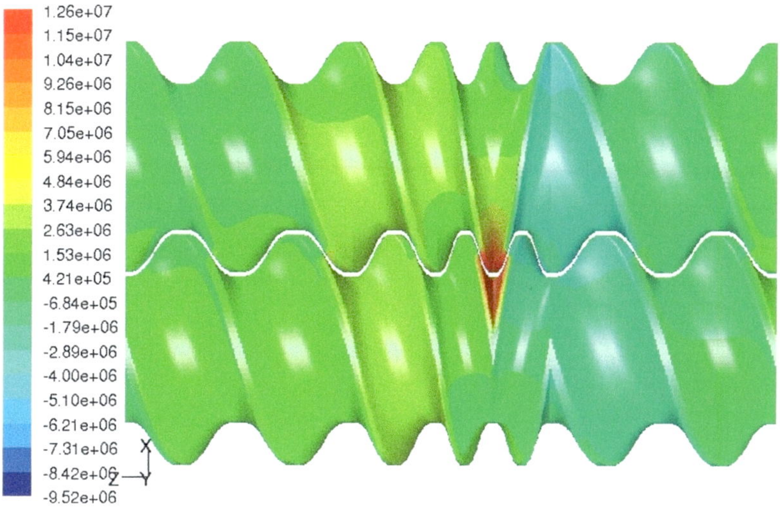

Figure 4.85 Pressure on the screw surface, front view, flow from left to right. The color scale ranges from −95 bar (blue) to +126 bar (red)

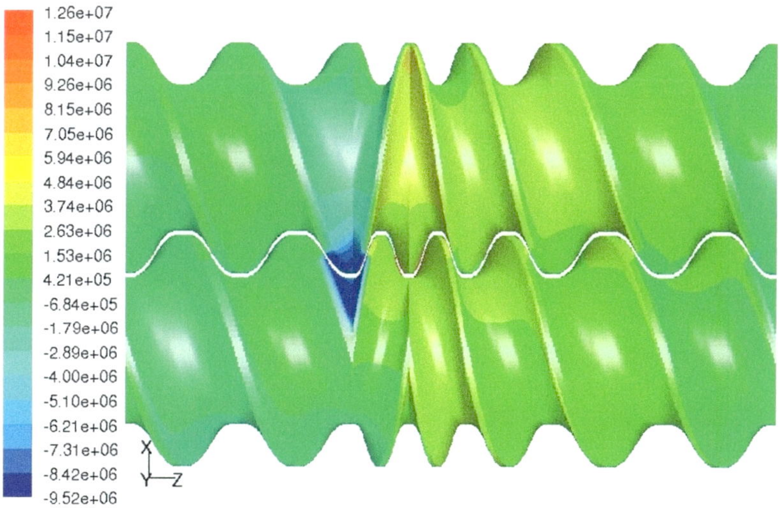

Figure 4.86 Pressure on the screw surface, rear view, flow from right to left. The color scale ranges from −95 bar (blue) to +126 bar (red)

4.7.2.3 Conclusion and Outlook

The two examples, deliberately chosen for their simplicity, show that computational fluid dynamics facilitate a more in-depth examination of the local flow behavior of twin screw extruders. Local peaks in the mechanical and thermal stresses can be easily identified. By changing the geometry, stresses can be reduced and the quality of the polymer can thereby be optimized. Another application focus is the rapid determination of the dimensionless axis intercepts for the pressure build-up $A_1$and A_2 and for the power requirement $B_1$and B_2. The significance of these parameters has already been discussed in detail in the two previous sections.

The grid program GiPEx 3d has been recently extended to mesh mixing and kneading elements. Figure 4.87 and Figure 4.88 show initial simulation results for a screw zone comprising a conveying element, spacer disc, mixing element, spacer disc, and kneading element at two different pressure gradients. The top part of the figure shows the pressure on the screw surface, while the bottom part shows the axial velocity at half the channel height.

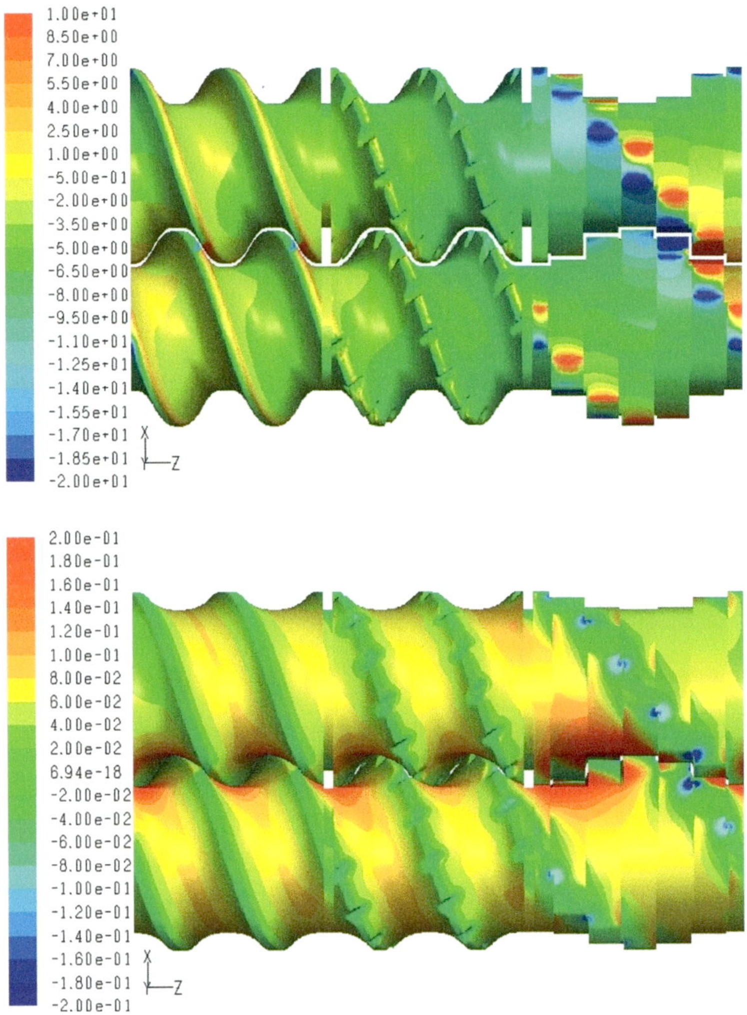

Figure 4.87 The screw zone comprises a conveying element – spacer disc – mixing element – spacer disc and kneading element at a pressure difference of 0 bar and melt flow of 0.11 kg/s. The top figure shows the pressure on the screw surface, while the bottom figure shows the axial velocity in the melt channel at half the channel height

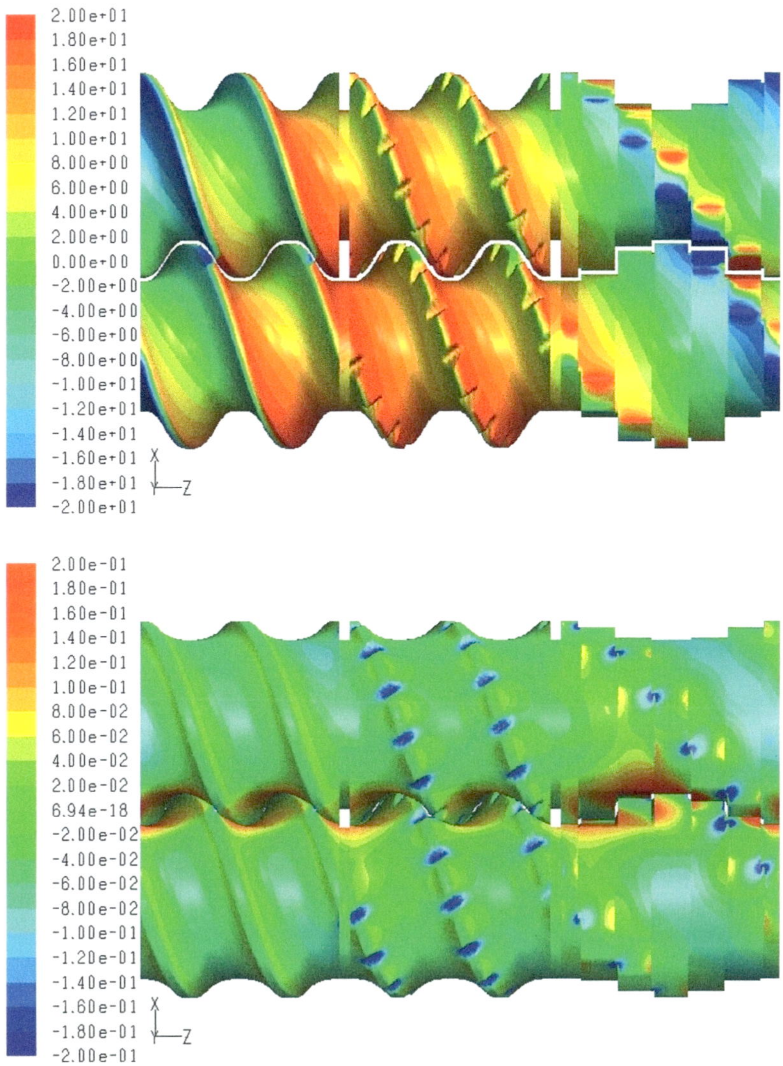

Figure 4.88 The screw zone comprises a conveying element - spacer disc - mixing element - spacer disc and kneading element at a pressure difference of 100 bar and melt flow of −0.01 kg/s. The top figure shows the pressure on the screw surface, while the bottom figure shows the axial velocity in the melt channel at half the channel height

The conveying element corresponds to the first conveying element from the second example. The two spacer discs have the same diameter as the core diameter of the other screw elements. The spacers are 3 mm thick. The mixing element is based on the geometry of the conveying element and is characterized by 10 V-grooves distributed evenly around the circumference, each groove having a depth of 10 mm. The grooves are left-hand rotating with a pitch of 240 mm. The kneading element comprises seven discs, each of which is offset by 30° rotating in a right-hand direction. The first and the last discs are 5 mm thick, while the remainder is 10 mm thick. The width of the groove between the kneading discs is 1 mm. Since the tip angle of the individual kneading discs is approximately 20°, an open groove with an opening angle of approximately 10° exists between them. The left and right boundaries of the geometry are linked together by periodic boundary conditions, as described in the two examples above. The physical properties and the operating conditions have been adopted from the second example.

In Figure 4.87, the pressure difference over the screw zone is 0 bar. The screw zone in question is operated at the maximum natural conveying. The pressure ranges between −50.2 bar and 39.9 bar and the pressure peaks occur at the kneading discs. The color scale for the pressure ranges from −20 bar (blue) to 10 bar (red). Capping the pressure peaks reveals the pressure distributions in the conveying and the mixing elements. The axial velocity ranges between −0.403 m/s and 0.598 m/s. Here again, the top velocities have been capped. The color scale ranges from −0.2 m/s (blue) to 0.2 m/s (red). In the conveying element area, we can see the pressure and axial velocity distributions noted in Figure 4.70 and Figure 4.71. We can see a slight pressure drop at the mixing element causing only a very small reflux via the grooves in the screw tips. The kneading element is largely neutral but exhibits very large pressure peaks upstream and downstream from the tips. This is due to the wide kneading discs compared to the small tip widths of the conveying and mixing elements which make it difficult for the polymer to evade. Additionally, we can see a reflux of the polymer melt in the grooves between the kneading tips.

In Figure 4.88, the pressure difference in the screw zone was set to 100 bar. The computation shows a resulting negative melt flow of −0.01 kg/s. The pressure ranges between −52.7 bar and 51.6 bar. The color scale ranges from −20 bar (blue) to 20 bar (red). The axial velocity ranges between −0.481 m/s and 0.536 m/s. The color scale once again ranges from −0.2 m/s (blue) to 0.2 m/s (red). The conveying element ensures the necessary pressure build-up and we can see a pressure drop in the mixing and kneading elements. With respect to the axial velocity, we can see positive material transport in the intermeshing region. There is a reflux via the grooves between the kneading tips as well as a significant reflux via the grooves of the mixing element.

We know from practical experience that the reflux through the grooves of the kneading and mixing elements leads to a good mixing of the polymer melt. Computational fluid dynamics allows for the investigation of these specific mixing processes. The results of the investigations will contribute to developing innovative and more efficient kneading and mixing elements.

4.7.3 Partly Filled Screw Sections

Twin screw extruders not only convey the melt to a pressure consumer, they also perform other process steps such as mixing and degassing volatile ingredients. For degassing it is necessary that the screw section is no longer completely filled with the melt. In addition to the liquid phase, there is a phase with gas (or vacuum). In fluid mechanics we speak of a two-phase flow, whereby the two phases are separated from each other by a free surface. Extensive experimental investigations e. g. in [2] show possible surface forms in rotating apparatus. Numerically, the simulation of such a flow is quite complex.

Different model approaches for the numerical simulation of two-phase flows are known in the literature [3]. In the mixing approach, the balance equations of mass and momentum are formulated globally for both phases. A locally variable volume fraction is defined which indicates the distribution of the two phases: $\alpha = 0$ for the gas phase and $\alpha = 1$ for the fluid phase. The fluid properties (density and viscosity) of the different phases in the balance equations are calculated from the volume fraction. The actually abrupt jump of the volume fraction at the phase boundary is modeled within a small space by a mixture of both phases. Here, the fluid properties change due to a linear interpolation, for example. In contrast, the Euler-Euler model [4] uses a separate momentum equation with constant fluid properties for each phase. At the phase boundary the model assumes a disperse distribution of the flow in the form of spheres. Gas bubbles in the fluid phase and fluid drops in the gas phase experience resistance and buoyancy forces due to the flow around them. Since these forces must be globally balanced, they can be used to couple the two momentum equations with each other. However, the mixture approach and the Euler-Euler model show problems if the viscosity difference is large. Here, the free surface model [5] has distinct advantages for the application of high viscosity polymer melts. Only the melt phase is calculated and a constant pressure is set at the surface as a boundary condition, see Figure 4.89. This procedure makes sense here because the shape of the free surface is only influenced by the melt flow due to the very high viscosity.

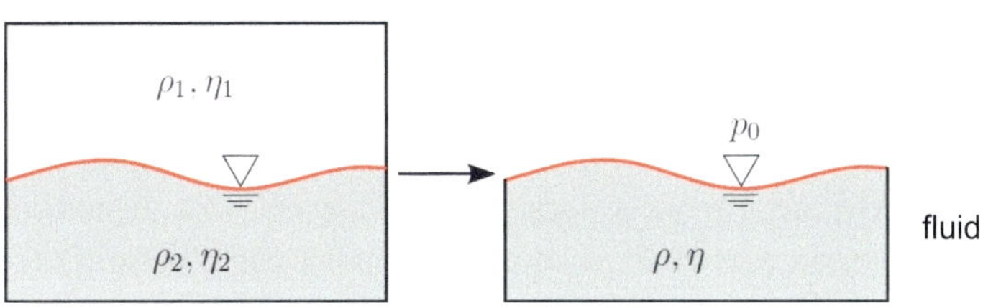

Figure 4.89 Modeling the two-phase flow with the free-surface model

To illustrate the advantages of this method, a simplified model of an extruder can be considered (Figure 4.90). There is also a rotating cylinder, the interior of which is divided into two chambers by a fixed beam. The chambers are half-filled with a highly viscous liquid and are subject to gravity. By moving the inner wall of the cylinder, the liquid is dragged upwards in a film in the right chamber, stripped from the cylinder wall by the beam and then flows downwards again on the beam supported by gravity. The higher the rotational speed, the thicker the film forms, up to a state of flow in which the entire liquid hangs in the upper area.

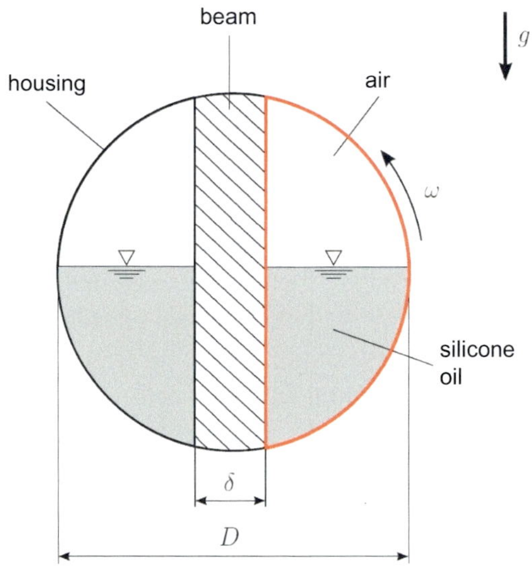

Figure 4.90 Simplified extruder model in the right chamber to illustrate the modeling approaches for two-phase flows

For a typical flow state, Figure 4.91 shows the results of the flow simulation after two complete revolutions with the different modeling approaches. In the mixing approach, a portion of the gas (blue) is transported into the liquid (red). This leads to a smearing of the phase boundary, which is clearly intensified during the calculation of further revolutions. The Euler-Euler model provides better results here. The phase boundary is clearly formed and remains stationary for further revolutions. The small gas bubbles on the beam still result from the initial conditions of the numerical simulation. They grow together in the course of the calculation and unite with the gas phase. Even a detail of the flow like the formation of the so-called cusp can be seen well [4, 6]. At this point on the surface, fluid particles move towards each other and immerse into the melt. However, the quality of the results depends strongly on the parameters of the model (e.g. the ball diameters in the disperse phase). Furthermore, numerical stability problems become visible especially for complex geometries such as the twin screw extruder. In the free-surface model, the phase boundary is also clearly visible, comparable to the Euler-Euler model. No gas bubbles can be detected, since only the flow of the melt is calculated during the free surface modeling. The amount of velocity in the melt is largely the same for all three modeling types. The gas velocity calculated with the mixture approach and the Euler-Euler model shows clear differences and is non-physical overall. This shows the weaknesses of the two models, which do not cope well with the high viscosity differences in particular.

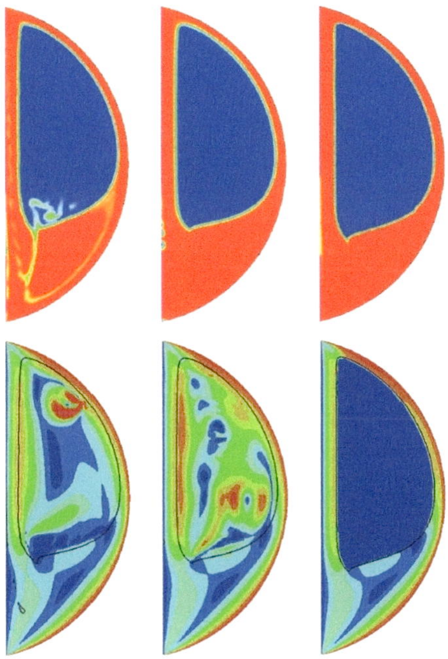

Figure 4.91 Simulation results in the simplified extruder model: phase distribution (top) and velocity magnitude (bottom). Left: mixing approach; middle: Euler-Euler model; right: free-surface model

The free-surface model can be implemented independently of the selected numerical method. In addition to the finite-volume method, which is very often used in flow simulation, it can also be used in particle methods, e. g. the Smoothed Particle Hydrodynamics (SPH) [7], which work without discretization of the calculation domain.

The performance of the model is to be demonstrated on a three-dimensional, double-flighted single-screw extruder. The geometry follows the Erdmenger profile with a pitch ratio $T/D = 1.5$ and a dimensionless gap (screw barrel clearance) of about $\delta/D = 0.01$, as it is also used in twin screw extruders. The calculation grid for the finite volume method used consists of approx. 300,000 cells and the filling degree of the extruder is 50%. Figure 4.92 shows the calculated phase distribution of a highly viscous melt for a given speed. The melt collects in a pool in front of the driving flank and conveys it forward. Since gravity is small compared to the friction forces under the selected operating conditions, the relative shape of the melt in the extruder does not change. The housing wall is covered with a thin melt film. The thickness of the film is approximately half the gap height between the screw tip and the housing.

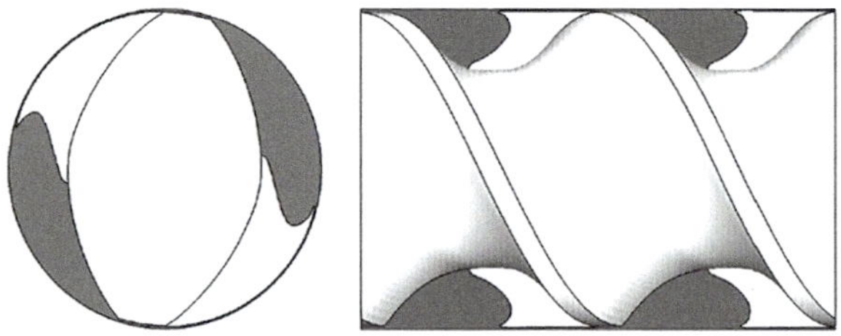

Figure 4.92 Calculated phase distribution of the melt of a single-screw extruder in cross-section (left) and in longitudinal section (right) for a filling degree of 50%

From the simulated phase distribution and the flow field, the pressure and friction stresses on the housing or shaft can be extracted. These are integrated over the surface and result in the power required to operate the extruder. In the fully filled screw section, the power for Newtonian melts and under creeping flow conditions is known to be a linear function of the volume flow. Applied dimensionless, see Figure 4.93, the characteristic axial section values B_1 and B_2 introduced in Section 4.2 can be read off. For partially filled screw sections, this course changes significantly. A partial filling can only be realized if the pressure between inlet and outlet does not change. In the fully filled state, this corresponds to point A_1 in Figure 4.93. Starting from this operating state, the output does not decrease linearly with decreasing filling degree, but with an S-stroke. At first, the volume flow remains largely constant, as the conveying properties are not yet influenced by the partial filling. Only when the filling level has dropped significantly the volume flow also decreases significantly. For very small filling grades of filling, essentially only the gap is filled with melt and the volumetric flow is close to zero. The power decreases again significantly and remains at a value caused by friction in the melt film between the screw tips and the barrel. This numerically calculated course of the performance curve on a single-screw extruder is confirmed by analytical considerations on an unwounded channel model [3, 8] and can also be qualitatively transferred to twin screw extruders.

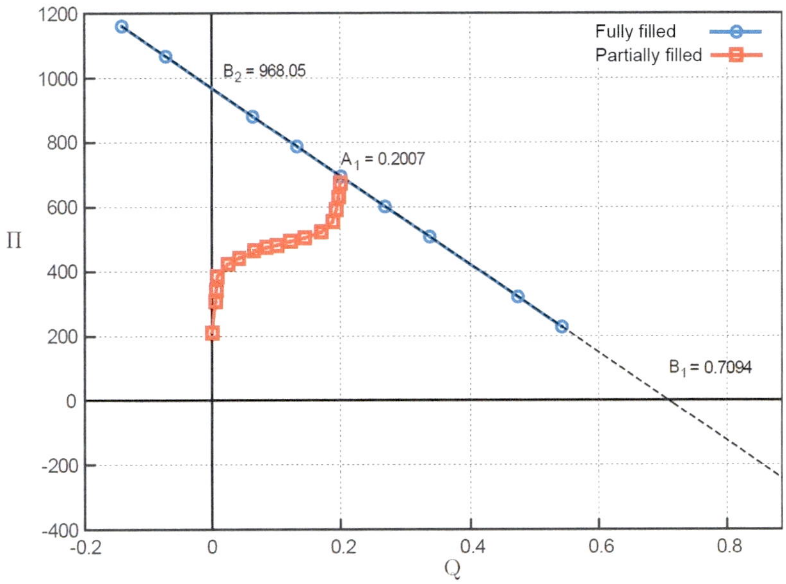

Figure 4.93 Dimensionless power characteristics depending on the dimensionless flow rate for a single-screw extruder: fully filled screw section (blue) and partially filled areas (red)

Although numerical simulation of the two-phase flow in twin screw extruders is much more difficult due to the complex geometry, it can basically be carried out using the above methods. Figure 4.94 shows an example of the distribution of the melt in a two-dimensional model of a twin-screw extruder for different points of time of a revolution, i. e. for different positions of the screws. The numerical simulation was carried out with a particle method instead of the previously used finite volume method. The shape of the free surface in the channels corresponds to the results in the single-screw extruder. In the intermeshing region the left screw scrapes a part of the melt from the right screw, a part remains on the right screw due to the gap between the two shafts (position 1 and 2). As a double-flighted configuration is considered here again, this process is repeated accordingly in positions 4 and 5.

Finally, Figure 4.95 illustrates a result of the phase distribution in a section of a three-dimensional real twin-screw extruder, also calculated numerically using a particle method. The geometry corresponds to a ZSK 32 from Coperion GmbH, Stuttgart with a filling degree of 50%. The distribution can be seen in the cross-section (bottom left), in a central longitudinal section (top) and in a three-dimensional view (bottom right). The color indicates the axial component of the velocity. The axial velocities are particularly high in the intermeshing region, where the main conveying also takes place during partial filling, while low velocities down to zero (green) prevail in the melt pool at the screw channels. Even with this partially filled state, which takes place globally without pressure, there are local pressure differences and thus backflows (blue). These details in the gap between the screw tip and the housing wall can also be verified numerically, see Figure 4.95 (bottom right).

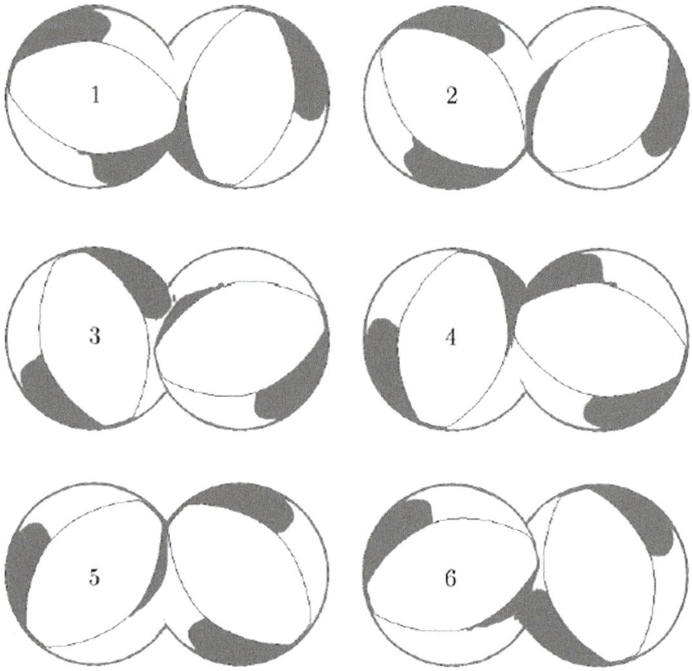

Figure 4.94 Calculated time-dependent phase distribution of the melt of a two-dimensional twin-screw extruder in cross-section for a filling degree of 50%

Figure 4.95 Phase distribution of the melt in a partially filled three-dimensional twin-screw extruder

Conclusion and Outlook

While the numerical simulation of the three-dimensional flow field in a completely filled screw section is now state of the art, the calculation of a partially filled section is currently still one of the exceptions in the design of twin screw extruders. The methodical bases for an efficient and exact calculation of the free surface in such complex geometries are intensively researched and tested in research and industry. Problems occurring in numerics such as instabilities and long computing times still speak against widespread application [9]. Extensive parameter studies are to be expected in the future only with the further developing hardware technology.

References for Section 4.7

[1] König, T., Brod, H., Bierdel, M., Temperaturspitzen in Schneckenmaschinen. *Tagungsband VDI-Aufbereitungstechnik*, Baden-Baden (2002)

[2] Zucht, A., Müller, A., Böhme G., Untersuchung der Phasengrenze eines fluiddynamischen Systems mit experimentellen und numerischen Methoden. *Forsch. Ingenieurwes.* (2008), 72(4), pp. 241–258

[3] Lübke, M., Wünsch, O., Modellierung teilgefüllter Schnecken. VDI Wissensforum GmbH, *Aufbereitungstechnik 2013*, VDI-Verlag GmbH, Düsseldorf (2013), pp. 79–88

[4] Müller, A., Theoretische Untersuchungen eines rotierenden fluiddynamischen Systems mit variierendem Füllgrad. Dissertation, Helmut-Schmidt-Universität Hamburg (2008)

[5] Wünsch, O., Lübke, M., Hirschfeld, S., IGF-Vorhaben Filmentgasung hochviskoser Polymere. Final report IGF project 17298N of the GVT (2015)

[6] Jeong, J.-T., Moffatt, H. K., Free-surface cusps associated with flow at low Reynolds number. *J. Fluid Mech.* (1992) 241, pp. 1–22

[7] Eitzlmayr, A., *Modeling of High-Viscous Flow and Mixing in Co-Rotating Twin-Screw Extruders*. Dissertation, TU Graz (2015)

[8] Lübke, M., Wünsch, O., Investigation of the free-surface flow in partially filled extruders. *Proc. Appl. Math. Mech.* (2013) 13, pp. 293–294

[9] Pokriefke, G., *Numerische Analyse reibungsbehafteter Strömungen in teilgefüllten Extrudern*. Dissertation, Helmut-Schmidt-Universität Hamburg (2005)

Index

F

G

H

I

N

O

P

R

S

Z